Lecture Notes in Computer Science 2025
Edited by G. Goos, J. Hartmanis and J. van Leeuwen

Springer
Berlin
Heidelberg
New York
Barcelona
Hong Kong
London
Milan
Paris
Singapore
Tokyo

Michael Kaufmann Dorothea Wagner (Eds.)

Drawing Graphs

Methods and Models

Series Editors

Gerhard Goos, Karlsruhe University, Germany
Juris Hartmanis, Cornell University, NY, USA
Jan van Leeuwen, Utrecht University, The Netherlands

Volume Editors

Michael Kaufmann
University of Tübingen, Wilhelm Schickard Institute for Computer Science
Sand 13, 72076 Tübingen, Germany
E-mail: mk@informatik.uni-tuebingen.de

Dorothea Wagner
University of Konstanz, Department of Computer & Information Science
Box D 188, 78457 Konstanz, Germany
E-mail: Dorothea.Wagner@uni-konstanz.de

Cataloging-in-Publication Data applied for

Die Deutsche Bibliothek - CIP-Einheitsaufnahme

Drawing graphs : methods and models / Michael Kaufmann ; Dorothea Wagner (ed.). - Berlin ; Heidelberg ; New York ; Barcelona ; Hong Kong ; London ; Milan ; Paris ; Singapore ; Tokyo : Springer, 2001
(Lecture notes in computer science ; 2025)
ISBN 3-540-42062-2

CR Subject Classification (1998): G.2, I.3, F.2

ISSN 0302-9743
ISBN 3-540-42062-2 Springer-Verlag Berlin Heidelberg New York

Springer-Verlag Berlin Heidelberg New York
is a part of Springer Science+Business Media

springeronline.com

Typesetting: Camera-ready by author, data conversion by Boller Mediendesign
Printed on acid-free paper SPIN: 10973851 06/3111 5 4 3 2 1

Preface

Drawings are an attractive and effective way of conveying information. *Graph Drawing* includes all aspects of visualizing structural relations between objects. The range of topics extends from graph algorithms, graph theory, geometry and topology, to visual languages, visual perception, and information visualization, and to computer-human interaction and graphics design. Clearly, the design of appropriate drawings is a complex and costly task where automation is urgently required.

The automated generation of graph drawings has important applications in many areas of computer science, such as compilers, data bases, software engineering, VLSI and network design, and graphical interfaces. Applications in other areas include graphical data analysis (e.g. in all fields of engineering, biology, or social sciences) and the visualization of information in general (e.g. by flow charts, schematic maps, or all kinds of diagrams).

The purpose of this book is to give an overview of the state of the art in graph drawing. It concentrates on algorithmic aspects, with an emphasis on interesting visualization problems with elegant solution methods. Each chapter provides a survey of some part of the field; in addition some selected results are described in more detail. This approach should make the book suitable for a first introduction as well as a good basis for an advanced course, where it may be supplemented by other sources. There is no claim of completeness – graph drawing is a very dynamic area – so the reader should be aware of the possibility that further progress might have been made since the publication of this book. There is also a chance that we may have failed to notice some subjects, since the necessity of drawing graphs arises in so many different areas.

The rapid growth of graph drawing as a field has caused some inconsistencies in terminology: terms like "drawing", "layout", "representation", or "model" are often used with different meanings. The authors have tried to achieve consistent notation; this has not always been possible without breaking with existing conventions, and we apologize for all remaining inconsistencies.

The book arose from a seminar for young computer scientists. The idea of the "GI Research Seminars" is to provide young researchers with the opportunity to gain insight into a new, relevant, and interesting area of computer science. The topics were chosen and prepared by the organizers. Lectures on the topics were then presented by the young researchers at the research seminar, which took place in April 1999 at Schloß Dagstuhl. Based on the presentations and discussions during the seminar, each chapter was elaborated by a team of authors, and carefully reviewed by other participants of the seminar. In addition, we obtained expert advice from Therese Biedl and David Wood. They provided detailed comments and improvements on several chapters in a preliminary version of the book.

The material covered in this book can be organized in various ways. However, a strict separation of topics according to, e.g., models, representations, or methods, would inevitably lead to an artificial distinction between topics that are in fact closely related with respect to some other aspect. Therefore, we decided to do without a strict partition into chapters under a single category of distinction, but rather provide cross-pointers where appropriate.

In the first chapter, Rudolf Fleischer and Colin Hirsch review some *application areas.* They discuss the traditional applications of graph drawing like ER-diagrams or software engineering, but also some of the areas that are less related to computer science like social networks or workflow. From those applications, different tasks and basic techniques are derived such that the reader can get some motivation to learn of the methods to solve the tasks.

We then start from a more graph-theoretic point of view. *Planarity* is a classical topic in graph theory and algorithms, and an important and central aspect in graph drawing as well. René Weiskircher reviews basic algorithms related to planar graphs (planarity tests, *st*-orderings, etc.). He then turns to simple and advanced drawing algorithms for straight-line and related representations of planar graphs.

In the following chapter, some further *special classes of graphs* are considered by Matthias Müller-Hannemann. Drawing algorithms for trees and series-parallel graphs typically use their recursive structure. A closely related issue is the drawing of order diagrams and lattices, which is summarized as well. Although at first sight, this topic seems to be only of special interest, the algorithms presented have direct applications since many structures to be visualized are just trees or series-parallel.

In the chapter on methods based on *physical analogies*, Ulrik Brandes gives an overview of the ideas behind the so-called spring embedder and its variants. These methods work by iterative improvement. Related methods to compute a minimum of the objective function directly are included as well. In the last subsection, the reader gets an idea of the wide field of possible applications of the underlying principle which ranges from clustering and 3D to dynamics and constraints.

A classical topic in graph drawing which cannot be missed in a review like this is an approach supporting *layered drawings.* This method is mainly used to display temporal structures like workflow or other unidirectional dependencies. Several interesting methodological questions like the maximal acyclic subgraph problem and the level-wise crossing minimization problem are discussed in this chapter by Oliver Bastert and Christian Matuszewski.

A long chapter is devoted to *orthogonal layouts*, written by Markus Eiglsperger, Sándor P. Fekete, and Gunnar W. Klau. Historically, the first methods for orthogonal layouts were developed in the context of VLSI routing and placement. Changing some of the criteria gave room for improvements and variations like visibility representations. After the review of some heuristics for planar and nonplanar graphs, a very elegant flow-based approach to solve the bend-minimization problem for planar graphs is considered in the second part of the chapter. The final part is devoted to compaction, i.e., the post-processing phase where the components of the layout are squeezed together to save area, edge length, or bends. Here again, VLSI methods are reviewed and new LP-based methods are presented that solve the compaction problem regarding some specific criteria to optimality.

The methods described in chapters 1–5 are fundamental. The following are more advanced and mostly arise from application demands such as labeling, clustering and hierarchies, dynamics and interactiveness, or 3D.

The generalization from drawing in the plane to *3D graph drawing* is discussed by Britta Landgraf. Various representations are covered. Starting from force-directed methods and layered approaches such as cone trees, she reviews the most important techniques of orthogonal routing in 3D in more detail and ends with the discussion on finding the best point from which to view a three-dimensional structure.

Handling large graphs is an important problem where some new methods are needed. Sabine Cornelsen and Ralf Brockenauer present some of the common *clustering techniques* like partitions and structural clusterings. Furthermore, hierarchical graphs are discussed with the focus on planar drawings and on the concept of compound graphs as well. Moreover, it is shown how force-directed methods can be applied to visualize clusters.

Dynamic graph drawing is a very recent and relevant topic. Jürgen Branke highlights many concepts in this area, like maintaining the mental map and support of dynamics in orthogonal structures and force-directed approaches.

At first glance, *labeling* is a side topic in graph drawing. However, labels arise in nearly all practical applications, and the labeling problem is highly nontrivial. Gabriele Neyer reviews different labeling models and most important methods. Most of them come from the fields of computational geometry, cartography, and optimization. The various aspects considered include the relation to satisfyability problems, sliding labels, and the combination with compaction.

In various aspects graph drawing is motivated by the relevance of visualizing relational data in many field of applications. Therefore, *software tools* for drawing graphs and *algorithms libraries* are an important issue in graph drawing. However, some difficulties about this topic prevented us from making it a regular chapter. Many graph drawing tools and libraries have been developed over the last few years, but easy access to the tools, stability, and support are often not guaranteed. Typically, software developed in academia cannot have the pretension of being "made for eternity". Moreover, it is hard to get an overview of the graph drawing tools available, or even reliable information about their actual abilities. On the other hand, we thought it would be useful to share what we knew of some existing systems. In the appendix, Thomas Willhalm provides an overview of some of the most common systems for graph drawing.

Finally, it is our pleasure to thank all the people whose contribution has made this book possible. First and foremost, these are the authors of the chapters, who not only put a huge amount of work into their own chapters, but also carefully reviewed other chapters. Special thanks go to Ulrik Brandes, who supported this project from the very beginning in many ways. Not only did he do a lot of administrative work in preparation of the research seminar, but he also inspired the choice of topics and supported the collection of relevant material. Last but not least, he handled most of the technical parts of the editing process. Finally, we would like to express our gratitude to the external experts Therese Biedl and David Wood, who did us the favor of reviewing several chapters in a preliminary version of the book.

January 2001

Michael Kaufmann
Dorothea Wagner

Table of Contents

List of Contributors

Editors

Michael Kaufmann
Universität Tübingen
Institut für Informatik
Sand 13
72076 Tübingen, Germany
mk@informatik.uni-tuebingen.de

Dorothea Wagner
Universität Konstanz
Fachbereich Informatik
& Informationswissenschaft
Fach D 188
78457 Konstanz, Germany
Dorothea.Wagner@uni-konstanz.de

Authors

Oliver Bastert
Technische Universität München
Zentrum Mathematik
80290 München, Germany
bastert@mathematik.tu-muenchen.de

Ralf Brockenauer
Algorithmic Solutions
Postfach 15 11 01
66041 Saarbrücken, Germany
Ralf.Brockenauer@algorithmic-solutions.com

Ulrik Brandes
Universität Konstanz
Fachbereich Informatik
& Informationswissenschaft
Fach D 188
78457 Konstanz, Germany
Ulrik.Brandes@uni-konstanz.de

Sabine Cornelsen
Universität Konstanz
Fachbereich Informatik
& Informationswissenschaft
Fach D 188
78457 Konstanz, Germany
Sabine.Cornelsen@uni-konstanz.de

Jürgen Branke
Universität Karlsruhe (TH)
Institut für Angewandte Informatik
und Formale Beschreibungsverfahren
76128 Karlsruhe, Germany
branke@aifb.uni-karlsruhe.de

Markus Eiglsperger
Universität Tübingen
Institut für Informatik
Sand 13
72076 Tübingen, Germany
eiglsper@informatik.uni-tuebingen.de

Sándor P. Fekete
Technische Universität Berlin
Fachbereich Mathematik
Straße des 17. Juni 136
10623 Berlin, Germany
fekete@math.tu-berlin.de

Rudolf Fleischer
The Hong Kong University
of Science and Technology
Department of Computer Science
Clear Water Bay, Kowloon
Hong Kong
rudolf@cs.ust.hk

Colin Hirsch
RWTH Aachen
Lehrstuhl für Informatik VII
52056 Aachen, Germany
hirsch@informatik.rwth-aachen.de

Gunnar W. Klau
Technische Universität Wien E186
Abteilung für Algorithmen
und Datenstrukturen
Favoritenstraße 9–11
1040 Wien, Austria
gunnar@ads.tuwien.ac.at

Britta Landgraf
Universität Bonn
Institut für Informatik I
Römerstraße 164
53117 Bonn, Germany
landgraf@informatik.uni-bonn.de

Christian Matuszewski
Universität Halle-Wittenberg
Institut für Informatik
06099 Halle/Saale, Germany
matuszew@informatik.uni-halle.de

Matthias Müller-Hannemann
Technische Universität Berlin
Fachbereich Mathematik
Straße des 17. Juni 136
10623 Berlin, Germany
mhannema@math.tu-berlin.de

Gabriele Neyer
ETH Zürich
Institut für Theoretische Informatik
CLW B2
8092 Zürich, Switzerland
neyer@inf.ethz.ch

René Weiskircher
Technische Universität Wien E186
Abteilung für Algorithmen
und Datenstrukturen
Favoritenstraße 9–11
1040 Wien, Austria
weiskircher@ads.tuwien.ac.at

Thomas Willhalm
Universität Konstanz
Fachbereich Informatik
& Informationswissenschaft
Fach D 188
78457 Konstanz, Germany
Thomas.Willhalm@uni-konstanz.de

1. Graph Drawing and Its Applications

Rudolf Fleischer and Colin Hirsch

1.1 Introduction

Graph drawing has emerged in recent years as a very lively area in computer science. We would like to explain in this chapter why this happened and why graph drawing is an important problem which should interest many people.

First of all, graphs are abstract mathematical objects (see Berge (1993); West (1996), for example), so why should it be important for non-mathematicians ever to draw one? Many probably would not even know a graph if they encountered one. One reason might be that people want to *visualize information*, i.e., objects and relations between objects. And even if they do not know it, this is exactly what graphs are designed for: describing relations between objects. The objects are called *nodes*, and relations between objects are called *edges*. If all relations are between two objects, we have an ordinary *graph*, otherwise we have a *hypergraph*. If the same set of objects is related in several different ways, we have *multiple* or *parallel* edges, and the resulting graph is a *multigraph*. A relation between an object and itself is a *self-loop*. There are a few more things you should know about graphs, when reading this book but you do not need to know more in this chapter.

Let's take an example from daily life. Living in a big city with a subway[1], we are confronted with a graph whenever we use the subway and try to find out which train to take to our destination. This information is usually displayed in form of a map, where stations are drawn as circles, and train lines are drawn as lines connecting the circles (see Figure 1.1). Stations are labeled by their name, and lines are either labeled by the train number, or different colors are used to distinguish between different trains. Such a map can naturally be interpreted as a graph. The circles (stations) are the nodes, and the lines (trains) connecting the stations are the edges. There are rarely selfloops, but there might be parallel edges if two train lines connect the same two stations and both connections are displayed by their own separate line.

We give a second example. People interested in genealogy try to find out as much as possible about their ancestors and then visualize this information by drawing a genealogy tree (see Figure 1.2)[2]. Note that trees are a subclass of graphs with a very simple structure: there are no cycles, so every two nodes of the graph are connected by exactly one path. In this sense, in a genealogy

[1] There is nothing special about subways here. Just replace 'subway' by 'bus' if your city hasn't got one yet.

[2] Actually, genealogy trees are not always trees because marriages can create cycles, for example.

M. Kaufmann and D. Wagner (Eds.): Drawing Graphs, LNCS 2025, pp. 1-22, 2001.

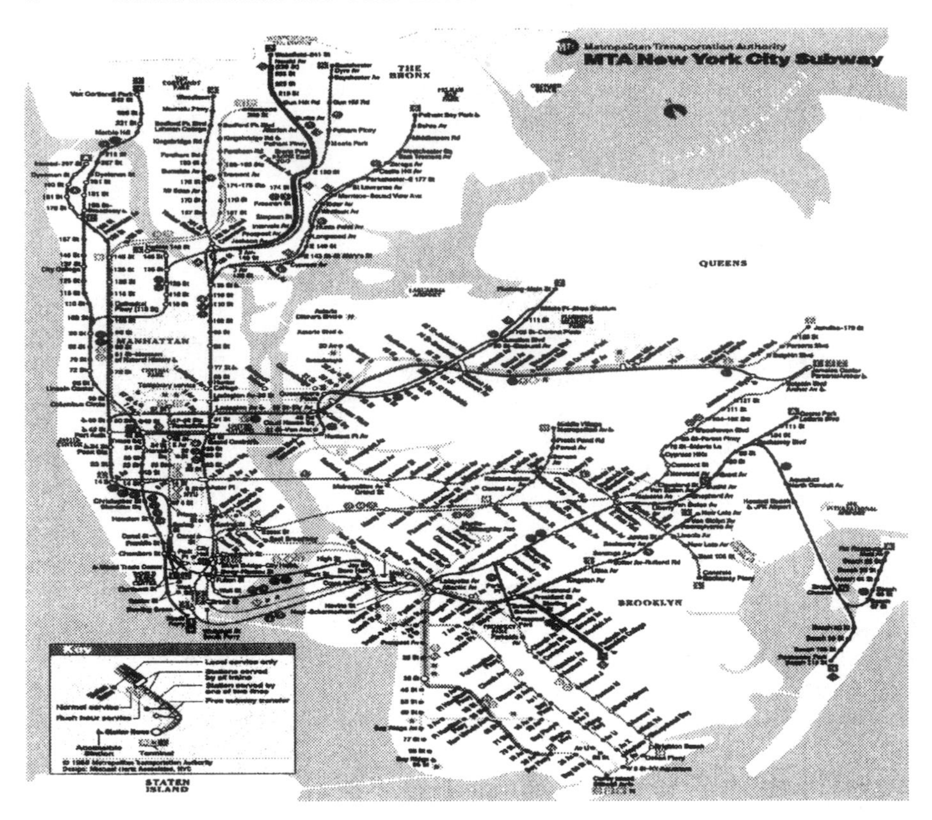

Fig. 1.1. Subway map, New York City (MTA, 1999).

tree each person is a node (labeled by the name of the person), and edges indicate direct decendency.

A natural question arises from these two examples: "What is the best way to represent a graph (and the information contained in it)?". Of course, we could describe a subway map by a list of sentences of the form "Train 1 runs from station A to station B via stations C, D, and E. At station D you can change to trains 3 and 11, etc." And a genealogy tree could easily be described by sentences of the form "A.B. C married D. E and they had two daughters F and G, and one son H, etc.". These verbal descriptions obviously contain all the information. However, this information is not easily accessible. A better way to present such data is by drawing a nice picture, i.e., a subway map or a genealogy tree (remember the proverb "A picture is worth a thousand words"). While reading the previous paragraphs you probably had a picture of your home town subway map or the genealogy tree of your family before your inner eye. This leads us to the problem of *graph drawing*. If we can describe our data as a graph, how should we draw the graph such as to best reveal all its information? In particular, this includes the question of which

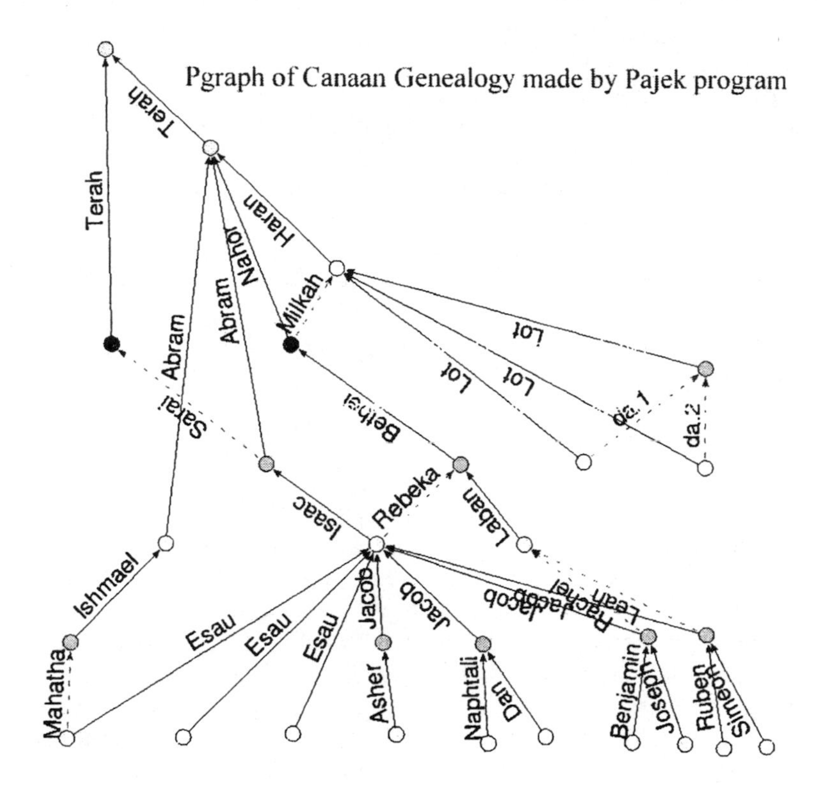

Fig. 1.2. Canaan Genealogy.

criteria determine whether a drawing is *good* or *bad.* We come back to this question in Section 1.3 after we have seen a few more applications of graph drawing in the next section.

1.2 Some Applications

In this section we will give some concrete examples for applications and their requirements from a user's point of view. Most applications inherently limit the types of graphs occuring to some intuitively or formally defined subclass, for example trees as in the genealogy example given above or graphs with a natural "layering". In some cases this can be a significant aid to an automatic layout algorithm trying to find a "good" drawing of a given graph. In more general cases, however, the lack of understanding of the intended structure and semantics of a graph seems to limit the usefulness of automatically drawn graphs. This does not even take into account that other measures for the quality of a graph drawing, like aesthetics, are difficult to quantify and even more difficult to realize in an implementation due to their very individual and subjective nature. Further demands are placed on the drawing algorithm in case of dynamic graph representations, where the graph to be drawn is modified over time, or in the case of very large graphs.

Most information in this section was gathered by talking to people at RWTH Aachen, Germany. In all cases graph drawing is not their main research subject. The drawing problem rather arises as by-product doing research on other subjects. Many thanks to Frank Huch, Stephan Kanthak, Ralf Molitor and Ansgar Schleicher for information, background and pictures on specification/verification, speech recognition, description logics and workflow diagrams, respectively.

1.2.1 Word-Graphs

One of the more promising, but also most demanding fields of research in computer science is speech processing and especially speech recognition. It is often seen in the context of building more user friendly computer systems capable of adapting to the human way of communication instead of the other way round. Of course the human brain and a common digital computer have entirely different approaches to information processing. This necessitates the use of complex mathematical methods and algorithms from stochastics and signal processing to handle natural language in form of speech.

Speech recognition is a process that involves several steps, each transforming the input information to a higher level of abstraction. First the audio signal is transformed into the frequency domain using cosine transformations and further normalizing steps. Next the building blocks of our speech, the phonemes[3] are identified. Then, omitting the intermediate step of syllables, the phonemes are put together to words and the words in turn form sentences.

Experiments have shown that a human listening to speech only recognizes about 70% of all phonemes correctly. However the higher-level capabilities of our brain including the so-called language model together with the context and semantics of what we hear enable us to understand whole sentences with next to no mistakes. Since speech recognition systems are built to imitate the behavior of the human brain they cannot be expected to, and indeed do not, perform better with respect to the phoneme error rate. Consequently, when transforming speech to text, such a system has to take into account that some phonemes are not correctly recognized. However, as some pairs of phonemes are more alike than others, a sane approach is to select a small set of phonemes which the system believes to have a high probability of faithfully representing the speach signal at each point in time. Using a dictionary it is then possible to eliminate most combinations and recognize the correct word with high probability.

Such a collection of hypothes finds a natural visualization in shape of a word-graph (Oerder and Ney, 1993). A word-graph is a layered graph, where

[3] Phonemes can be thought of as the atoms of speech such as "ah", "sh", "mhm".

consecutive layers represent consecutive time intervals. Each layer consists of a set of possible phonemes, augmented by additional data such as the probability of each phoneme. The possible words can be shown as paths going from layer to layer and so selecting one phoneme out of each layer.

This visualization is not necessary when using speech recognition as end-user or application developer; however, it is an invaluable debugging tool for the speech recognition system developer. Currently word-graphs are used in the VERBMOBIL project (German Research Center for Artificial Intelligence GmbH, 1999; Warnke et al., 1997; Amtrup and Jost, 1996).

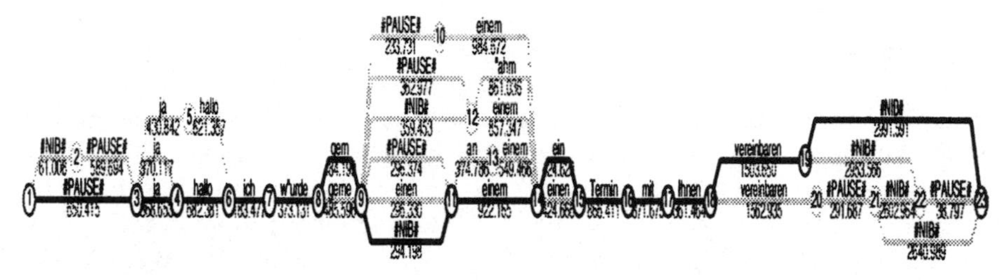

Fig. 1.3. Word graph.

Figures 1.3 and 1.4 show two typical word graphs. The horizontal positions of nodes are given by their points in time. Each edge represents one hypothesis about the words recognized during the interval given by the two incident nodes. The most probable path from left to right of the word graph is emphasized, showing what the system ultimately believes was spoken.

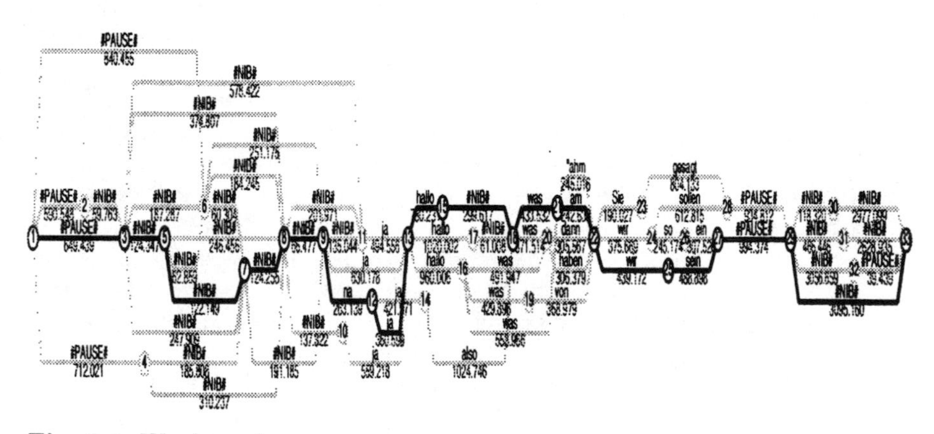

Fig. 1.4. Word graph.

Since the horizontal coordinate of each node is given, the problem of finding a suitable layout is greatly reduced. This strong limitation of the possible node and edge arrangements is the main reason for the graph drawing problem associated with this application to be solved in practice. That is, drawing algorithms currently in use produce layouts faithfully representing the intended meaning of the graph.

The layout algorithm here directly uses the position on the timeline for the horizontal placement of nodes. The freedom for the vertical positioning is used to minimize heuristically the number of edge crossings. The labels are added at the center of the corresponding nodes and edges, respectively. More on layered graphs can be found in Chapter 5. The placement of labels in general is considered in Chapter 10.

1.2.2 Specification and Verification

A software system running on a set of distinct machines with the capability to exchange data and synchronization events via some means of communication is called a distributed system. The added complexity in designing (the software for) a distributed system requires more advanced support techniques compared to dealing with single computing nodes. Various methods to model and/or verify properties of distributed systems are available.

The basic abstraction underlying these methods looks upon distributed systems as (finite) transition systems. A transition system in turn can be defined using so-called process algebras. A process algebra is a formalism used to describe inductively transition systems and is similar to a grammar used to describe formal languages. Since states of a distributed system have to contain the state of each component, the state space grows exponentially in the number of components.

Therefore a specification for a distributed system is typically broken down into specifications for each component. These component-wise specifications contain, besides a set of states and actions or transitions leading from one state to another, information on which transitions are to be synchronized with transitions in other components. Given such a specification of each component, the transition system modeling the complete distributed system can be generated automatically. These formal specifications are used for several purposes.

When writing the software intended to implement a specification, it is necessary to place procedure calls to communicate with remote components in the program text whenever an action has to be executed synchronously at two or more components. Using the formal approach of process algebras this step can be automated — the specification is in a format that can easily be machine processed and contains all required information. The benefits of this approach are twofold. For one the time and consequently the costs of the programing phase are reduced. Furthermore, one possible cause of errors in the implementation is eliminated, once the tool converting specifications into the basic application source code is debugged.

The downside is that using process algebras in the specification process is rather unintuitive. Before proceeding with further steps it has to be verified that the written specification is correct with respect to the ideas of the developers. To overcome this problem, graph drawing is used to give a more intuitive representation of the transition systems (Indermark et al., 1999).

Module: talkHandler

Global Knowledge: FromName,FromPid

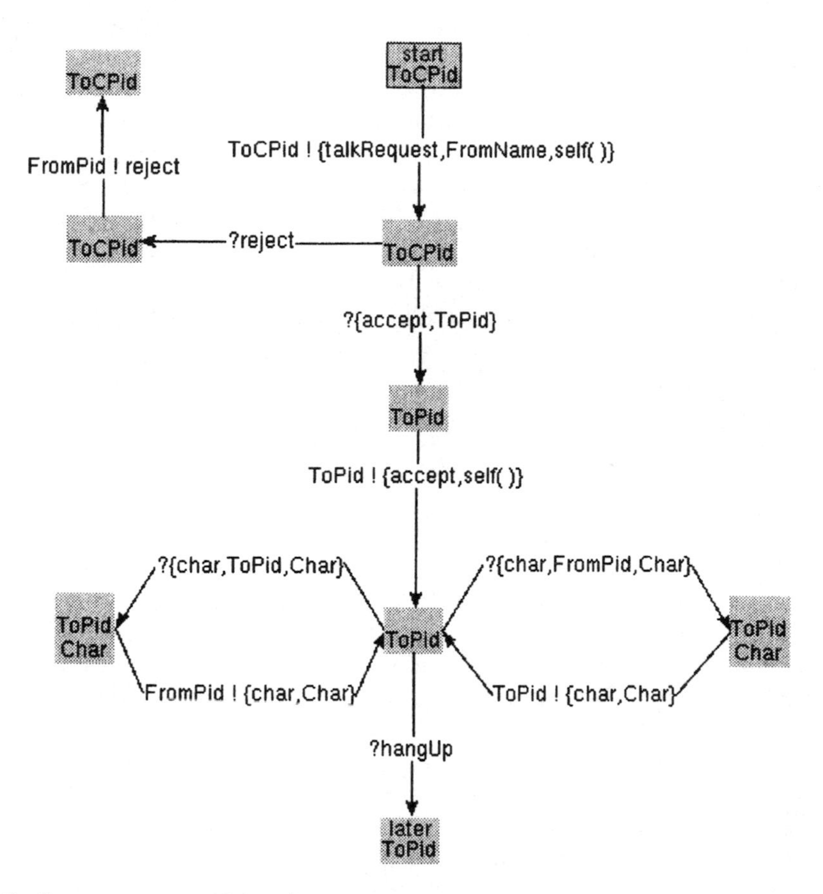

Fig. 1.5. Component transition diagram.

Figure 1.5 gives an example of a hand-drawn transition system belonging to one component of a distributed system. This particular example is used in testing a tool taking a different approach. Here the specification is entered graphically by the user and machine-translated into an algebraic specification later. However, the availability of this technology does not obsolete the earlier approach and the necessity to automatically layout these graphs, since there are cases when it is not feasible to use this method, especially when the specification was machine generated from some other form of description and creating the graph would add more work.

As can be seen, these graphs come with node *and* edge labels, giving information about the states and the actions associated with the transitions, respectively. Hence a layout algorithm for these graphs should be able to cope with edge labels. However the greater problem is to layout according to the semantics of the graph. In this example the top five nodes form a directed rooted tree, inducing a natural layering on these nodes. The layout chosen obviously disregards this fact, rather using the semantics of the graph for a differing layering.

Regarding the lower half of the graph it does not seem clear, from an algorithmic point of view, why the strongly connected nodes necessarily belong to one layer. The question arises how much additional information an algorithm needs in order to obtain a layout of similar quality? Furthermore, is it necessary for this information to refer explicitly to the layout (e.g., "nodes A and B are on the same level"), or is it sufficient to formalize some of the semantics (e.g., "node C is a rejecting end node"). In favor of flexibility and intuitivity the second choice would be preferable. However it is unclear how to formalize or process this kind of data.

Taking the leap from the transition system of an individual component to that of the whole distributed system most noticably comes with a large increase in size. Figure 1.6 gives a typical, if rather simple example of a transition system. Apart from the greater size and complexity in comparison to the single components, additional requirements arise due to the way the users work with these graphs.

Besides the behaviour of the single components, it has to be verified that the whole system satisfies certain requirements. Traditionally these requirements are formalized using some temporal logic. Characteristic for these logics is that they check whether any or all paths in the transition system satisfy some properties. Hence, especially in the case that the automatic verification fails, the user needs to trace manually along paths in order to find the error and adapt the specification.

As can be seen, the layout shown in Figure 1.6 is not well suited. Many edges cover large distances and as such are difficult to follow. This problem is further amplified if the graph is too large to fit onto the display and the user has to scroll the image. A layout algorithm is needed that tries to minimize the length of the edges. Furthermore it is desirable to be able to select paths and emphasize them in the drawing. However, optimizing the layout for the selected path should not completely change the rest of the drawing. After working with a graph for some time, the so-called "mental map" forms in the mind of the user. If the layout completely changes each time a path is examined, the user has to orientate herself anew. Isolating a path and

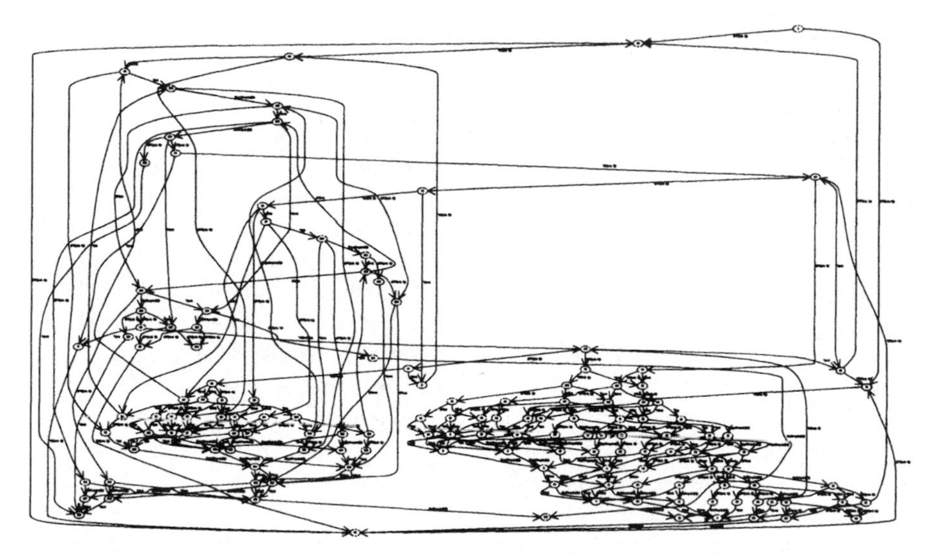

Fig. 1.6. Full transition diagram.

rendering it into a seperate window would make it difficult to retain the surrounding for orientation. Dynamic layout and mental maps are considered in more detail in Chapter 9.

1.2.3 Workflow

Workflow technology has emerged as an important tool in managing various types of business processes. This includes purely administrative tasks such as keeping track of employee holiday requests or processing orders from customers to more technical development and production processes such as quality assurance or organizing the steps of a chemical reprocessing plant. In all cases the expected benefits of using a workflow system are manifold (Abbott and Sarin, 1994; Georgakopoulos et al., 1995; Carpano, 1980b).

Our task is to record and store knowledge about details of the processes used in a company on an everyday basis. In many cases no documentation for these processes has been created. Rather the people on the job have an implicit knowledge of what to do when, which is later passed on to their successors. However the infrastructure of a company consists not only of material values. The corporate identity and resources to be found in the experience of the workforce constitutes a significant fraction of the overall value – especially in technology oriented branches. Without documentation, the loss of one or more employees in a position crucial to a process, be it due to staffing a new company branch, illness or retirement, can lead to severe problems when training new people.

Another aspect is to gain further insight into the processes themselves. In our highly competitive economy each production process has to be streamlined in order to reduce costs and enable faster delivery of service to the customer. The data gathered in the workflow management system makes it possible to identify, and, equally important, to prioritize places of inefficiency. These can then be handled one at a time starting with the most dire cases. Furthermore, only when the processes which form the activities of a company are fully understood is it possible to make accurate predictions on the impact of making changes to the structure, or even noticing changes happening by themselves over time.

To achieve these goals the data collected by a workflow system is as diverse as the fields of application. The temporal aspects have to be considered as well as resource usage and information flow. Resources are funds, raw materials, employees, machinery, etc. For each step of the process the resources and information it requires and the output have to be made explicit.

In order to comprehend the nature of a process once all the relevant information has been entered into a workflow management system, the only feasible approach is to give the user a graphical view of the process graph. This graph is obtained by creating nodes for the individual steps of the process. Edges depict temporal dependencies as well as the flow of information and/or objects. Figure 1.7 gives a simple example of a graph that was laid out manually.

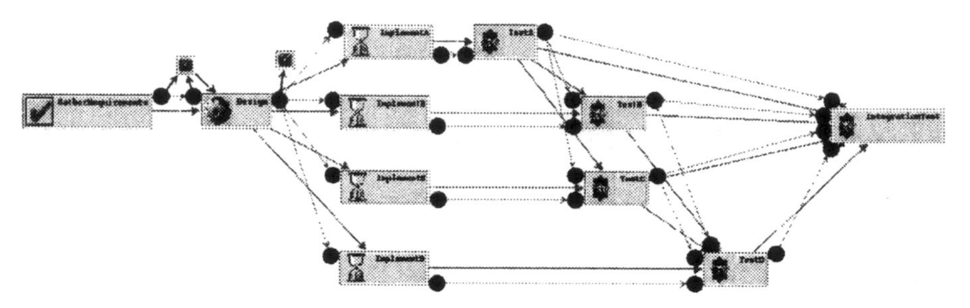

Fig. 1.7. Process graph, laid out manually.

Such a graph often consists of several types of nodes and edges. The main nodes represent stages or steps of the process. In the graph of Figure 1.7 smaller, attached nodes rendered as black discs are used to group and expose certain types of adjacent edges. A layout algorithm has to take some constraints about the relative positions of nodes into account, in order to place these extra nodes near the principal nodes. Such weight and constraint based methods are presented in more detail in Chapter 8.

Different kinds of edges are used to represent different kinds of dependencies and/or flows. This can be used to seperate information or other non-physical goods from physical objects being passed along or to differentiate information and meta-information. In some cases one kind of edge is more important than others, so the layout should be optimized accordingly. Consider planning the arrangement of machinery in an industrial plant. In this case the layout of the process graph can be made to follow mainly one criteria, e.g. the flow of information.

This application is especially sensitive to the quality of the layout in terms of correct logical groupings and aesthetics. Process graphs can be used for visual presentations when trying to convince customers or superiors of new ideas. If a production process is implemented and something goes wrong it is crucial to understand quickly where the problem is and which other parts of the process could be affected.

1.2.4 Data Modeling

One of the areas in computer science and information technology enjoying a rapid evolution and growth of importance are databases and data warehouses. This development is fueled by the wide availability of cheap powerful networking infrastructure making it feasible to store centrally large amounts of data and allowing concurrent remote access.

An important step in the creation of a database is the design phase. The information to be stored has to be categorized and typed; relationships between different types have to be found and conflicting types resolved. Depending on the underlying database paradigm this process can look rather differently.

One of the most important database models is the relational database model (Ullman, 1989; Vossen, 1991). Here, data elements are stored as lines in tables, or, mathematically speaking, tuples in relations. Each table represents a set of objects. The attributes of the object are the components of the tuples. Relationships and dependencies between objects are again written into tables.

Entity-relationship diagrams are the most common method to aid structuring large volumes of data by defining attributes on and relations between the data. Both objects, or entities, and their attributes are modeled as nodes of a graph. Edges linking an attribute to an entity express that said attribute is indeed an attribute of said entity. Annotated edges linking entities depict a relation with cardinality constraints between these entities. Figure 1.8 gives an example of an entity-relationship diagram.

The most noticeable feature is the possibility of edges being incident to more than two nodes, i.e., entity-relationship diagrams are actually hypergraphs. A layout algorithm can either directly handle these cases or introduce new nodes to join the different parts of such an edge. Furthermore, as opposed to many other graph drawing applications, no prior statement about the structure of the entity-relationship diagram can be made. The underlying graph can contain circles or be a tree, although typically without root or any other "special" element that can be taken as starting point for the layout.

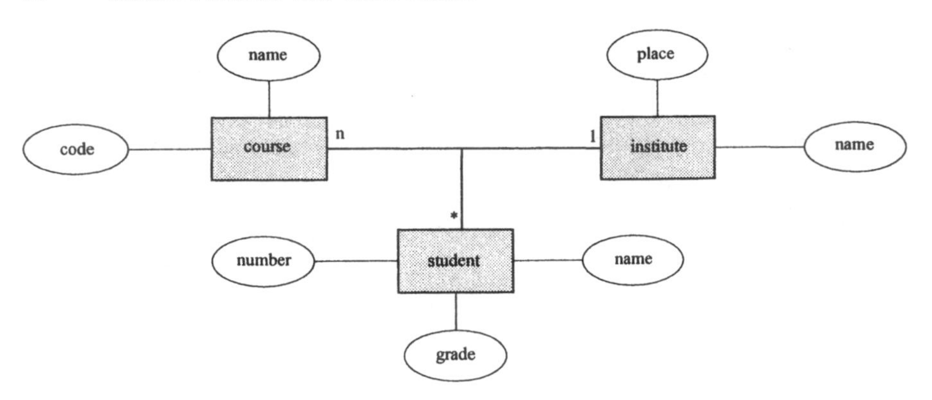

Fig. 1.8. Entity-relationship diagram.

Hierarchical Models In contrast to the relational model, which proposes a flat view of the different object types, many other database paradigms have a hierarchical type system. The most prominent version at the moment are object oriented databases (Hughes, 1993). The idea is similar to that of the relational database in that each entity, here called an object, belongs to a certain table or type, here called a class, sharing the same set of attributes. However, classes may inherit attribute sets from other classes. For example, first define a class `building` with attributes such as address, owner and colour, then define `hotel` as `building` with additional attributes name, rating ...

This results in a directed acyclic graph representing the "inherits" relation between all classes. Visualizing this graph is an important issue during the design phase where one tries to find mistakenly introduced cyclic dependencies.

A kind of inclusion dependency between different types resulting in similar graphs arises in the context of terminological knowledge representation systems or description logics, as used in the CLASSIC structural data model (Borgida et al., 1989). However, there is a huge difference in how the graph is obtained. In case of an object oriented database, it is explicitly stated which class inherits the properties of what other classes. Hence the user starts with a fairly accurate idea about the class hierarchy.

In the case of description logics the knowledge base consists of a set of formulae defining inclusions and other constraints about the types. Further inclusions can follow from this set of formulae. Hence a visualization is necessary to give the user an idea about what the complete hierarchy looks like. In that case an automated method is used to compute this hierarchy.

Figure 1.9 gives a partial example of such a hierarchy. As can be seen the graph is fairly tree-like, with few edges to the contrary. Since trees are both planar and naturally layered, finding satisfiable layout algorithms is comparatively easy. On the other hand the subsumption hierarchy of a real world

Fig. 1.9. Subsumption hierarchy.

example can easily have several thousands of nodes, severely complicating matters.

The user would like to be able to view parts of the hierarchy without losing all edges adjacent to nodes outside of the current drawing. Also the layout should take into account that the simple way of drawing trees as used in Figure 1.9 produces rather long edges in the higher levels of the tree. Of course not all applications generate clean trees, and the general case of directed acyclic graphs should be handled appropriately.

Clustering methods and hierarchical layouts as described in Chapter 8 can help giving an overview and navigating in larger graphs.

1.2.5 Social Networks

An interesting interdisciplinary field of research using methods from group theory, statistics, discrete combinatorics and graph-drawing for sociological studies are social networks (Freeman, 1999a; McGrath and Borgatti, 1999; Blythe et al., 1996; Kenis, 1999; Brandes et al., 1999). Simply speaking a social network is obtained by taking a group of people as nodes and inserting edges based on some abstract relationship between these people. This research is motivated by the belief that visualizing social structures and detecting patterns gives insight into how a society works, how individuals interact with society and even why certain societies or individuals are more successful than others. Aspects taken into account can be sociological, economical, demographical, ethnical, or even medical, and are used for various purposes on all scales. Examples are analyses of economic growth in third-world countries, changes in social relationships of married couples over time or the social networks of ethnic minorities and their function.

Figure 1.10 shows a small hand-drawn social network. Each node represents a person, each edge represents a relationship such as acquainted (dotted line) and marriage (triple line). Most nodes and edges are annotated with ages and dates adding a temporal aspect to the picture.

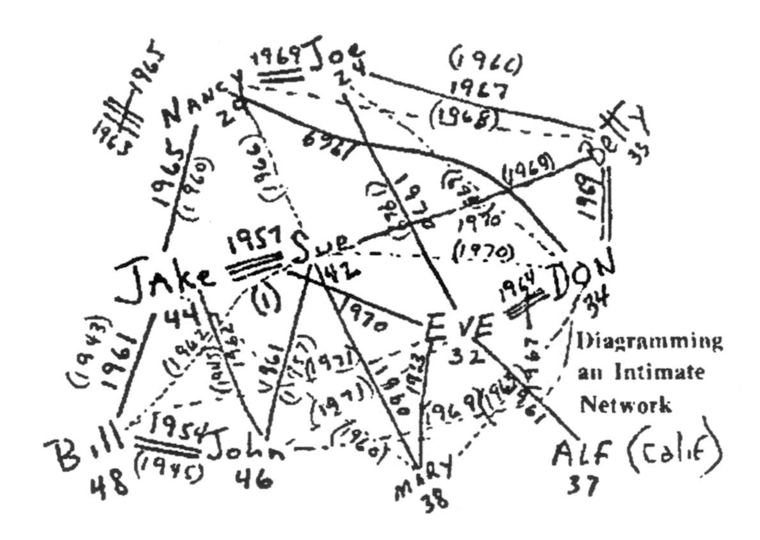

Fig. 1.10. Hand drawn social network.

What makes the graph-drawing problem interesting here is that there is no fixed a-priori idea of what is a good, or even "optimal" layout. The patterns implicitly contained in the description of a workflow diagram for example, which can be used by a human for drawing the graph manually or evaluating the quality of an automatic layout, are not yet known. Indeed, the drawing of the network is intended to aid the user in finding these patterns. Hence improving and judging the quality of layout algorithms is an especially demanding problem.

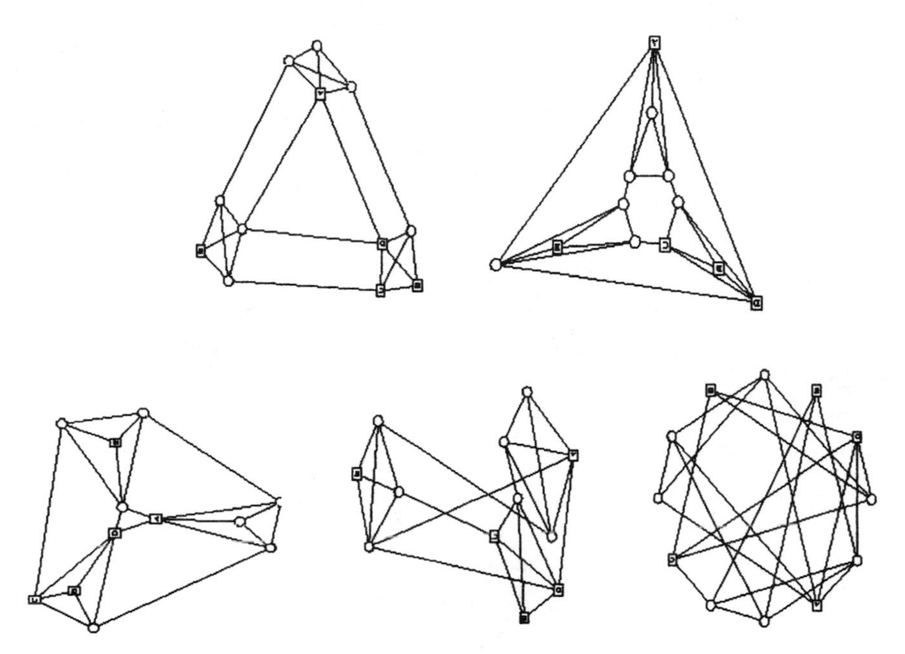

Fig. 1.11. Five drawings of a single network (Blythe et al., 1996).

Figure 1.11 shows five different layouts of the same graph of a social network. Regarding each of the drawings by itself leads to completely different conclusions: the upper left image suggests the existance of three groups of four individuals, whereas the one in the lower right looks rather tigthly interwoven.

It would be desirable to find one layout expressing the key features and patterns of the given graph. When dealing with larger sets of individuals a further help would be to identify dense regions and fold these into aggregate nodes. Such a clustering allows for an analysis of the macroscopic structure without getting lost in the details.

Another idea examined at the moment applies algorithms from molecular visualiziations to social networks (Humphrey et al., 1996; Hermansson and Ojamae, 1994; Freeman, 1999b). Implementations used in chemistry have been available for some time. Besides the capability of dealing with coloured nodes, the most exciting benefit is the ability to visualize dynamically data in form of animations. The application of this feature is immediate, since social networks naturally change over time.

1.2.6 Data Structures

When writing new software typically less than half the required time is used for actually writing new lines of code. Most of the time is spent looking for bugs and errors in the finished code. Therefore special software has been developed to aid this process by tracking runs of other software.

This greatly simplifies testing, since the programmer does not need to trace manually the state of his software. Rather it is possible to output and analyze the contents of data storage during runtime. This is especially important when using error-prone dynamic data structures. It lies in the nature of dynamic data structures that many small objects reside in the storage of the machine. These data elements are linked together via references to other data elements. We thus obtain a directed graph where each data element is a node connected to the data elements it refers to.

Interactive debuggers give the user the opportunity to browse these data structures (Zeller and Lütkehaus, 1996; Isoda et al., 1987). Figure 1.12 shows the output of such a debugger.

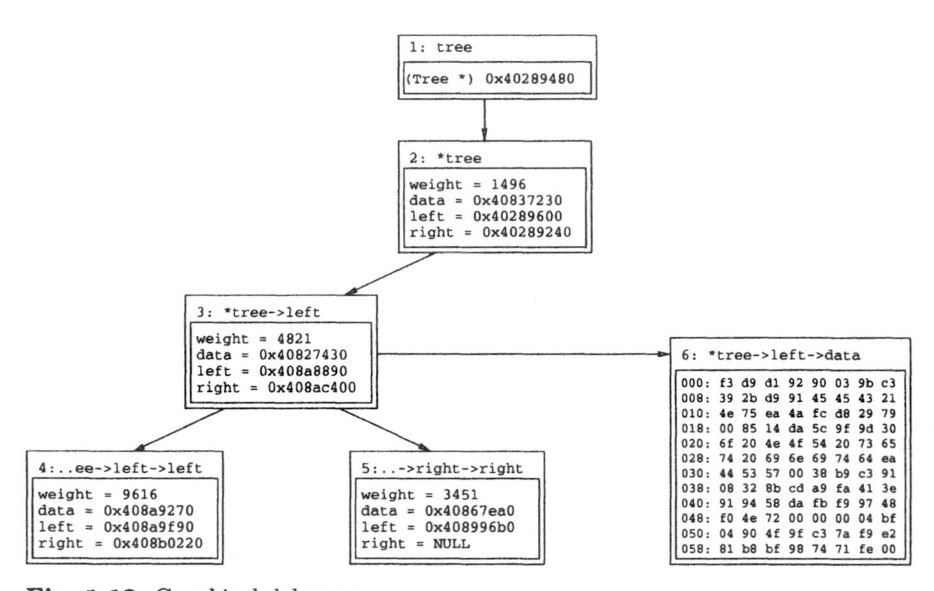

Fig. 1.12. Graphical debugger.

The first problem for any graph layout algorithm is the vastly different sizes of the nodes. Data structures of completely different type and size may refer to each other. Also the structure of the graph is completely arbitrary.

Most demanding are the interactive and the dynamic aspect of the data, the program execution and the user input. The two phases of letting the software run and examining the results alternate throughout the debugging process. In the first phase new data elements can be created and others deleted. During the second phase the user takes the currently displayed subgraph as starting point and continues to fold and unfold parts of the graph along references. The size of the complete graph can easily be in the range of millions, hence drawing the whole thing is not feasible and this kind of incremental navigation essential.

Depending on the number of nodes already visible it is again desirable to retain key features of the current layout when adding or removing nodes to let the user build a mental map. Unfortunately, the more nodes are present, the more important this effect becomes, but also the more difficult to take into account. Various methods have been proposed that either reduce the necessary changes to the layout or limit the adverse effects on the mental map (see Chapter 9).

1.3 How to Draw a Graph

1.3.1 Graph Representations

In the previous section we have seen many examples of graph drawings which differ widely in their appearance. Depending on the application, the basic features of a graph should be drawn in different ways. Nodes may be drawn as dots (as in Figures 1.1 and 1.2), circles (as in Figures 1.3, 1.4, and 1.6), boxes (as in Figures 1.5 and 1.12), a mixture of styles (as in Figures 1.7, 1.8 and 1.11), or not at all as in Figures 1.9 and 1.10 where nodes are represented implicitly by their name labels. Edges may be drawn as straight lines (as in Figures 1.2, 1.5, 1.7, 1.9 1.11 and 1.12), orthogonal polygonal paths (as in Figure 1.8), arbitrary polygonal paths (as in Figures 1.3 and 1.4), or arbitrary curves (as in Figures 1.1, 1.6 and 1.10). The information corresponding to the nodes and edges can be visualized using text labels at various positions in or next to a graph object, different colors (as on a subway map), or other visual elements such as thickness of lines, size of boxes, etc. A graph may be drawn in the plane or in three dimensions. It may be drawn completely, partially, or hierarchically, i.e., clusters are shrunken to a single node which can be expanded on request. We call these (and other) drawing style considerations the *representation* of a graph.

A special case of graphs are *planar graphs*, i.e., graphs that can be drawn in the plane without edge crossings. Planar graphs arise in algorithm animation, CAD systems, circuit schematics, information systems design, and VLSI schematics, for example. We have seen planar drawings of graphs in Figures 1.3, 1.4, 1.5, 1.8 and 1.12. Algorithms for drawing planar graphs are given in Chapter 2.

A subclass of planar graphs are trees (see Figures 1.8 and 1.12, for example). They can be found in algorithm animation, circuit design, visualization of class hierarchies, flowcharts, project management diagrams, and syntax trees. We deal with trees in Chapter 3.

Planar graphs are often drawn orthogonally as in Figure 1.8 because orthogonal drawings usually look much tidier than drawings with arbitrarily curved edges (note that the near-orthogonal drawings in Figures 1.3, 1.4, 1.5 and 1.9 are not bad either). Other applications for orthogonal drawings include architectural floorplan design, network visualization, data base schemas, flow diagrams, entity relationship diagrams, molecular structure diagrams, project management charts, software engineering diagrams, VLSI schematics, and workflow visualization. Algorithms for orthogonal drawings can be found in Chapter 6.

Three-dimensional drawings are suitable to display large and dense graphs such as file system graphs or WWW structure graphs. They are also used in algorithm animation, business graphics, database design, visualization of multimedia documents, software engineering tools, and VLSI schematics. Chapter 7 describes techniques for three-dimensional drawings.

1.3.2 Aesthetics Criteria

Once we have decided upon the representation we face the question of how to actually draw the nodes and edges of a graph. A drawing, or better a *layout*, is a mapping of the nodes and edges into the plane (or into R^3 for three-dimensional drawings). But what distinguishes a good layout from a bad one? The two examples in the Introduction indicate that different applications may require different criteria. For example, drawing a genealogy tree with an algorithm designed to draw subway maps (trees are a subclass of subway graphs, so this should be possible) will most likely produce a highly unsatisfactory picture. That is because an algorithm for drawing a subway graph also uses some general knowledge on subways. For example, customers would be highly confused if stations were drawn at random locations on the map instead of at locations (approximately) corresponding to the locations were they would be found on a real map of the city. On the other hand, genealogists would expect the drawing to reflect the chronology of events; so genealogy trees are usually drawn with nodes ordered by time (left-right or top-down), and the nodes corresponding to a married couple are usually drawn next to each other.

So for every particular graph drawing problem (subway maps, genealogy trees, etc.) we need specially customized algorithms which look beyond the structural properties of the given graphs and also use additional knowledge about the semantics behind the graph structure. This seems to imply that we would need a great number of different graph drawing algorithms, one for each application (in a recent survey of the graph drawing literature, we counted more than 130 different applications). Fortunately, this is not the case. There are a few general concepts and techniques which are powerful enough to cover a wide range of graph drawing applications. Finding a good layout is thus reduced to optimizing one or a few criteria. In applications where the main goal is to produce layouts for human consumption these criteria are appropriately called *aesthetics criteria*, however in applications where graphs are drawn for other purposes as in VLSI schematics, for example, technical criteria such as wire length might be more important than aesthetics criteria. Also, in some cases traditional drawing styles must be honored (it would be very unusual to draw an electrical circuit non-orthogonal, for example).

We now list some of the commonly used aesthetics criteria together with examples of their practical importance. Note that these criteria depend entirely or mostly on the structure of the graph, so algorithms for optimizing these criteria can be devised easily and plugged in as extension to improve the output of a graph drawing system.

Unfortunately, there is no clear ranking among these criteria which would be valid for all possible applications. Figure 1.11 shows that emphasizing some criteria over others can result in very different layouts of the same graph. And each layout has its own virtues, depending on the semantics of the graph it is supposed to depict.

Crossing minimization. If too many edges cross each other, the human eye can not easily find out which nodes are connected by an edge. If a graph can be drawn without edge crossings (such graphs are called *planar*), then this is very often preferable to a drawing with edge crossings. To compute a planar layout is not too difficult, even if we restrict ourselves to straight line edges (see Chapter 2). Crossing minimization is also an important technical criterion. In circuit schematics, wire crossings should be avoided as much as possible to reduce the number of layers.

Bend minimization. This is an important aesthetics criterion for orthogonal layouts because the human eye can much more easily follow an edge with none or only a few bends than an edge wildly zig-zagging through the picture. In VLSI production, bends in wires are potential spots of trouble, so minimizing bends is also an important technical criterion.

Area minimization. Minimizing the area of a layout is again crucial for VLSI schematics, but it is also a general aesthetics criterion: a picture looks much better if the nodes and edges fill the space with homogenous density. There may also be more profane reasons for area minimization, e.g. when producing pocket size maps of a bus network (see Herdeg (1981, page 137, fig. 263)).

Angle maximization. This aesthetics criterion becomes more important nowadays. If a graph is displayed on a video screen with low resolution, it is important that edges are as far apart as possible. In numerics, simulations using finite element nets behave better if the net embeddings have large angles.

Length minimization. In VLSI schematics, edges correspond to wires which carry information from one point on the chip to another. To do this fast, wires should be short.

Symmetries. If a graph contains symmetrical information then it is important to reflect this symmetry in its layout. Technical drawings often contain hidden symmetries. Unfortunately, displaying symmetries is not an easy task.

Clustering. When drawing social networks, parse trees, graphs in CASE tools, or large graphs such as WWW graphs, large networks, or graphs arising in VLSI schematics, then it is necessary to cluster the nodes to reveal some of the graph's structure.

Layered drawings. Organizational charts, ER diagrams, flow diagrams, or graphs in CASE tools usually require a layered layout where node positions are restricted to distinct layers.

1.4 Algorithmic Approaches to Graph Drawing

Having decided on the representation and the right mix of aesthetics criteria, we usually face two problems. Firstly, many of the criteria cannot be optimized efficiently, so we must retreat to approximation algorithms or heuristics. An example is crossing minimization without fixed embedding. And secondly, if we need to optimize several criteria at the same time, we might find this task impossible because the criteria might contradict each other. For example, there are graphs whose optimal orthogonal layout needs an edge of length $\Omega(n^2)$ with $\Omega(n)$ bends (Tamassia et al., 1991).

There are a few powerful techniques which can be used to attack these optimizing problems. They are described in detail in other chapters of this book, so we give here only a short summary.

Planarization. Planar layouts are usually much more appealing than non-planar layouts. Also, in circuit schematics these planarization techniques are important for layer minimization. Unfortunately, in practice many graphs are non-planar. Then one can try to make it planar by removing as few edges as possible (this is an $\mathcal{NP}$-complete problem) or by removing those edges whose insertion would afterwards create the least number of crossings. The problem of crossing minimization is in general NP-hard, but some heuristics for planarization yield acceptable results. These techniques are discussed in Chapter 2.

Force-directed methods. In Chapter 4 we describe energy based layout algorithms. These algorithm interpret a graph as a physical system with forces between the nodes and then try to minimize the energy of the system to obtain a nice drawing. Such algorithms are used for drawing arbitrary (sparse) networks such as flow charts, program planning graphs, telephone call graphs, etc. They can also be applied to clustered layouts.

Sugiyama-like methods. The most widely used algorithms for drawing layered graphs are the Sugiyama type algorithms (see Chapter 5). They produce layered layouts while also trying to minimize the number of crossings or the area of the layout.

Flow methods. Bend minimization can efficiently be solved by reduction to a network flow problem (see Chapter 6), at least if the topology of the embedding is fixed. The same techniques can be used to maximize angles between edges (see Chapter 6).

Interactive drawings. The methods above are good for drawing static graphs. However, interactive applications such as the visualization of debugging tools, document retrieval, entity relationship modules, VLSI schematics, and WWW graphs require the display of graphs which change over time. Techniques for interactive graph drawing are described in Chapter 9.

Labeling. Another important aspect of graph drawing, for example in drawing maps, state diagrams, or engineering diagrams, is *labeling*, i.e., naming nodes and edges in the drawing. Methods for labeling are discussed in Chapter 10.

These aesthetics and efficiency criteria stand in contrast to more intuitive criteria concerning the semantics and intended meanings of graphs. As the example of Figure 1.5 shows, the semantics and the structure of a graph can give very different hints for the layout. However, completely disregarding aspects such as the length of edges, a purely technical term, can lead to rather unbalanced and difficult to trace layouts as in Figure 1.6.

It can be speculated that the lack of layout algorithms respecting the semantics of graphs and therefore being more capable of creating a drawing that is informative as well as "favourable to the eye" lies in the nature of the problem. The manually arranged drawings were all created by individuals with an intimate knowledge of the semantics. Disregarding the aesthetic aspect concerning aesthetics as in artwork, it remains a challenge to identify and, more importantly, formalize aesthetics criteria taking the semantic aspects into account. The word graphs in Figures 1.3 and 1.4 using the additional information "time" is but a small step in this direction.

1.5 Conclusion

Graph drawing applications are so manifold that we could only show a few examples in this chapter. They mainly come from applications within computer science — not too surprising since both authors work in that field. We found these examples just by talking to the people in the offices around the corner. This shows that graph drawing problems appear in many places, indicating that its study is an important task.

The examples also show that graph drawing is not a single well-defined problem but an art, namely the art of describing what a *nice* drawing of a graph means in the context of a particular application. In Section 1.3.2 we have seen a list (which should not be understood to be complete) of aesthetics criteria whose optimization can lead to acceptable layouts. Most of the following chapters of this book give algorithms for optimizing these criteria.

However, these algorithms should not be considered as the final solution to the graph drawing problem, i.e., even if we put them all together in a tool box we cannot expect to always find at least one algorithm perfectly suited for a particular application problem at hand. More often than not, the output from an automatic graph drawing tool can 'easily' be improved manually (so far, the winning drawings in the annual graph drawing competition (Eades and Marks, 1995, 1996; Eades et al., 1996, 1997c, 1998) have never been produced fully automatically, except when explicitly demanded). The reason is that graphs in an application usually have semantics known to the people working with the graph but not known to the graph drawing tool (because it was designed to draw graphs according to some aesthetics criteria, not according to some particular graph semantics). Therefore, for most applications the standard algorithms for optimizing standard aesthetics criteria (these algorithms are covered by this book) must be refined to include problem specific aesthetics criteria (as given by the graph semantics). On the other hand, there are many commercial and non-commercial graph drawing tools available, so chances are good that for a given application one might find some tool which produces at least reasonable drawings, but one should not expect miracles. In Appendix A, we give a short summary of some of the better (in our opinion) tools.

2. Drawing Planar Graphs

René Weiskircher

2.1 Introduction

When we want to draw a graph to make the information contained in its structure easily accessible, it is highly desirable to have a drawing with as few edge crossings as possible (Purchase et al., 1997; Purchase, 1997). The class of graphs that can be drawn with no crossings at all is the class of *planar graphs*. Algorithms for drawing planar graphs are the main subject of this chapter.

We first give some necessary definitions and basic properties of planar graphs. In section 2.3, we take a closer look at two linear time algorithms for testing if a graph is planar. When a graph is not planar but we want to apply an algorithm for drawing planar graphs, we can transform the graph into a similar planar graph; Section 2.4 gives an overview of methods to do so.

Most drawing algorithms presented in this chapter require a 2-connected planar graph as input. If a planar graph does not have this property, we can add edges to make it 2-connected and planar. Section 2.5 describes ways to accomplish this. The following sections describe drawing algorithms for planar graphs. Section 2.6 treats the generation of convex straight-line representations, while section 2.7 gives an overview of some algorithms that use a special ordering of the vertices of a graph called a *canonical ordering.*

2.2 What Is a Planar Graph?

To define what we mean by the term *planar graph* we first have to define what is meant by the term *planar representation.*

Definition 2.1 (Planar Representation). *A* planar representation *D of a graph $G = (V, E)$ is a mapping of the vertices in V to distinct points in the plane and of the edges in E to open Jordan curves with the following properties:*

- *For all edges $e \in E$, the representation of edge $e = (v_1, v_2)$ connects the representation of v_1 with the representation of v_2.*
- *The representations of two disjoint edges $e_1 = (v_1, v_2)$ and $e_2 = (v_3, v_4)$ have no common points except at common endpoints.*
- *The representation of edge $e = (v_1, v_2)$ does not contain the representation of $v_3 \in V$ with $v_3 \notin \{v_1, v_2\}$.*

M. Kaufmann and D. Wagner (Eds.): Drawing Graphs, LNCS 2025, pp. 23-45, 2001.

If D is a planar representation, the set $\mathbb{R}^2 - D$ is open and its regions are called the *faces* of D. Since D is bounded, exactly one of the faces of D is unbounded. This face is called the *outer face* of D.

Using the definition of planar representations, it is easy to define the term *planar graph.*

Definition 2.2 (Planar Graph). *A graph G is planar if and only if there exists a planar representation of G.*

There is an infinite number of different planar representations of a planar graph. We can define a finite number of equivalence classes of planar representations of the same graph using the term *planar embedding*.

Definition 2.3 (Planar Embedding). *Two representations D_1 and D_2 of a planar graph G realize the same planar embedding of G, if and only if the following two conditions hold:*

- *The cycles of G that bound the faces of D_1 are the same cycles that bound the faces of D_2.*
- *The outer face in D_1 is bounded by the same cycle of G as in D_2.*

The definition of a planar graph above is very simple but it is a geometric definition. Since the set of all planar representations of a graph is infinite and uncountable, it is not immediately clear how to test a graph for planarity. Kuratowski found a combinatorial description of planar graphs but to present this description, we have to define the *subdivision* of a graph.

Definition 2.4 (Subdivision). *A* subdivision *of a graph $G = (V, E)$ is a graph $G' = (V', E')$ that can be obtained from G by a sequence of split operations where we insert a new vertex u and replace an edge $e = (v_1, v_2)$ by the two edges $e_1 = (v_1, u)$ and $e_2 = (u, v_2)$.*

Thus, a subdivision of a graph is another graph where some edges of the original graph have been replaced by paths. Planar graphs are now characterized by the following:

Theorem 2.5. *A graph G is planar if and only if it does not contain a subdivision of K_5 (the complete graph with 5 vertices, see Figure 2.1(a)) or $K_{3,3}$ (the complete bipartite graph with 3 vertices in each set, see Figure 2.1(b)).*

If a graph G is directed (each edge is an ordered pair of vertices), we can define a more restricted class of planar graphs, the *upward planar graphs*. First we define the term *upward representation.*

Definition 2.6 (Upward Representation). *Let $G = (V, E)$ with $E \subseteq V \times V$ be a directed graph. A representation of G is called* upward *if the representation of every edge (u, v) is monotonically nondecreasing in the y-direction when traced from u to v.*

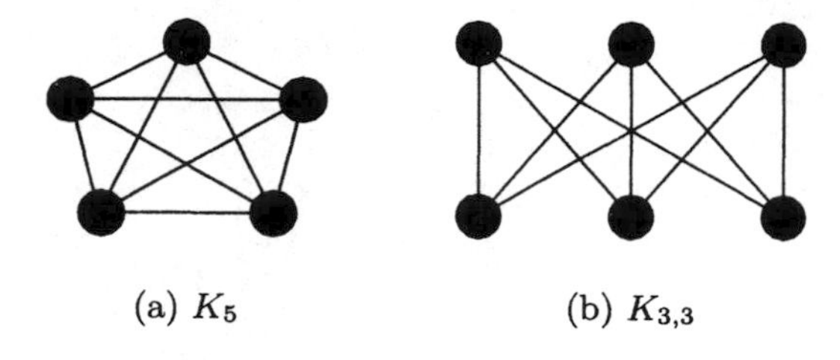

Fig. 2.1. The basic non-planar graphs.

We use this definition to define *upward graphs* and *upward planar graphs.*

Definition 2.7 (Upward Graph). *A directed graph is* upward *if and only if it admits an upward representation.*

Definition 2.8 (Upward Planar Graph). *A directed graph is* upward planar *if and only if it admits an upward and planar representation.*

It is possible to test in linear time whether a directed graph admits an upward representation (because only acyclic graphs admit such a representation) and, as we will see in the next section, we can test in linear time, whether a graph admits a planar representation. But testing whether a graph admits an upward planar representation is $\mathcal{NP}$-complete for general graphs (Garg and Tamassia, 1995a).

A survey about upward planarity testing can be found in Garg and Tamassia (1995a). We will not treat the topic in this chapter, but algorithms for drawing upward graphs can be found in Di Battista and Tamassia (1988).

2.3 Planarity Testing

The first algorithm for testing whether a given graph is planar was developed by Auslander and Parter (1961) and Goldstein (1963). Hopcroft and Tarjan (1974) improved this result to linear running time. Another linear time algorithm for planarity testing was developed by Lempel, Even and Cederbaum (Lempel et al., 1967) and Booth and Lueker (1976). We will only give a short overview of the two linear time algorithms.

2.3.1 The Algorithm of Hopcroft and Tarjan

This overview of the algorithm follows that of Mutzel (1994). In principle, the algorithm works as follows: Search for a cycle C whose removal disconnects the graph. Then check recursively whether the graphs that are constructed by merging the connected components of $G - C$ and the cycle C are planar. In a second step, combine the computed embeddings for the components to get a planar embedding of the whole graph, if possible.

The algorithm needs a depth first search tree $G' = (V, T, B)$, where V is the set of DFS numbers of the vertices in G, T is the set of *tree edges* of the depth first search tree and B the set of *back edges* (for DFS trees, see Mehlhorn (1984)). We assume that G is 2-connected (this is not a restriction, because a graph is planar if and only if all its 2-connected components are planar).

Let C be a *spine cycle* of G, which is a cycle consisting of a path of tree edges starting at the root (vertex 1) of the DFS tree followed by a single back edge back to the root vertex. Because G is 2-connected, such a cycle must exist. We assume that removing all edges of C splits G into the subgraphs $G_1, G_2, \ldots, G_k$. For $1 \leq i \leq k$, we define the graph G'_i as the graph G_i together with the cycle C and all edges in G between a vertex in G_i and a vertex on C. First, we recursively check whether each G'_i is planar and compute a planar embedding for it. Planar embeddings are equivalence classes of planar representations that describe the topology of the representation but not the length and shape of edges or the position of vertices (see definition 2.3).

The planar embeddings of the G'_i must have all edges and vertices of C on the outer face. Now we assume that we have found a suitable embedding for each G'_i. We must test whether we can combine these embeddings to a planar embedding of G. The reason why this may fail is that each G_i shares at least two vertices with C. Figure 2.2 shows how this fact can make it impossible to embed two graphs G_i and G_j on the same side of C. We say that the two graphs *interlace*.

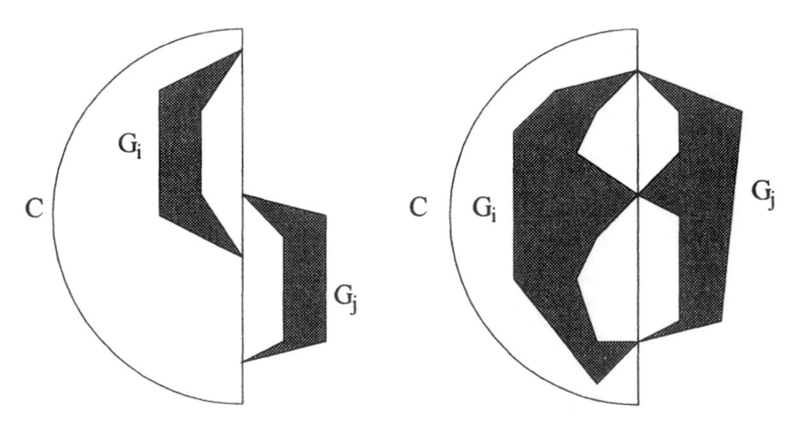

Fig. 2.2. Interlacing graphs G_i and G_j that can't be embedded on the same side of C.

To test whether there is an assignment of the G_i's to the two sides of C so that the resulting representation is planar, we build the *interlace graph* I_G. This graph has one vertex for each G_i and two vertices are adjacent if and only if they interlace. We can only draw G planar if I_G is bipartite. If there is an embedding with the necessary properties for each G_i' and the interlace graph is bipartite, we know that G is planar and we can construct a planar embedding for it.

2.3.2 The Algorithm of Lempel, Even, and Cederbaum

A vertex-based method for planarity testing is the test developed by Lempel et al. (1967); Even (1979). We say that this test is vertex-based because we add the vertices one by one to a special data structure and check after each step if the information seen so far proves that the graph is non-planar. This test can be implemented in linear time (Booth and Lueker, 1976), like the algorithm of Hopcroft and Tarjan discussed above.

The input of the algorithm is again a 2-connected graph $G = (V, E)$. We assume that $V = \{v_1, v_2, \ldots, v_n\}$ where the numbering of the vertices is an st-numbering as defined below.

Definition 2.9 (st-numbering). *Given an edge $\{s, t\}$ in a graph $G = (V, E)$ with n vertices, an st-numbering is a function $g : V \to \{1, \ldots, n\}$, such that*

- $g(s) = 1,\ g(t) = n$
- $\forall v \in V \setminus \{s, t\}\ \exists u, w \in V \quad (\{u, v\}, \{v, w\} \in E \ \wedge\ g(u) < g(v) < g(w))$

Lempel, Even and Cederbaum showed that for every edge $\{s, t\}$ in a graph G, there exists an st-numbering if and only if G is 2-connected. A linear time algorithm to find it is given in Even (1979).

We define G_k as the subgraph of G induced by $\{v_1, \ldots, v_k\}$. This graph is extended to a graph B_k as follows. For each edge $(u, v) \in E$ with u in G_k and v not in G_k, the graph B_k has a new *virtual vertex* and an edge connecting u to this vertex. So there may be several virtual vertices in B_k that correspond to the same vertex in G. The idea of the algorithm is to check whether we can identify the virtual vertices corresponding to the same vertex in G without losing the planarity property.

If G is planar, B_k has a planar embedding where each vertex v_i for $1 \leq i \leq k$ is drawn on y-coordinate i, all virtual vertices are placed on y-coordinate $k+1$ and all edges are disjoint y-monotone curves (which means that they are only intersected at most once by any horizontal line). Such a representation is called a *bush form*. Figure 2.3 shows an example for a bush form.

Let v_i be a vertex in a bush form. If the removal of v_i disconnects the bush form, we call it a *cut vertex*. Let B' be the bush-form after the removal of v_i. The *split-components* of v_i are those connected components of B', where the indices of all vertices are greater than i. Now consider the bush form in

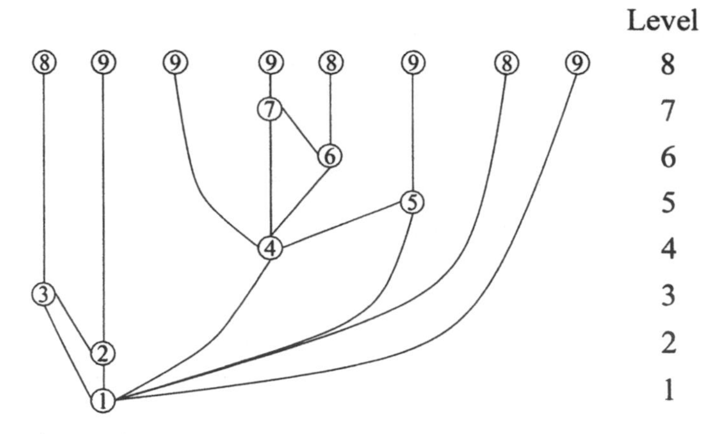

Fig. 2.3. A bush form.

Figure 2.3. Since the labels of the vertices are their *st*-numbers, this is bush form B_7. When we want to draw B_8, we must first transform B_7 so that all virtual vertices with label 8 form a consecutive sequence on level 8. This can be done by flipping around the split component of vertex 1 which includes the vertices 2 and 3, so that the virtual vertices labeled 8 and 9 in the split component swap their positions. We also have to move the virtual vertex labeled 9 adjacent to vertex 4 to the right and flip the split component of vertex 4 with the vertices 6 and 7. The resulting graph is shown in Figure 2.4.

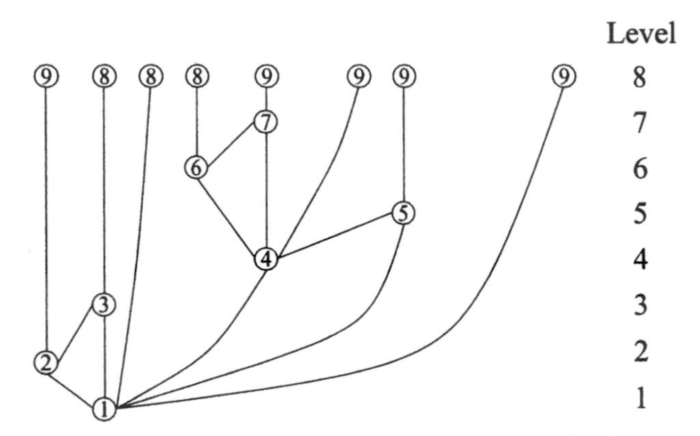

Fig. 2.4. The bush form from Figure 2.3 has been transformed so that all vertices labeled 8 form a consecutive sequence.

If v is a split vertex of a bush form (which means that removing v disconnects the bush form), then we can freely permute the split components which have vertices with higher *st*-number than v and we can flip each individual component. There may be several possible ways of producing a consecutive sequence of the vertices labeled $k+1$ and since not all may eventually lead to a planar representation of G, we have to keep track of all of them. This can be done in linear time using a data structure called *PQ-tree* proposed by Booth and Lueker (1976). If it is not possible to make the vertices labeled $k+1$ consecutive, we know that the graph is not planar. Otherwise, the algorithm will produce a planar embedding of the graph. In Mehlhorn and Näher (1999) a detailed description of the complete algorithm can be found.

2.4 How to Make a Graph Planar

There are many popular algorithms for drawing planar graphs and they produce a great variety of styles of representations. Therefore, it makes sense to transform a non-planar graph into a similar planar graph, apply a graph drawing algorithm for planar graphs to the result and then modify the resulting representation so that it becomes a representation of the original non-planar graph. A survey of methods for doing this can be found in Liebers (1996).

A quite drastic way of making a graph planar is to delete vertices. This method is not used very much in graph drawing, because deleting vertices changes a graph considerably. The problem of deciding for an integer k if we can make a non-planar graph planar by deleting at most k vertices is $\mathcal{NP}$-complete (Lewis and Yannakakis, 1980).

Another way of making a graph planar is to split vertices. This is a rather complex operation, so we will give the formal definition from Liebers (1996).

Definition 2.10 (Vertex Splitting). *Let $G = (V, E)$ and $G' = (V', E')$ be two graphs. Then we say G' has been obtained by splitting vertex v of G into the vertices v_1 and v_2 if the following conditions are satisfied:*

$$\begin{aligned} V &= (V' \backslash \{v_1, v_2\}) \cup \{v\} \\ E &= (E' \backslash \{uv_i \,|\, u \in V' \text{ and } uv_i \in E' \text{ for } i \in \{1,2\}\} \\ &\quad \cup \{uv \,|\, u \in V \backslash \{v\} \text{ and } (uv_1 \in E' \text{ or } uv_2 \in E')\} \end{aligned}$$

Splitting a vertex is also a drastic operation and is not commonly used in graph drawing to planarize graphs. Testing whether a non-planar graph can be made planar by at most k vertex-splitting operations is $\mathcal{NP}$-complete (Faria et al., 1998).

Two more commonly used ways of transforming a non-planar graph into a planar graph are the insertion of new vertices and the deletion of edges.

2.4.1 Inserting Vertices

Assume we have a non-planar graph G and a representation D of G with k crossings. Then we can transform G into a planar graph G' in the following way: Let $e = (u, v)$ and $f = (x, y)$ be two edges that cross in D. Then we can add a new vertex v_c to G, remove the edges e and f from G and insert the four new edges $e_1 = (u, v_c)$, $e_2 = (v_c, v)$, $f_1 = (x, v_c)$ and $f_2 = (v_c, y)$. This is equivalent to replacing the crossing in D between e and f by the new vertex v_c. If we do this for every pair of crossing edges, we will transform G into a planar graph G' and D into a planar representation D' of G'.

Since the graph G' is planar, we can draw it by using any algorithm for drawing planar graphs. If D'' is the resulting representation, we can transform this representation into a representation of the original non-planar graph G by replacing all introduced vertices by crossings again. Since we want to have as few crossings as possible in the resulting representation, we want to introduce as few new vertices as possible.

The minimum number of vertices we have to insert is equal to the minimum number of crossings in any representation of G. But the problem of deciding for a graph G whether it can be drawn with at most k crossings is $\mathcal{NP}$-complete (Garey and Johnson, 1983). The only known heuristics for inserting few vertices to construct a planar graph are the algorithms for drawing non-planar graphs. By inserting vertices at every crossing of the representation produced we get a planar graph.

2.4.2 Deleting Edges

If G is a non-planar graph, there is a non-empty subgraph of G that is planar. In particualr, each spanning tree of G is planar, since every graph without cycles is planar. So we can derive a planar graph from a non-planar graph by deleting a subset of its edges. But the problem of deciding for a non-planar graph $G = (V, E)$ and a number $k < |E|$ if there is a planar subgraph with at least k edges is $\mathcal{NP}$-complete. This was independently shown by Liu and Geldmacher (1977), Yannakakis (1978) and Watanabe et al. (1983). The associated $\mathcal{NP}$-hard maximization problem is to find a planar subgraph of a G with the property that there exists no other planar subgraph that has more edges. This problem is called the *maximum planar subgraph problem*. The problem of finding a planar subgraph, which is not a proper subgraph of another planar subgraph of G is called the *maximal planar subgraph problem* and is solvable in polynomial time.

Definition 2.11 (Maximal Planar Subgraph). *A maximal planar subgraph of a graph $G = (V, E)$ is a subgraph $G' = (V, E')$ of G in which there exists no edge in $E - E'$ that can be added to G' without losing planarity.*

One approach to solving this problem is to start with the subgraph $G_1 = (V, \emptyset)$ of G and to test for each edge if we can add it to the current solution without losing planarity. If we can do that, we add the edge and proceed to the next edge. Since we have to perform a planarity test for each edge of the graph and such a test can be implemented in linear time, this algorithm has a running time of $O(n \cdot m)$ where n is the number of vertices in the graph and m the number of edges.

Di Battista and Tamassia developed a data structure called SPQR-tree, which can be used for decomposing a planar 2-connected graph into its 3-connected components and for fast online planarity testing (Di Battista and Tamassia, 1989; Di Battista and Tamassia, 1990; Di Battista and Tamassia, 1996). Using this data structure, they were able to develop an algorithm for finding a maximal planar subgraph in $O(m \log n)$ running time. There is also an algorithm with the same asymptotic running time developed by Cai et al. (1993) which is based on the planarity testing algorithm in Hopcroft and Tarjan (1974).

La Poutré (Poutré, 1994) proposed an algorithm for incremental planarity testing yielding an algorithm for the maximal planar subgraph problem running in time $O(n + m \cdot \alpha(m, n))$ where $\alpha(m, n)$ is the inverse of the Ackermann function and grows very slowly. There are even linear time algorithms for the problem, by Djidjev (1995) and by Hsu (1995), which has the best asymptotic running time possible for solving the maximal planar subgraph problem.

A heuristic for the maximum planar subgraph problem is the Deltahedron heuristic (Foulds and Robinson, 1978; Foulds et al., 1985). This heuristic starts with the complete graph on 4 vertices (tetrahedron) as the initial planar subgraph and then places the remaining vertices into the faces of the current planar subgraph. The sequence of the vertices depends on a chosen weight function. Leung (1992) proposed a generalization of this method. The current planar subgraph has only triangular faces and in each step, we add a single vertex and 3 edges or we add 3 vertices and 9 edges. A list of other heuristics can be found in Liebers (1996).

Jünger and Mutzel (Mutzel, 1994; Jünger and Mutzel, 1996) proposed a branch and cut algorithm for solving the maximum planar subgraph problem based on an integer linear program that excludes the presence of subdivisions of $K_{3,3}$ and K_5 in the solution graph. The advantage of a branch and cut algorithm is that it either finds an optimum solution together with a proof of optimality or finds a solution together with an upper bound on the value of the optimum solution. For problems of moderate size (about 50 vertices), their approach finds an optimal solution in most cases.

2.5 How to Make a Planar Graph 2-Connected Planar

Many graph drawing algorithms only work for 2-connected or 3-connected graphs. This is true for most algorithms presented in this chapter. Therefore if we want to draw a graph which does not have the necessary connectivity property for applying a specific graph drawing algorithm, we can increase its connectivity by adding new edges (*Augmentation*). After a representation of the augmented graph has been computed, we remove the representations of the additional edges to get a representation of the original graph. Since we do not want to change the graph too much, we want to add a minimum number of edges in the augmentation step.

The planar augmentation problem is the problem of adding a minimum number of edges to a given planar graph so that the resulting graph is 2-connected and planar. Kant and Bodlaender (1991) introduced this problem and showed that it is $\mathcal{NP}$-hard. They have also given a 2-approximation algorithm running in time $O(n \log n)$ and a $\frac{3}{2}$-approximation algorithm with running time $O(n^2 \log n)$, where n is the number of vertices in the graph. However, the $\frac{3}{2}$-approximation algorithm is not correct because there are problem instances where it computes only a 2-approximation.[1]

Fialko and Mutzel developed a $\frac{5}{3}$-approximation algorithm (Fialko and Mutzel, 1998). The running time of the algorithm is $O(n^2 T)$ where T is the amortized time bound per insertion operation in incremental planarity testing. Using the algorithm in Poutré (1994), a running time of $O(n^2 \alpha(k,n))$ can be achieved where α is the inverse Ackermann function and k is $O(n^2)$. Recently, the algorithm has been improved by Mutzel (private communication) to guarantee a $\frac{3}{2}$-approximation.

The $\frac{5}{3}$-approximation algorithm works on the *block tree* of the graph we want to make 2-connected. The block tree has two types of vertices: The *b-vertices* correspond to the maximal 2-connected components of the graph and the *c-vertices* to the cut vertices (as already mentioned, the removal of a cut vertex disconnects the graph). We have an edge between a c-vertex and a b-vertex if and only if the corresponding cut vertex belongs to the 2-connected component represented by the b-vertex. The idea is now to insert edges, merging paths of the block tree into single blocks until the tree has only one vertex and is thus 2-connected.

A crucial role in the algorithm is played by the *pendants* of the block tree, which are b-vertices with degree one. The algorithm connects pendants via edges if possible and otherwise connects pendants to non-pendant blocks. To achieve the approximation ratio, pendants are combined to form larger structures that are called *labels.* The algorithm looks at these labels in the order of decreasing number of pendants and tries to connect the pendants of two labels by introducing new edges. Inserting edges that connect the pendants of two labels is called a *label matching.*

The algorithm prefers certain matchings, but because the resulting graph has to be planar, not all of the preferred label matchings can be realized. Some labels cannot be matched at all and so the algorithm introduces edges that connect pendants of the same label and an additional edge from one of the pendants to a non-pendant vertex outside the label.

The approximation guarantee of the algorithm is tight which means that graphs exist for which the number of added edges is $\frac{5}{3}$ of the optimum number. On realistic instances, the algorithm performs very well and very often

[1] Goos Kant, personal communication (1994).

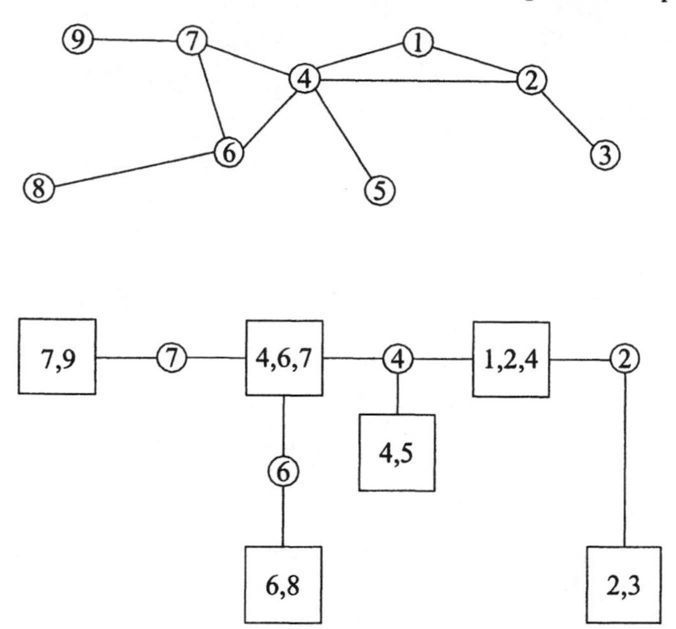

Fig. 2.5. A graph and its block tree.

finds a solution that uses at most one edge more than the optimum solution. This has been tested using a branch and cut algorithm for the planar augmentation problem developed by Mutzel (1995) which is able to solve instances of realistic size optimally.

2.6 Convex Representations

Some planar graphs can be drawn in such a way that all cycles that bound faces are drawn as convex polygons. An example for such a drawing is given in Figure 2.6. Such a representation is only possible if all face boundaries of the graph are simple cycles. Thus, a graph that is not 2-connected cannot have a convex representation. It has been shown that such a convex representation exists for all 3-connected graphs (Tutte, 1960) and Tutte gave an algorithm for producing representations of 3-connected graphs which involves solving $O(n)$ linear equations, where n is the number of vertices in the graph (Tutte, 1963).

Nishizeki and Chiba (1988) developed an algorithm for producing a convex representation of a 2-connected planar graph (if it admits a convex representation) in linear time. The drawing algorithm is based on the proof of Tutte's result given by Thomassen (1980). The testing algorithm works by dividing a 2-connected planar graph into 3-connected components as described

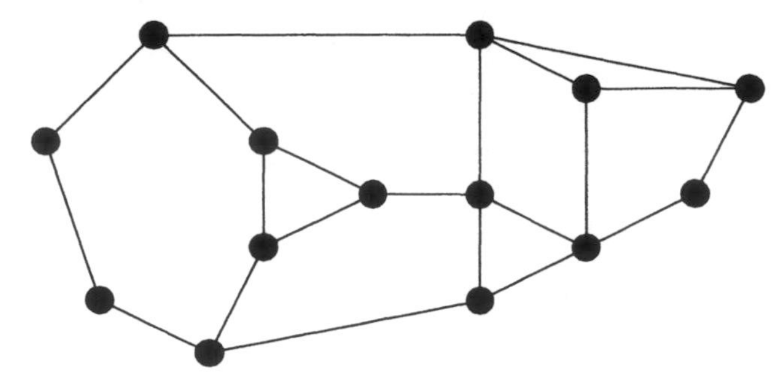

Fig. 2.6. A convex drawing of a graph.

in Hopcroft and Tarjan (1973) and testing planarity of a special graph constructed from the original graph using the algorithm described in Hopcroft and Tarjan (1974).

To give a short outline of the drawing algorithm, we have to define what we mean by the term *extendible polygonal representation* of a *face cycle* of a graph G. A face cycle is a cycle in the graph that is the boundary of a face (region) of a planar representation of the graph. A convex representation S^* of a face cycle S is a convex polygon in which all vertices of S are drawn on the boundary of S^* and each apex of S^* is occupied by the representation of a vertex on S. The polygonal representation S^* of S is called extendible if there is a convex representation of G, in which S^* is the outer face of the representation.

Thomassen (1980) showed that the polygonal representation S^* of S is extendible if and only if the following conditions hold.

1. For each vertex v of G not on S, there are three vertex disjoint paths from v to vertices on S.
2. There are no connected components C in $G - S$, in which all vertices in S adjacent to a vertex in C are located on the same straight segment P of S^*.
3. There is no edge that connects two vertices on a straight segment of S^*.
4. Any cycle in G that does not share an edge with S has at least three vertices with degree greater than 2.

If the conditions above are satisfied, the following algorithm will correctly compute a convex representation of G.

The input of the algorithm `convex-draw` is a triple consisting of the graph G, a face cycle S of G and an extendible polygonal representation S^* of S.

Algorithm `convex-draw` (G, S, S^*):

1. We assume that G has more than 3 vertices, and some of them do not belong to S, otherwise, our problem is already solved. Select an arbitrary

apex vertex v of S^* and set $G' = G - v$. Divide G' into the blocks $B_1, \ldots, B_p$ as shown in Figure 2.7 according to the cut vertices on S^*.

2. Draw each B_i convex applying the following procedure:
   a) Let v_i and v_{i+1} be the cut vertices that split B_i from the rest of G'. Then these two vertices have already a fixed position, because they belong to S. These vertices also belong to the outer facial cycle S_i of B_i. We now draw all the vertices of S_i that do not belong to S on a convex polygon S_i^* inside the triangle given by the vertices v, v_i and v_{i+1}. Each apex of the polygon is occupied by a vertex of S_i which is in G adjacent to v. The other vertices of S_i are drawn on the straight line segments of S_i^*.
   b) Recursively call the procedure `convex-draw` for all blocks with the arguments (B_i, S_i, S_i^*).

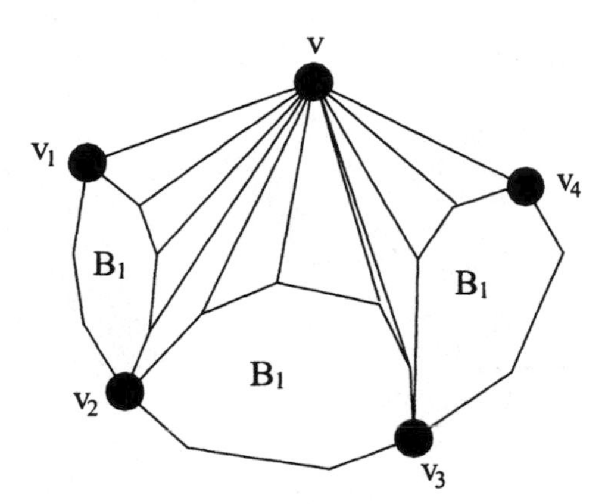

Fig. 2.7. Recursive computation of a convex representation.

The algorithm for testing whether a 2-connected planar graph has a convex representation relies on determining the *separation pairs* of the graph. A separation pair is a pair of vertices whose removal disconnects the graph.

Definition 2.12 (Separation Pair). *A separation pair of a graph is a pair of vertices $\{x, y\} \subset V$ so that there exist two subgraphs $G_1 = (V_1, E_1)$ and $G_2 = (V_2, E_2)$ which satisfy the following conditions:*

1. $V = V_1 \cup V_2$, $V_1 \cap V_2 = \{x, y\}$
2. $E = E_1 \cup E_2$, $E_1 \cap E_2 = \emptyset$, $|E_1| \geq 2$, $|E_2| \geq 2$.

A separation pair is called *prime separation pair* if at least one of the graphs G_1 and G_2 is either 2-connected or is a subdivision of an edge joining two vertices with degree greater than two.

In the algorithm for testing convex planarity, the *forbidden separation pairs (FSPs)* and the *critical separation pairs* (CSPs) play a crucial role.

Definition 2.13 (Forbidden Separation Pair). *A prime separation pair is called a* forbidden separation pair *(FSP) if it has at least four split components or three split components none of which is a path.*

If a graph has an FSP, then it has no convex representation of the graph. Figure 2.8 shows two examples of FSPs. Neither of these graphs has a convex drawing.

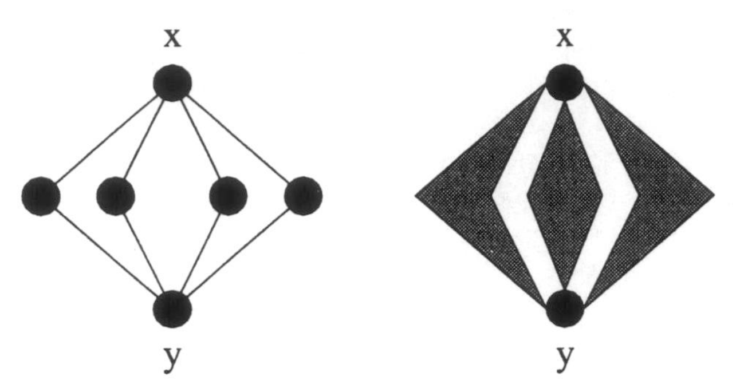

Fig. 2.8. Two examples of FSPs $\{x, y\}$. The shaded regions in the drawing on the right are subgraphs.

Definition 2.14 (Critical Separation Pair). *A prime separation pair is a critical separation pair (CSP) if it has 3 split components of which at least one is a path or if it has two split components of which none is a path.*

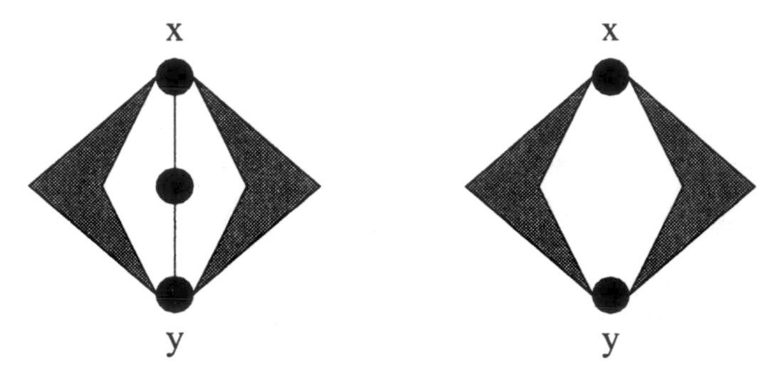

Fig. 2.9. Two examples for CSPs $\{x, y\}$. The shaded regions in the drawings are subgraphs.

The algorithm `convex-test` works as follows:

1. Find all separation pairs of G by the linear time algorithm described in Hopcroft and Tarjan (1973) for finding 3-connected components. Determine the set F of FSPs and the set C of CSPs.

2. If $F \neq \emptyset$, then there is no convex representation of G. If both F and C are empty, we can produce a convex representation by choosing any face cycle of G as the cycle S that starts the computation. If there is exactly one pair in C, we choose S as a cycle with the CSP on it, depending on the structure of the split components. If there is more than one pair in the set C, we go to the next step.
3. We transform each CSP with three split components by removing one component that is a path. Then we connect all vertices of all pairs in C to a new vertex v_S and check if the resulting graph G' is planar. If this is not the case, we know that there is no convex representation of G. Otherwise, let Z be any planar representation of G'. Let S be the face cycle that surrounds v_S in Z after deleting all edges incident to v_S. Then we know that there is a convex representation of G if we choose S as the start cycle for the recursive computation of the algorithm `convex-draw`. This is the case because all CSPs belong to S.

2.7 Methods Based on Canonical Orderings

There are several methods for drawing a planar graph that rely on a special ordering of the vertices which is often called the *canonical ordering*. The vertices are ordered and successively added in this special order to a data structure that describes a representation of the graph. In some of these algorithms, the vertices are added one by one while in others a set of vertices can be added in one step. Before the execution of each step, the data structure always describes a representation of the subgraph induced by the vertices that have already been added.

The vertex orderings used in all these algorithms and the algorithms themselves have several common properties:

1. The ordering is defined by some embedding of the graph.
2. The ordering of the vertices defines an ordered partition $V_1, V_2, \ldots, V_k$ of the vertices in the vertex set V of the graph. The union of the V_i is V, each V_i has at least one vertex and the V_i are pairwise disjoint.
3. In step i of the algorithm, the vertices in V_i together with the edges that connect them to the vertices in $V_1 \cup V_2 \cup \ldots \cup V_{i-1}$ and the edges between the vertices in V_i are added to the data structure that defines the representation.
4. The set V_1 has at least 2 elements and there is at least one edge in the subgraph induced by V_1 which is on the outer face of every representation D_i.
5. Let S_i be the data structure after inserting the vertices in V_i and let D_i be the corresponding representation. Then D_i is the representation of a 2-connected graph where all vertices adjacent to vertices in $V_{i+1} \cup \ldots \cup V_k$ are on the outer face of the representation.

The last point is not true for the algorithm proposed by Schnyder (1990) because in this algorithm, the vertices are inserted inside the triangle given by the three vertices in V_1. This algorithm (described in subsection 2.7.2) is quite different from all others treated in this section, because it computes three barycentric coordinates for each vertex before computing the x- and y-coordinates.

2.7.1 The Algorithm of De Fraysseix, Pach, and Pollack

The first algorithm using a canonical ordering for drawing planar graphs with straight edges using polynomial area was described by de Fraysseix et al. (1990). The algorithm draws a planar triangulated graph on a grid of size $(2n - 4) \times (n - 2)$, where n is the number of vertices in the graph. The running time of the algorithm is $O(n \log n)$. In the same paper, the authors give a linear time and space algorithm for adding edges to a planar connected graph to produce a planar triangulated graph. The outer face of the representation is always a triangle. This result was later improved by Kant (1996), but his algorithm is very similar to the one described in de Fraysseix et al. (1990).

Let $G = (V, E)$ be a triangulated graph with a planar representation D where $(u, v) \in E$ is on the outer face. Let $\pi = (v_1, \dots, v_n)$ be a numbering of the vertices in V with $v_1 = u$ and $v_2 = v$. We define G_i as the subgraph induced by the vertex set $\{v_1, \dots, v_i\}$. The face C_i is the outer face of the representation D_i of G_i that we get by removing all representations of vertices and edges from D that do not belong to G_i.

Then π is a canonical ordering if and only if the following conditions hold for all $4 \leq i \leq n$:

- The subgraph G_{i-1} is 2-connected and C_{i-1} contains the edge (v_1, v_2).
- In the representation D, the vertex v_i is in the outer face of G_{i-1} and its neighbors in G_{i-1} form a subinterval of the path C_{i-1} with at least two elements.

Such a canonical ordering exists for any triangulated planar graph and can be computed in linear time by starting with the representation D and successively removing single vertices from the outer face that are not incident to any chords of the outer face. It is easy to show that such a vertex always exists for a triangulated planar graph.

The invariants of the actual drawing algorithm are that after step i (inserting the vertex v_i and the necessary edges), the following conditions hold:

- The vertex v_1 is at position $(0, 0)$ and v_2 at position $(2i - 4, 0)$.
- If the sequence of the vertices on the outer face is $c_1, c_2, \dots, c_k$ with $c_1 = v_1$ and $c_k = v_2$, then we have $x(c_j) < x(c_{j+1})$ for $1 \leq j < k$.
- The edge (c_j, c_{j+1}) has slope $+1$ or -1 for $1 \leq j < k$.

To describe the idea of the drawing algorithm, we define the *left-vertex* c_l of vertex v_i as the leftmost vertex on C_{i-1} that is adjacent to v_i. By leftmost we mean that the vertex comes first on the path from v_1 via C_{i-1} to v_2 that does not use the edge (v_1, v_2). The *right-vertex* c_r of v_i is defined as the rightmost vertex on C_{i-1} adjacent to v_i. From now on we will refer to the vertex c_{l+1} on C_{i-1} as the vertex directly right of c_l on C_{i-1}.

When we want to add the vertex v_i, we move the vertices c_{l+1} to c_{r-1} one unit to the right while we move the vertices c_r to c_k two units to the right. We also have to move some inner vertices of the representation to the right to make sure that the representation remains planar. This is achieved by storing for every vertex v on C_i a set of *dependent vertices* that have to be moved in parallel with v. When v vanishes from the outer cycle, we add v to its own list of dependent vertices and make this updated list the set of dependent vertices of the new vertex on the outer cycle.

We place v_i at the intersection of the line with slope $+1$ starting at c_l and the line with slope -1 starting at c_r. Figure 2.10 shows an example for the construction of such a representation. This approach can also be applied to non-triangulated graphs by first adding edges to make the graph triangulated (augmentation), applying the algorithm, and deleting the additional edges in the computed representation.

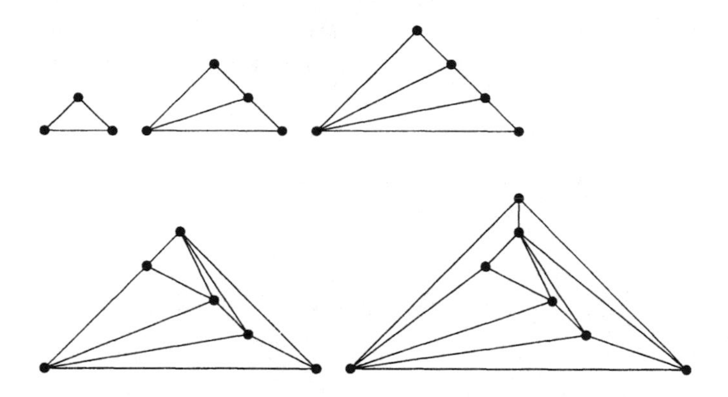

Fig. 2.10. An example for the straight-line algorithm of de Fraysseix, Pach and Pollack.

2.7.2 The Barycentric Algorithm of Schnyder

In the same year, Schnyder described an algorithm for solving the same task in time $O(n)$ using a grid of size $(n-2) \times (n-2)$ (Schnyder, 1990). This algorithm computes three coordinates for each vertex in the sequence given by the same canonical ordering as used by de Fraysseix, Pach and Pollack. In a second step, it computes the actual grid coordinates for the vertices using the barycentric coordinates.

The vertex positions are defined using a *barycentric representation* of the input graph G.

Definition 2.15 (Barycentric Representation). *A barycentric representation of G is an injective function*

$$v \in V \rightarrow (v_1, v_2, v_3) \in \mathbb{R}^3$$

satisfying the following conditions:

1. $v_1 + v_2 + v_3 = 1$ *for all vertices* v.
2. *For each edge* $\{u, v\}$ *and each vertex* $w \notin \{u, v\}$ *there is some* $k \in \{1, 2, 3\}$ *such that* $u_k < w_k$ *and* $v_k < w_k$.

A barycentric representation of the input graph is computed by first constructing a *normal labeling* of the angles of the faces of the input graph. Since the input graph is triangulated, every face has exactly three angles. The angles of each face are numbered 1, 2 and 3 so that the numbers appear in counterclockwise order around the face and for each interior vertex, the angles around it in counterclockwise order form a nonempty sequence of 1's followed by a nonempty sequence of 2's followed by a nonempty sequence of 3's. Such a labeling can be constructed in linear time.

For each normal labeling, every edge has two different labels on one end while the labels on both sides of the other end are the same. We call the repeated label the label of the edge. Thus, each normal labeling defines a *realizer* of the graph.

Definition 2.16 (Realizer). *A realizer of a triangular graph G is a partition of the interior edges of G into three sets T_1, T_2 and T_3 of directed edges so that for each interior vertex v the following conditions are satisfied:*

1. *The vertex v has outdegree 1 in T_1, T_2 and T_3.*
2. *The counterclockwise order of the edges around v is: leaving in T_1, entering in T_3, leaving in T_2, entering in T_1, leaving in T_3, entering in T_2.*

Every normal labeling has the following property: For each number in $\{1, 2, 3\}$ there is exactly one vertex on the outer face where every adjacent angle is labeled i. For each interior vertex, there is exactly one path leaving the vertex where all edges are labeled i for $i \in \{1, 2, 3\}$. This path ends in the vertex of the outer face where all adjacent edges are labeled i. These 3 paths leaving each interior vertex define 3 regions of the graph and the number of faces in each of these regions are the 3 barycentric coordinates of the vertex.

If we have 3 arbitrary non-collinear points α, β and γ in the plane and vertex v has the barycentric coordinates (v_1, v_2, v_3), then drawing every vertex v at position $v_1\alpha + v_2\beta + v_3\gamma$ will result in a planar straight-line embedding of the graph.

2.7.3 The Straight-Line Algorithm of Kant

Kant used the canonical ordering approach to develop several drawing algorithms (Kant, 1996). The first one also produces straight-line representations, but in contrast to the algorithms mentioned before, it guarantees that the inner regions are convex for 3-connected graphs, even if it is not the case that every face of the graph is bounded by 3 edges. This is not necessarily the case for the algorithms mentioned before, because if we want to apply them to non-triangulated graphs, we first have to augment the graph by adding edges to produce a second graph where every face is a triangle, then produce a representation for this graph and finally delete the added edges from the final representation. Thus it might happen that not every inner face of the representation is convex. The algorithm of Kant has a maximum grid-size of $(2n - 4) \times (n - 2)$ and runs in $O(n)$ time. Chrobak and Kant (1997) later improved this algorithm so that it only uses an area of $(n - 2) \times (n - 2)$.

Since this algorithm is an improved version of the algorithm of de Fraysseix et al. (1990), we will only give an overview of the differences. The algorithm of Kant can also handle 3-connected graphs that are not triangulated. This is achieved by defining the canonical ordering not as an ordering of the vertices but rather as an ordered partition of the vertices. Let $G = (V, E)$ be a 3-connected graph with a planar representation D where $v_1 \in V$ is on the outer face. Let $\pi = (V_1, \dots, V_k)$ be a partition of V and G_i the subgraph of G induced by $V_1 \cup V_2 \cup \dots \cup V_i$. The face C_i is the outer face of the representation D_i of G_i that we get by removing all representations of vertices and edges from D that do not belong to G_i.

Then π is a canonical ordering if and only if the following conditions hold:

- $V_1 = \{v_1, v_2\}$, v_1 and v_2 both lie on the outer face of D and $(v_1, v_2) \in E$.
- $V_k = v_n$ and v_n lies on the outer face of D with $(v_1, v_n) \in E$ and $v_n \neq v_2$.
- Each C_i for $i > 1$ is a cycle containing (v_1, v_2).
- Each G_i is 2-connected and internally 3-connected (removing any two inner vertices will not disconnect the graph).
- For each $i \in \{2, \dots, k - 1\}$ one of the following conditions holds:
  1. V_i is a single vertex z belonging to C_i and having at least one neighbor in $G - G_i$.
  2. The vertices in V_i form a *chain* (a path where all inner vertices have degree 2) $(z_1, \dots, z_l)$ on C_i where each z_j has at least one neighbor in $G - G_i$. The vertices z_1 and z_l each have exactly one neighbor in C_{i-1}, and these are the only neighbors of the vertices in V_i.

This canonical ordering can be computed in linear time by starting with the representation D and successively removing chains or single vertices from the outer face so that the resulting graph G' is 2-connected. To do this in linear time, we have to store and update for each face the number of its vertices and edges on the outer face and for each vertex the number of adjacent faces having a separation pair.

To compute the actual representation, the canonical ordering is first transformed into a *leftmost canonical ordering* which can be computed in linear time from a canonical ordering and is necessary for achieving linear running time. The invariants of the drawing algorithm after step i (inserting the vertices of the set V_i and the necessary edges) are:

- v_1 is at position $(0,0)$ and v_2 at position $(2i-4,0)$.
- If the sequence of the vertices on the outer face is $c_1, c_2, \ldots, c_k$ with $c_1 = v_1$ and $c_k = v_2$, then we have $x(c_j) < x(c_{j+1})$ for $1 \leq j < k$.

The only difference in the actual drawing algorithm compared to the algorithm of de Fraysseix, Pach and Pollack is that we can now insert several vertices at once. These vertices form a chain and we give them the same y-coordinate. Figure 2.11 shows an example for the construction of such a representation.

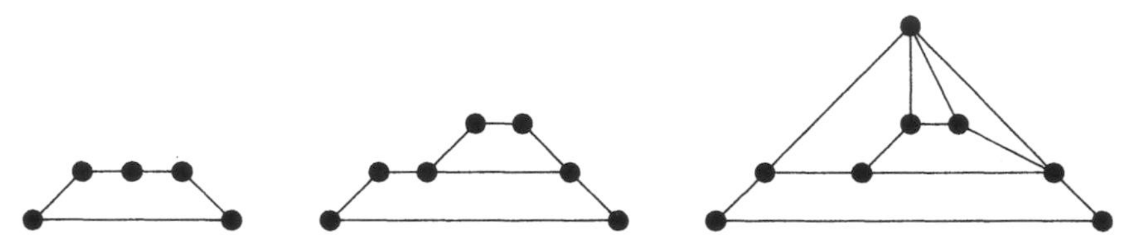

Fig. 2.11. An example for the straight-line algorithm of Kant.

2.7.4 The Orthogonal Algorithms of Kant

In the same paper (Kant, 1996), Kant also gives two algorithms for producing *orthogonal representations* of planar graphs. In an orthogonal representation, all edges consist only of horizontal and vertical segments. If every vertex is drawn as a point, such a representation can only be used for planar graphs in which every vertex has at most degree 4 (*4-planar graphs*). Since orthogonal drawing algorithms are explicitly treated in Chapter 6, we will only give a short overview of the two algorithms. The first algorithm draws 3-connected 4-planar graphs on an $n \times n$ grid with at most $\lceil \frac{3}{2}n \rceil + 4$ bends so that each edge has at most two bends. The second algorithm produces an orthogonal representation for planar graphs with maximum degree 3 having at most $\lfloor \frac{n}{2} \rfloor + 1$ bends on a grid of size $\lfloor \frac{n}{2} \rfloor \times \lfloor \frac{n}{2} \rfloor$. The running time of both algorithms is linear.

The algorithm for producing orthogonal representations of 3-connected 4-planar graphs given in Kant (1996) also uses the canonical ordering. Since the edges only consist of vertical and horizontal segments, there are exactly 4 directions from which an edge can attach to a vertex v. They are called *up(v)*, *down(v)*, *left(v)* and *right(v)*. One of these directions is called *free* if we have not yet attached an edge to it. The idea of the algorithm is to add each vertex v in the canonical ordering to the subgraph that is already placed so that *down(v)* is not free and *up(v)* is free. The algorithm works in two phases. In the first phase, we assign the 4 directions of each vertex to the incident edges and give the vertices the y-coordinates. We also store for each vertex and bend a pointer to its column. During the algorithm, we may have to add new columns. In the second phase, we assign x-coordinates to the columns and thus indirectly to the vertices and bends of the representation.

The algorithm in Kant (1996) for drawing planar graphs with maximum degree 3 is based on an algorithm for 3-connected graphs with maximum degree 3. This algorithm is similar to the algorithm of the last paragraph, but we can place all vertices of the same partition of the canonical ordering on the same y-coordinate. The algorithm is generalized for working on 2-connected 3-planar graphs using SPQR-trees. We recursively use the algorithm for drawing 3-connected 3-planar graphs and then merge the representations into a representation for the whole graph. This method is again generalized to connected 3-planar graphs by drawing every 2-connected component so that the cut vertex is in the upper-left corner and then merging the representations into a representation of the whole graph without introducing new bends.

2.7.5 The Mixed Model

Kant also introduces a new method for drawing 3-connected planar graphs called the *Mixed Model.* In this model, each edge is a poly-line which may have at most three bends. Each edge consists of at most four parts. The parts connected to the vertices may be diagonal, while the two middle parts of each edges are vertical and horizontal. The principle of the algorithm is to define a set of points around each vertex where the orthogonal edges coming from other vertices connect. These points define the boundary of the *bounding box* of the vertex. The points are then connected by straight lines to the vertex itself. Each edge consists of a straight line segment between the start vertex and a point on the boundary of the bounding box, an orthogonal part with at most one bend from the bounding box of the start-vertex to the bounding box of the target-vertex and another straight part from the boundary of the bounding box of the target-vertex to the target-vertex itself.

The grid size for this algorithm is $(2n-6) \times (3n-9)$ and the number of bends is at most $5n-15$. An important property of the algorithm is that it guarantees that the angle between two edges emanating from the same vertex is larger than $2/d$ radians where d is the degree of the vertex. The minimum angle of two edges emanating from the same vertex in a representation is called the *angular resolution* of the representation. Having a large angular resolution improves the readability of a drawing.

Gutwenger and Mutzel have improved Kant's algorithm for the Mixed Model to achieve a grid size of $(2n-5) \times (\frac{3}{2}n - \frac{7}{2})$ (Gutwenger and Mutzel, 1998). They also have improved the angular resolution for graphs which are not 3-connected. Since Kant's algorithm only works for 3-connected graphs, graphs that are not 3-connected have to be augmented by adding additional edges before the algorithm is applied and afterwards the additional edges have to be deleted from the representation. This can lead to an angular resolution of $\frac{4}{3d+7}$ where d is the maximum degree in the original graph. Since the algorithm in Gutwenger and Mutzel (1998) can be applied directly to 2-connected graphs, an angular resolution of $2/d$ can be guaranteed for any planar graph. The running time for both algorithms is linear.

The algorithm for drawing graph G works in three phases:

1. If the graph is not 2-connected, edges are added to produce a planar 2-connected graph G'.
2. A suitable canonical ordering for G' is computed.
3. This ordering is used to draw the original graph G.

For each vertex, we define a set of *inpoints* and *outpoints*. The inpoints are the points where the edges from vertices that have already been placed arrive and the outpoints are the points where the edges to vertices that still have to be placed leave. The inpoints and outpoints of each vertex are located on the boundary of a roughly diamond shaped bounding box and will be placed on grid coordinates. Figure 2.12 shows two examples of bounding boxes.

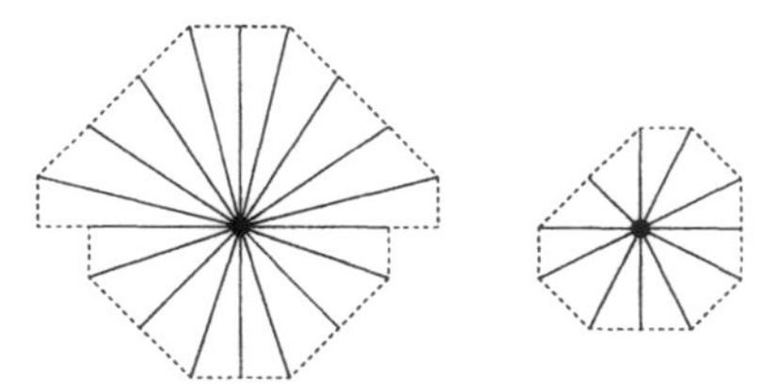

Fig. 2.12. Two examples of bounding boxes.

The point straight down from a vertex as well as the two points straight to the right and to the left are inpoints, while the point straight above the vertex is always an outpoint. Thus, a vertex with an indegree not greater than 3 and an outdegree of at most 1 will have no adjacent diagonal edges. These inpoints together with the edges that connect them to the vertex form a cross. We denote the four sectors defined by this cross NW, NE, SE and SW, as on a compass.

If there are more than three incoming edges, we distribute the remaining inpoints evenly among the sectors SE and SW. If there are at least two outgoing edges, they are distributed evenly between the sectors NE and NW. If the remaining number of edges is not even, we get an asymmetric configuration like in the right picture of Figure 2.12.

When a vertex is placed, we have to avoid overlapping bounding boxes except if we can identify the outpoint of an adjacent vertex with the vertex we want to place. If the set V_i of vertices in the canonical order we want to add in step i has only one element v, we place this vertex directly above the adjacent vertex which is connected by the inedge going straight down. We choose the y-coordinate so that the minimum vertical distance between the bounding box of an adjacent vertex and the bounding box of v is 1. We may have to shift the adjacent vertices already placed and their dependent sets to the right to make room for the edges. If V_i has more than one element, all the vertices in the set will get the same y-coordinate. Figure 2.13 shows an example of a drawing produced with Kant's original algorithm. Figure 2.14 shows two drawings computed with the algorithm of Gutwenger and Mutzel (1998).

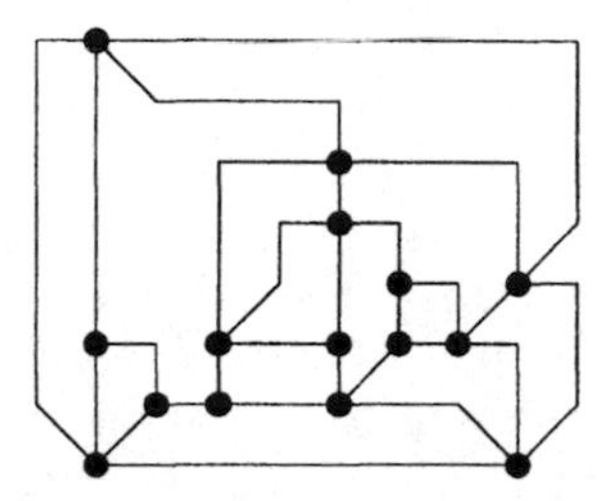

Fig. 2.13. A drawing produced by the Mixed Model algorithm of Kant.

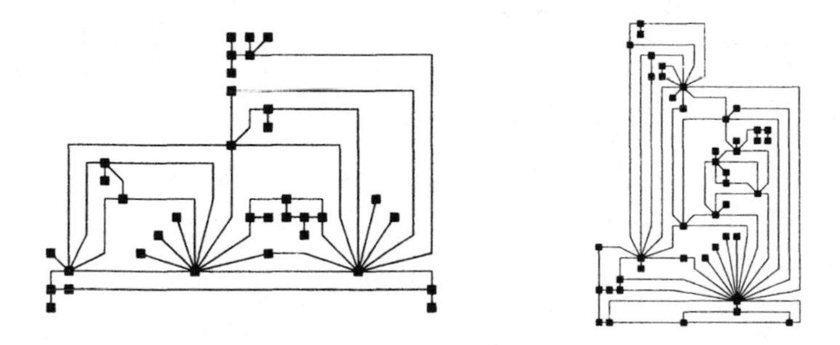

Fig. 2.14. Example drawings produced by the algorithm of Gutwenger and Mutzel.

3. Drawing Trees, Series-Parallel Digraphs, and Lattices

Matthias Müller-Hannemann*

In many applications of graph drawing which have been described in Chapter 1 one faces graph classes with a special structure. First of all, this means that specific layout criteria become possible. Second, the structural properties of these classes allow the development of more powerful algorithms with respect to running time and layout space requirements. Among the many special graph classes which exist, probably most attention has been paid to trees and planar graphs. Algorithms and methods for planar graphs in general have been given in Chapter 2. This chapter concentrates on three different graph classes, namely on trees and series-parallel digraphs (which are, of course, even more specialized planar graphs), but also on graphs arising from lattices, more precisely, on covering digraphs of lattices.

In the corresponding sections, we first discuss the required terminology, explain the specific layout styles and criteria, and then give a (partial) survey of the main results in these areas. The main part of each section presents some typical highlight in more detail. In Section 3.1, we describe recent work of Chan (1999) on the drawing of ordered binary trees with near-linear area bounds. For the drawing of series-parallel digraphs, we also selected some very recent work, namely that of Hong et al. (1998) devoted to the display of symmetries. Finally, in our section on the drawing of lattice diagrams our presentation is application-driven: we sketch the field of formal concept analysis as introduced by Wille and co-workers.

3.1 Trees

Trees are widely used as a data structure and representation of hierarchies. Applications requiring a suitable visualization include among others organization charts of companies, search trees, parse trees of computer programs or family trees in genealogy.

3.1.1 Rooted Trees

We adopt the following standard terminology for trees. A *tree* is a connected acyclic graph. A *rooted tree* T is a tree with a special vertex $r \in T$, the so-called *root* of T. For rooted trees, it is common to orient the edges "away

* The author was partially supported by the special program "Efficient Algorithms for Discrete Problems and Their Applications" of the Deutsche Forschungsgemeinschaft (DFG) under grant Mo 446/2-3.

M. Kaufmann and D. Wagner (Eds.): Drawing Graphs, LNCS 2025, pp. 46-70, 2001.

from the root" such that the root is the only vertex with no incoming edge, but every other vertex has exactly one incoming edge. In particular, there is a unique path from the root to each vertex. For a directed edge (u, v) in a rooted tree, the vertex u is the *parent* of v and v is a *child* of u. In a rooted tree, the *depth* of a vertex v is the number of edges of the path from v to the root. A vertex is called *leaf* if it has no child. A *binary tree* is a rooted tree where each vertex has no more than two children.

An *ordered tree* is a rooted tree with a given ordering of the children of each vertex. In an ordered binary tree, the first child of a vertex with two children is called the *left* and the second one is called the *right* child. If v is the vertex of some rooted tree T, then the *subtree* rooted at v is the subgraph induced by all vertices reachable on directed paths starting at v. In case of an ordered binary tree and a vertex v with two children, the subtree rooted at the left or right child of v is the *left subtree* and *right subtree* of v, respectively.

Free trees are trees without a prespecified root. However, after selecting some vertex as a fictitious root they can be handled like rooted trees. A typical choice for a root of a free tree is a *center*, that is a vertex such that the height of the resulting rooted tree is minimized.

Typical requirements for "ideal drawings" of rooted trees involve representations, layout models and constraints of the following kind.

1. *Planar drawings*: No two edges cross.
2. *Grid drawings*: Vertices have integer coordinates.
3. *Straight-line drawings*: Each edge is a straight-line segment, whereas in a *polyline drawing* each edge is a polygonal chain.
4. *(Strictly) upward drawings*: A child should be placed (strictly) below its parent in the y-direction.
5. *Strongly order-preserving drawings*: The line segments from the parent to the leftmost child is monotone decreasing in the x-direction, whereas the line segment to the rightmost child is monotone increasing, and the line segments of all children from a vertex are sorted by angle from left to right.

A *layered drawing* of a tree is a drawing where a vertex v of depth i has as y-coordinate the negative of its depth, that is $y(v) = -i$. Hence, a *layer* is formed by the set of vertices of the same depth. In *radial drawings*, the layers are mapped to concentric circles. In an *orthogonal drawing* each edge is a chain of alternating horizontal and vertical segments. An *hv-drawings* is a planar straight-line orthogonal and upward drawing where additionally for every vertex the smallest bounding rectangles of its subtrees do not intersect. See Figure 3.1 for examples of a layered, a *hv*- and a radial drawing.

3.1.2 Area Bounds

The *area of a drawing* of a tree is, as usual, defined as the area of the smallest rectangular box with horizontal and vertical sides covering the area under

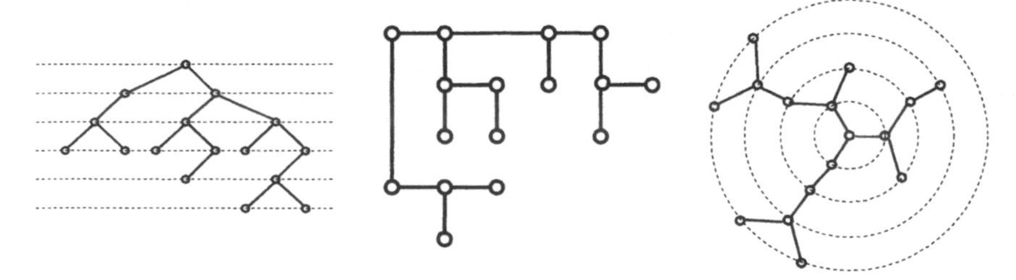

Fig. 3.1. Examples of a layered, a *hv*-, and a radial tree drawing.

some resolution convention (for example, that any two vertices have minimum distance one). Similarly, the *aspect ratio* of the drawing is the ratio of the length of the longest side to the length of the shortest side of the smallest enclosing box. In many cases, there is a trade-off between the required area or the achievable aspect ratio and some of the mentioned aesthetic requirements. Table 3.1 gives an overview of existing algorithms, the criteria they meet and the achieved bounds on the area and the aspect ratio.

Layered tree-drawings with several nice additional features (straight-line, grid, order-preserving, isomorphic subtrees have congruent drawings) require $O(n^2)$ area and can be found in linear time (Reingold and Tilford, 1981). A layered tree-drawing of a binary tree with minimum width can be solved in polynomial time by means of linear programming, but if a grid drawing is required, the width minimization problem becomes $\mathcal{NP}$-hard (Supowit and Reingold, 1983).

Table 3.1. Upward grid drawings of rooted trees. The columns indicate whether the drawings are strictly upward (2), straight-line (3), orthogonal (4), or (strongly) order preserving (5), and give bounds for the area (6) and aspect ratio (7).

tree	str. upw.	straight-line	orthogonal	ord. pres. (strongly)	area	ratio	reference
rooted	+	+	–	+	$O(n^2)$	$O(n)$	Reingold and Tilford (1981)
rooted	+	+	–	–	$\Theta(n \log n)$	$O(\frac{n}{\log n})$	Shiloach (1976), Crescenzi et al. (1992)
binary	–	+	+	–	$O(n \log n)$	$O(1)$	Chan et al. (1996)
binary	–	–	+	–	$\Theta(n \log \log n)$	$O(\frac{n \log \log n}{\log^2 n})$	Garg et al. (1996)
binary	+	+	–	+ (+)	$O(n^{1+\epsilon})$	$O(n)$	Chan (1999)
Fibonacci	+	+	–	–	$\Theta(n)$	$O(1)$	Crescenzi et al. (1992)
AVL	+	+	–	–	$\Theta(n)$	$O(\frac{n}{\log n})$	Crescenzi and Piperno (1995)
degree–$O(n^a)$, $0 \leq a < 1$	–	–	–	–	$\Theta(n)$	$O(n^a)$	Garg et al. (1996)

An area bound of $O(n \log n)$ is possible if the property of being order-preserving is dropped (Shiloach, 1976; Crescenzi et al., 1992). This area bound is tight, that is, there is a class of binary trees which requires $\Theta(n \log n)$ area in any strictly upward planar grid drawing.

For binary trees, upward, order-preserving polyline drawings also achieve the bound $\Theta(n \log n)$ (Garg et al., 1996), whereas it is interesting to note that upward orthogonal polyline grid drawings (but not order-preserving ones) allow even tight bounds of $\Theta(n \log \log n)$ (Garg et al., 1996).

Linear-area drawing algorithms are available for AVL-trees (Crescenzi and Piperno, 1995) and Fibonacci trees (Crescenzi et al., 1992). Moreover, for any rooted bounded-degree tree one can construct a planar upward grid polyline drawing with $O(n)$ area but not preserving a given order (Garg et al., 1996).

We note that these theoretical results usually assume that vertices consume the same space. However, the drawing area required to display vertex labels may be quite different.

Chan's Binary Tree Drawing. In the following, we will present recent methods of Chan (1999) for binary trees achieving a near-linear area bound for planar, straight-line, strictly upward and strongly order-preserving drawings. The analysis of these methods may highlight the typical kind of reasoning necessary for similar results on area bounds.

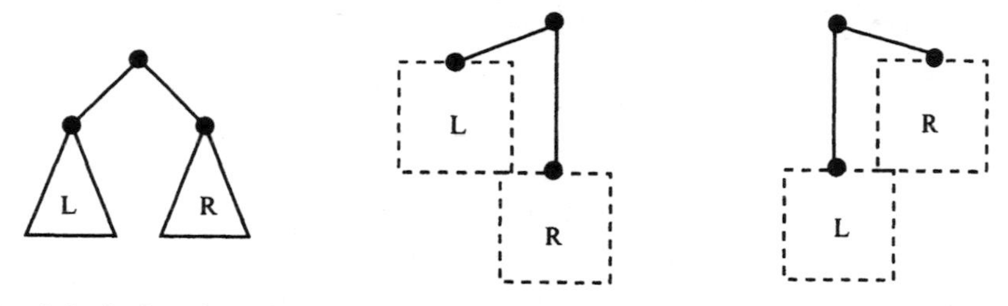

Fig. 3.2. Left-right-rules.

Let us consider a recursive procedure for drawing a given ordered binary tree T with root v. If we have constructed drawings of the left subtree L and the right subtree R, these partial drawings can be combined by two rules as follows. In the *left rule*, we vertically align v with the root of R, place the right upper corner of the bounding box of L one unit below and one unit to the left of v, and place the upper border of the bounding box of R on the same line as the lower border of the bounding box of L. Symmetrically, we define the *right rule*, see Figure 3.2. If we specify which rule to use, we immediately get a drawing algorithm. A first and very simple method is given in Algorithm 1.

The validity of the claimed properties of the output specification should be obvious. It is also immediate that the height of such a drawing is at most $n = |T|$; a bound on the width, however, is not obvious.

Algorithm 1: Simple binary tree drawing.

input : an ordered binary tree T with root v

output : a straight-line planar, strictly upward and strongly order preserving drawing of the binary tree T

```
begin
    if |T| ≤ 1 then return trivial drawing;
    L ← left subtree of T;
    R ← right subtree of T;
    draw L and R;
    if |L| ≤ |R| then
        combine drawings using the left rule;
    else
        combine drawings using the right rule;
end
```

A reformulation of this algorithm in a slightly different way makes the analysis of the required width more accessible.

Algorithm 2: Generic binary tree drawing.

input : an ordered binary tree T with root v

output : a straight-line planar, strictly upward and strongly order preserving drawing of the binary tree T

```
begin
    if |T| ≤ 1 then return trivial drawing;
    determine a path P = (v_0 = r, v_1, ..., v_k) from the root to some [illegible];
    for i = 0 to k do
        draw the subtree rooted at the sibling of v_i;
    combine the drawings by applying the left and right rules at nodes of P
    such that all nodes of P are vertically aligned;
end
```

Algorithm 2 is a generic version which is identical to Algorithm 1 if the *greedy path* P is chosen as follows: Let v_i, L_i, R_i be the root, left and right subtree of T_i, respectively. If $|R_i| \leq |L_i|$, then set $T_{i+1} = L_i$, and set $T_{i+1} = R_i$, otherwise. See Figure 3.3 for an illustration.

Lemma 3.1. *For any two different subtrees α and β of the greedy path, either (i) $|\alpha| \leq n/2$ and $|\beta| \leq (n - |\alpha|)/2$ or (ii) $|\beta| \leq n/2$ and $|\alpha| \leq (n - |\beta|)/2$.*

If we denote the width of a drawing for T by $W(T)$, then

$$W(T) = W(\alpha) + W(\beta) + 2$$

for some left subtree α and some right subtree β of the path P. With Lemma 3.1 we get the following recurrence on the maximum width $W(n)$ for trees of size n:

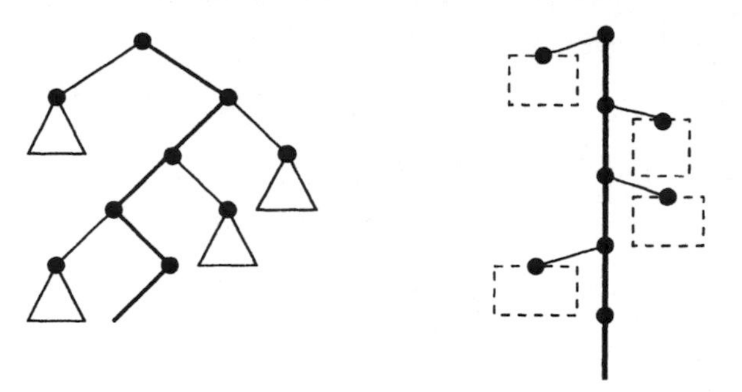

Fig. 3.3. Chan's generic algorithm (with the greedy path).

$$W(n) \leq \max_{n_1 \leq n/2,\; n_2 \leq (n-n_1)/2} (W(n_1) + W(n_2) + 2).$$

It can be shown that this recurrence solves to $W(n) = O(n^{0.695})$.

Near-Linear Drawings. A further improvement on the area bounds can be achieved by extending the left and right rules. The *extended left rule* applied at the root v of T translates the bounding box of the right subtree R horizontally by an arbitrary amount, as long as the x-coordinate of the root of R is not less than the x-coordinate of v. The *extended right rule* is defined symmetrically. Clearly both rules still guarantee straight-line planar, strictly upward and strongly order preserving drawings.

Consider the greedy path $P = (v_0, v_1, \ldots)$ and subtrees T_i rooted at v_i as defined above. For some fixed parameter A to be chosen later, let k be the largest index such that $|T_k| \geq n - A$. Suppose that v_k is a left (right) child. Let P' be the subpath $(v_0, v_1, \ldots, v_k)$ and P'' be the leftmost (rightmost) path from v_{k+1} to a leaf. The generic Algorithm 2 applied on the concatenation of the two paths P' and P'' leads to the improved Algorithm 3, shown in Figure 3.4.

For the analysis we may assume that v_k is a left child. Then the width $W(T)$ is given by

$$W(T) = \max\{W(\alpha) + W(\beta) + 2, W(\gamma) + 1\},$$

where α and β are left and right subtrees of P', respectively, and γ is a right subtree of P''.

Lemma 3.2. *Let P' and P'' be defined as above. Then for any subtree α of a vertex of P', $|\alpha| \leq A$. For any subtree γ of a vertex of P'', $|\gamma| \leq n - A$.*

Hence, for any choice of A we get the following recurrence on the maximum width $W(n)$ for trees of size n:

$$W(n) \;\leq\; \max\{2W(A) + 2,\; W(n - A) + 1\}.$$

From that, one can obtain

$$W(n) \leq 2W(A) + O(n/A).$$

If we set the parameter $A = n/2^{1/\epsilon}$ for a fixed $\epsilon > 0$, the recurrence solves to $W(n) = O(2^{1/\epsilon}n^{\epsilon})$. This implies an area bound of $O(n^{1+\epsilon})$. However, the best result is achieved for a non-constant value $\epsilon = 1/\sqrt{\log_2 n}$ which leads to the following theorem.

Theorem 3.3 (Chan 1999). *Any binary tree of size n admits a straight-line planar, strictly upward and strongly order-preserving drawing of height at most $n-1$ and width $O(4^{\sqrt{\log_2 n}})$.*

A similar result is also possible for ordered trees of arbitrary degree:

Theorem 3.4 (Chan 1999). *Any ordered tree of size n admits a straight-line planar, strictly upward and strongly order-preserving drawing with an area of $O(n4^{\sqrt{2\log_2 n}})$.*

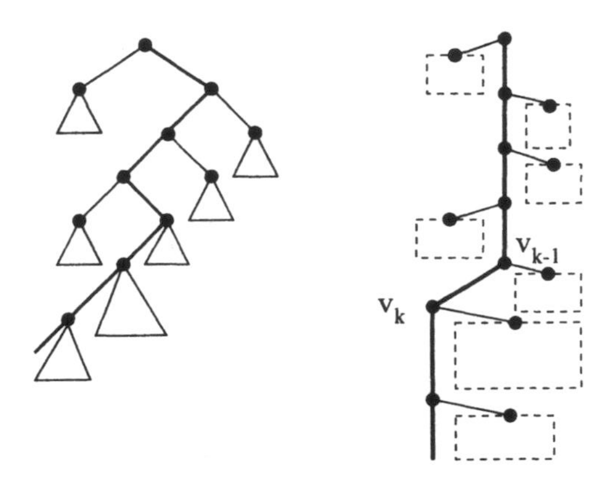

Fig. 3.4. Chan's improved algorithm (extended left-right-rule).

3.2 Series-Parallel Digraphs

Series-parallel digraphs arise in the analysis of electrical networks, and they appear in flow diagrams, dependency charts, and PERT networks. This section first reviews basic facts about such graphs, and then describes algorithms for upward straight-line drawings. In the main part we sketch a recent method for symmetry drawings.

Algorithm 3: Improved binary tree drawing.

input : an ordered binary tree T with root v, a parameter A
output : a straight-line planar, strictly upward and strongly order preserving drawing of the binary tree T

begin
  if $|T| \leq 1$ **then** return trivial drawing;
  determine the greedy path $P = (v_0 = r, v_1, \ldots, v_\ell)$ from the root to some leaf v_ℓ;
  let k be the largest index such that $|T_k| \geq n - A$ for the subtree rooted at v_k;
  if *v_k is a left child* **then**
    let $P' = (v_0, v_1, \ldots, v_k)$;
    let P'' the leftmost path from v_{k+1} to a leaf;
    draw the subtrees of P' and P'';
    combine the drawings by applying the left and right rules at $v_0, v_1, \ldots, v_{k-2}$ such that $v_0, v_1, \ldots, v_{k-1}$ are vertically aligned;
    apply the left rule at nodes on P'' and v_k such that these nodes are vertically aligned as well;
    apply the extended left rule at v_{k-1} such that v_k is aligned with the left side of the bounding box of the entire drawing of T;
  else
    "symmetric case", choosing the rightmost path from v_k and replacing right for left;
end

3.2.1 Terminology and Basic Facts

Series-parallel digraphs (more precisely, sometimes also called *two-terminal series-parallel multidigraphs*) are defined recursively as follows. A digraph consisting of two vertices, a *source* s and a *sink* t, joined by a single edge is a series-parallel digraph. If G_1 and G_2 are series-parallel digraphs, so are the digraphs constructed by each of the following operations:

1. The *parallel composition*: Identify the source of G_1 with the source of G_2 and the sink of G_1 with the sink of G_2.
2. The *series composition*: Identify the sink of G_1 with the source of G_2.

There exist other notions of series-parallel digraphs (see e.g. Valdes et al. 1982), but to our knowledge only the given one has been studied for specialized drawing algorithms. Note that every series-parallel digraph is acyclic and planar. Drawing algorithms for series-parallel digraphs usually assume that the given graphs are simple.

Decomposition and Recognition. Given an arbitrary multidigraph G, a *series reduction* is an operation which can be applied to the arcs $(u, v), (v, w)$ if v has in-degree and out-degree one. In such a case this operation deletes v and both incident arcs from G and reinserts a new arc (u, w). In a *parallel reduction*, exactly one arc of a pair of parallel arcs is deleted.

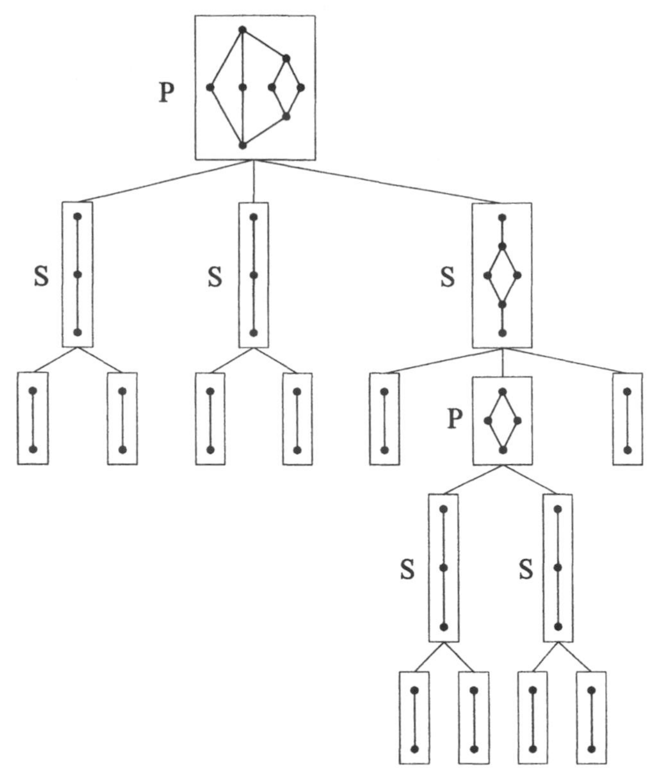

Fig. 3.5. Example of a canonical decomposition tree for a series-parallel digraph.

Based on reductions, series-parallel digraphs can also be characterized by the following lemma.

Lemma 3.5. *A graph is a series-parallel digraph if and only if it can be reduced to the one-edge series-parallel digraph by a sequence of series and parallel reductions.*

Using this lemma, one obtains an efficient algorithm for the recognition of series-parallel digraphs. Given a graph G, one repeatedly applies series and parallel reductions until no reduction is possible. It is a nice property of series-parallel digraphs that the result of such a reduction sequence is independent of the order in which the specific reductions are applied.

A series-parallel digraph G can be represented in a natural way as a *binary decomposition tree* T, which is obtained as a by-product of such a reduction sequence. The decomposition tree contains S-nodes, P-nodes and Q-nodes and is recursively defined as follows. If G is a single edge, then T consists of a single Q-node. If G is a series composition of G_1 and G_2 with decomposition trees T_1 and T_2 and roots r_1 and r_2, respectively, then T consists of an S-node root with left child r_1 and right child r_2. Similarly, if G is a parallel composition of G_1 and G_2 with decomposition trees T_1 and T_2 and roots r_1 and r_2, respectively, then T consists of a P-node root with children r_1 and r_2 (in an arbitrary order).

Valdes et al. (1982) describe how to recognize and to build up a binary decomposition tree of a series-parallel digraph in linear time. It is straightforward to get a *canonical decomposition tree* (no longer binary) by contracting each connected group of S-nodes and each connected group of P-nodes into a single node. Such a tree is unique up to reordering the children of each P-node. See Figure 3.5 for an example.

In the following we assume that the decomposition tree T of a series-parallel graph G is given as part of the input for the drawing algorithm.

3.2.2 Upward Straight-Line Drawings

Fixed Embedding Requires Exponential Area. We start with the negative result that upward straight-line drawings of series-parallel digraphs which preserve a given embedding may require exponential area. Consider the recursively defined class G_n of series-parallel digraphs as shown in Figure 3.6.

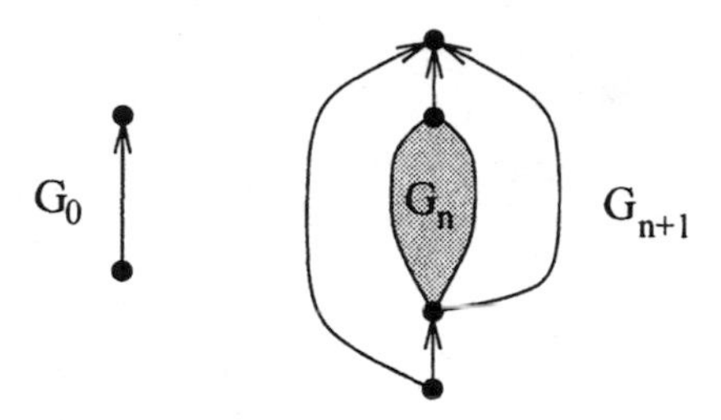

Fig. 3.6. The class G_n of series-parallel digraphs leading to exponential area bounds.

Lemma 3.6 (Bertolazzi et al. 1994a). *There exist embedded series-parallel digraphs such that any upward straight-line drawing that preserves the embedding requires exponential area, namely $\Omega(4^n)$ under any resolution rule for the class G_n.*

Linear-Time Drawings with $O(n^2)$ Area. However, if one allows small changes in the embedding much better area bounds are possible. In a so-called *right-pushed embedding*, a single edge which forms a component of a parallel composition is always embedded on the right side. (Since we assume that the series-parallel digraphs are simple, there is at most one single edge component in a parallel composition.)

Bertolazzi et al. (1994a) describe an algorithm for right-pushed embeddings. The *Δ-drawing* Γ of a series-parallel digraph is inductively defined inside a bounding triangle $\Delta(\Gamma)$ that is isosceles and right angled. The hypotenuse of $\Delta(\Gamma)$ is a vertical segment, and the other two sides are on its left. See Figure 3.7 for a sketch of the construction for the base case, a series composition and for parallel compositions (from left to right) in the corresponding Algorithm Δ-SP-Draw. More details can be found in the book of Di Battista et al. (1999).

Theorem 3.7 (Bertolazzi et al. 1994a). *Let G be a series-parallel digraph with n vertices. The algorithm Δ-SP-Draw yields a strictly upward planar straight-line grid drawing of G with $O(n^2)$ area such that isomorphic components of G have drawings congruent up to a translation. This algorithm can be implemented to run in linear time.*

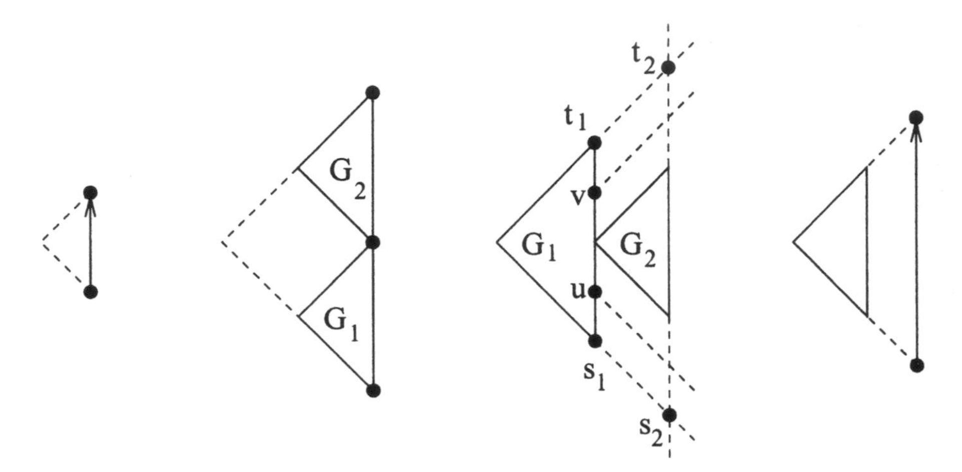

Fig. 3.7. Construction steps in Algorithm Δ-SP-Draw (sketch): base case, series composition, parallel composition (general case; the edges (s_1, u) and (v, t_1) are the rightmost edges incident on the source and sink of G_1), and parallel composition with a right-pushed single edge (from left to right).

Extension to Dynamic Drawings. We also mention extensions to dynamic drawings. For more information on dynamic drawings in general see Chapter 9.

Cohen et al. (1995) consider a framework for dynamic graph drawings where an implicit representation of the drawing of a graph is maintained such that the following operations can efficiently be performed.

- *Update operations*, i.e., insertion and deletion of vertices and edges or replacement of an edge by a graph.
- *Drawing queries*, which return the drawing of a subgraph S of a graph G consistent with the drawing of G.

- *Point location queries*, which return the vertex, edge or face containing a given point p in the subdivision of the plane induced by the drawing of G.
- *Window queries*, which return the portion of the drawing inside a query rectangle.

Within this framework the algorithm Δ-SP-Draw can be modified such that updates take $O(\log n)$ time and $O(n)$ memory space, drawing queries take time $O(k+\log n)$ for a series-parallel subgraph and $O(k \log n)$ for an arbitrary subgraph of size k, point location queries take $O(\log n)$ time, and window queries $O(k \log^2 n)$ time.

3.2.3 Display of Symmetries

Very recent work studies how to draw series-parallel digraphs with as much symmetry as possible (Hong et al., 1998, 1999a).

An *automorphism* of an undirected graph is a permutation of the vertex set which preserves adjacency of vertices. For a directed graph $G = (V, E)$, we will consider two kind of automorphisms. A *direction preserving automorphism* is a permutation p of the vertex set V such that $(u, v) \in E$ if and only if $(p(u), p(v)) \in E$, whereas in a *direction reversing automorphism* we require that $(u, v) \in E$ if and only if $(p(v), p(u)) \in E$. The set of all automorphisms (direction preserving and reversing) forms the *automorphism group* of G.

In general, the problem of finding an automorphism group of a graph is *isomorphism complete*, i.e., as hard as testing whether two graphs are isomorphic. The exact complexity status of this problem is open, namely, it is neither known to be $\mathcal{NP}$-complete nor are polynomial algorithms available.

In our context, we are only interested in those automorphisms which can be represented geometrically as a symmetry of an upward planar drawing. The corresponding groups are called *upward planar automorphism groups.* Based on earlier work of Manning (1990) and Lin (1992), Hong et al. show that only a few different automorphism groups occur for upward planar drawings. Examples are shown in Figure 3.8.

Lemma 3.8 (Hong et al. 1999a). *An upward planar automorphism group of a series-parallel digraph is either*

- *trivial, or*
- $\{1, p\}$ *where p is either vertical, horizontal, or a rotation of 180 degrees, or*
- $\{1, p, q, r\}$ *where p is of type vertical, q of type horizontal, and r a rotation of 180 degrees.*

The detection of upward planar automorphisms involves the following steps:

1. Construct the *canonical decomposition tree.*
2. Check for existence of *horizontal* automorphisms.
3. Check for existence of *vertical* automorphisms.

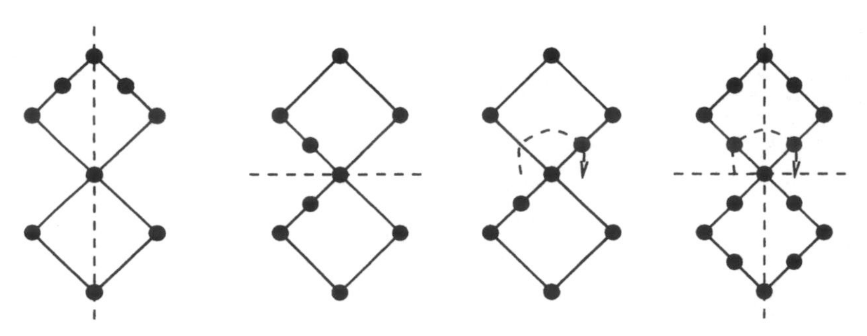

Fig. 3.8. Examples of automorphism groups: vertical, horizontal, rotational, and group of size 4 (from left to right).

4. Check for existence of *rotational* automorphisms.
5. Compute the *maximum* upward planar automorphism group.

Detection of Vertical Automorphisms. We will only sketch the detection of vertical automorphisms. The detection of horizontal and rotational automorphisms is similar, but slightly more complicated. For details we refer to Hong et al. (1999a).

Lemma 3.9 (Hong et al. 1999a). *Suppose that G is a series-parallel digraph, where the children of the root in the canonical decomposition tree represent $G_1, \dots, G_k$, and α is a vertical automorphism. If G is a series composition, then α fixes each one of $G_1, \dots, G_k$. If G is a parallel composition, then α fixes at most one of $G_1, \dots, G_k$.*

An *automorphism partition* of a set $\mathcal{G}$ of graphs is a partition of $\mathcal{G}$ into subsets $\mathcal{G}_1, \mathcal{G}_2, \dots, \mathcal{G}_m$ such that two graphs are in the same subset if and only if they are isomorphic. The sets $\mathcal{G}_i$ are called *isomorphism classes.* The partition can be expressed by assigning an integer label $code(G)$ to every graph $G \in \mathcal{G}$ such that, for each $G, G' \in \mathcal{G}$, $code(G) = code(G')$ if and only if G is isomorphic to G'.

An adaption of a tree isomorphism algorithm (Aho et al., 1974), applied to the canonical decomposition tree of a series-parallel digraph, allows an efficient labeling procedure (Algorithm 4) which yields the following lemma.

Lemma 3.10 (Hong et al. 1999a). *Suppose that u, v are nodes on the same level in the canonical decomposition tree of a series-parallel digraph G. Then the component represented by u is isomorphic to the component represented by v if and only if $code(u) = code(v)$.*

See Figure 3.9 for an example of such a labeling of a canonical decomposition tree.

Theorem 3.11 (Hong et al. 1999a). *Suppose that G is a series-parallel digraph.*

Algorithm 4: Vertical labeling of a canonical decomposition tree.

input : a canonical decomposition tree T of a series-parallel digraph
output : a vertical labeling of T

begin
initialize the tuples for each leaf u of T with $tuple(u) = (0)$;
for *each level i, from the maximum level to the root level* **do**
  for *each internal node u of T at level i* **do**
    set $tuple(u) = (code(v_1), code(v_2), \ldots, code(v_k))$, where the children of u are $v_1, v_2, \ldots, v_k$, from left to right;
  if *u is a P-node* **then**
    sort $tuple(u)$;
  let S be the sequence of tuples for the nodes on level i. Sort S lexicographically;
  for *each node u at level i* **do**
    set $code(u) = j$ if u is represented by the j-th distinct tuple of the sorted sequence S;
end

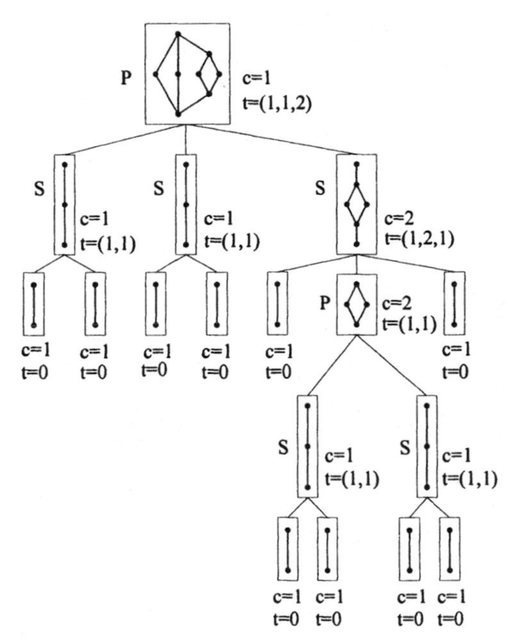

Fig. 3.9. Example of the vertical labeling of a canonical decomposition tree (the code of a component is abbreviated with c, an auxiliary tuple with t).

1. *If G is a* series composition *of $G_1, \ldots, G_k$, then G has a vertical automorphism if and only if each one of $G_1, \ldots, G_k$ has a vertical automorphism.*
2. *Suppose G is a* parallel composition *of $G_1, \ldots, G_k$. Consider the automorphism partition of $G_1, \ldots, G_k$.*
   a) *If there is more than one isomorphism class with an odd number of elements, then G has no vertical automorphism.*
   b) *If all automorphism classes have an even number of elements, then G has a vertical automorphism.*
   c) *If one isomorphism class has an odd number of elements, then G has a vertical automorphism if and only if the component of the odd size automorphism class has a vertical automorphism.*

Proof. Just look at Figure 3.10. Note that for an isomorphic pair G_i, G_j, we can construct drawings D_i of G_i and D_j of G_j such that D_i is a mirror image of D_j. By applying a "croissant-shape"–transformation (see Figure 3.11) to both drawings, isomorphic pairs can be arranged symmetrically on the opposite sides of a vertical line as in Figure 3.10.

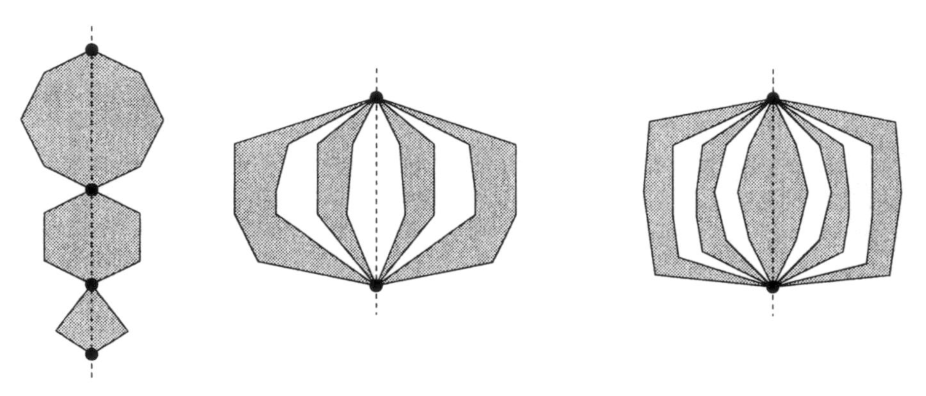

Fig. 3.10. Vertical arrangements for the different cases in Theorem 3.11.

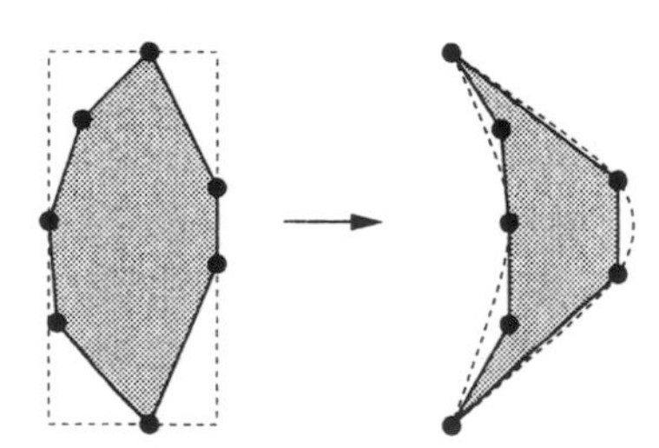

Fig. 3.11. The "croissant" transformation.

From Theorem 3.11 a recursive checking algorithm for vertical automorphisms is immediate, see Algorithm 5.

Algorithm 5: Checking for a vertical automorphism.

```
input  : a canonical decomposition tree T of a series-parallel digraph with
         root u and a vertical labeling
output : TRUE if T has a vertical automorphism, and FALSE otherwise
begin
    if u is a Q-node then
        return TRUE;
    if u is a S-node then
        if vertical-check(v) == TRUE for every child v of u then
            return TRUE;
        else
            return FALSE;
    if u is a P-node then
        partition the children of u into classes with equal values of code;
        if all the sizes of the classes are even then
            return TRUE;
        if more than one class has odd size then
            return FALSE;
        if only one class has odd size then
            choose some node v in this odd sized class;
            return vertical-check(v);
end
```

Note that a topological embedding of a series-parallel digraph is defined by the order of the P-nodes in the canonical decomposition tree. It is straightforward to adjust the subroutines of the checking algorithm for the maximum upward planar automorphism group such that it reorders the P-nodes corresponding to that automorphism.

It remains to explain how to construct a symmetric drawing based on such an embedding. We will sketch the construction for two layout styles, namely *visibility drawings* and *bus-orthogonal drawings*. In a *visibility drawing*, each vertex is mapped to a horizontal and each edge to a vertical line segment. For a series-parallel digraph, we may also require that the vertical line segment for the source is a horizontal translation of the vertical line segment for the sink. The drawing for a graph with a single edge is obvious. In general, the visibility drawing is constructed recursively by series and parallel compositions, as illustrated in Figure 3.12.

The principle of bus-orthogonal drawings is to connect neighboring vertices via a so-called bus. A *bus* is a horizontal line segment just below or above a vertex. In a *bus-orthogonal drawing* of a series-parallel digraph, the

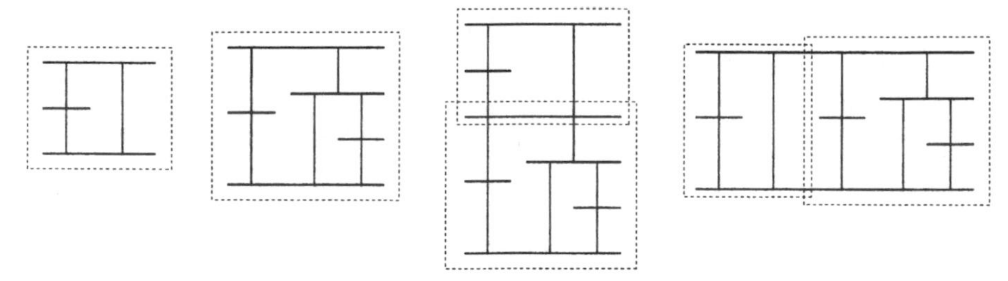

Fig. 3.12. Recursive construction of visibility drawings for series-parallel digraphs: Given two visibility representations (first two figures), a representation for a series composition (third figure) is obtained by "stretching" the narrower and identifying the sink of the first with the source of the second, whereas a representation (right figure) for a parallel composition is constructed by "stretching" the shorter representation and identifying the sources and sinks.

source s has a bus just above, the sink t has a bus just below, and all other vertices have exactly one bus above and one below. Each vertex is connected to its bus(ses) by vertical line segments, and neighboring vertices share a bus. An easy transformation yields bus-orthogonal drawings from visibility drawings, see Figure 3.13.

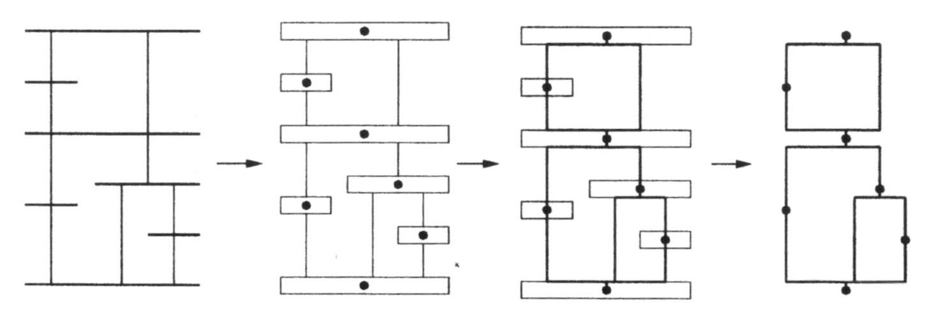

Fig. 3.13. Transformation from visibility drawings to bus-orthogonal drawings.

3.2.4 Three-Dimensional Drawings

Finally, we report on an algorithm for a bus-orthogonal drawing in three dimensions which minimizes the footprint (Hong et al., 1999b). The *footprint* of a three-dimensional drawing is its projection into the xy–plane, its *size* is measured by the minimum enclosing rectangle. A layout has *minimal footprint* if it has size $X \times Y$, and there is no layout with footprint of size $X' \times Y'$ where $X' \leq X, Y' \leq Y$, and $(X', Y') \neq (X, Y)$.

Hong et al. (1999b) developed a dynamic programming approach which yields a minimum size footprint layout. The basic idea is that for each parallel node in the canonical decomposition tree we have the freedom either to align all its children with the x-axis or with the y-axis. By a rotation of a component by 90° at the z-axis, it might be possible to reduce the footprint, see Figure 3.14. In contrast, the extent in the z-axis is fixed by the height of the canonical decomposition tree. Hence, in order to minimize the footprint one has to choose for each parallel composition the alignment to either the x- or y-axis. The mentioned dynamic programming algorithm traverses the canonical decomposition tree in a bottom-up fashion and computes the minimal layouts for each node of the decomposition tree from the minimal layouts of its children. For the details, we refer to Hong et al. (1999b).

Theorem 3.12 (Hong et al. 1999b). *There is a dynamic programming algorithm which computes a minimum size footprint layout of a series-parallel digraph in time $O(n^2)$.*

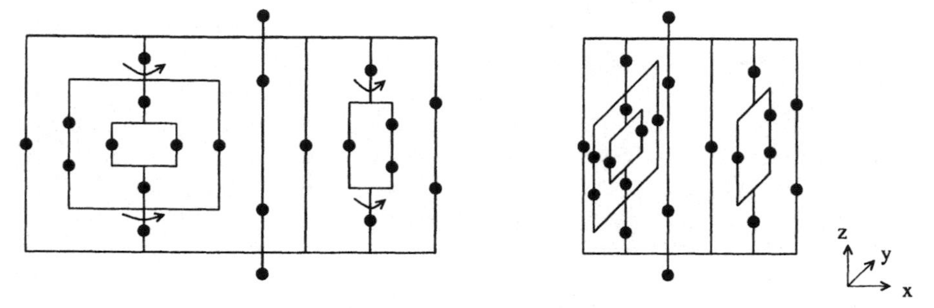

Fig. 3.14. Two- and three-dimensional bus-orthogonal drawings, the latter after rotations as indicated on the left side.

3.3 Lattices

The theory of ordered sets and lattices has become a fundamental discipline in modern mathematics. In many cases a diagram generated from an abstract representation of a lattice (or an ordered set) is an important aid for the understanding of its structure. Therefore, researchers quite often use diagrams to gain structural insights about lattices. Besides these inner-mathematical applications, the practical need for drawings arises in scheduling, in graphical analysis of statistical data, and formal concept analysis, where "the diagrams should not only reflect the structure of a concept lattice but also unfold views for interpreting the data" (Wille, 1989).

Lattices are usually represented by hierarchically layered drawings. A complete chapter of this book (Chapter 5) is devoted to general methods for layered drawings. These methods usually take arbitrary digraphs as input, but are, of course, well-suited for acyclic digraphs related to ordered sets. In this section we will only introduce the general concept of a diagram for ordered sets but then specialize our treatment to results for lattices. In particular, we discuss in more detail the relationship of lattices and planarity in Subsection 3.3.2, and the application of lattices in formal concept analysis in Subsection 3.3.3.

3.3.1 Order Diagrams

Let $(P, \leq)$ be a *partially ordered set (poset)* with ground set P and order relation $\leq$ which is a reflexive, antisymmetric and transitive binary relation on P. Distinct elements $a, b \in P$ are *comparable* when either $a < b$ or $b < a$. For two elements $a, b \in P$, we say b *covers* a and a *is covered by* b (written as $a \prec b$) if $a < b$ and $a \leq c < b$ implies $c = a$. We also call a a *lower cover* of b, and b an *upper cover* of a. In addition, (a, b) is called a *covering pair.*

The *line diagram (Hasse diagram)* or simply *diagram* of a poset draws the elements of P as small circles (vertices) in the plane such that if $a, b \in P$, and $a < b$ then a is drawn with smaller y-coordinate than b. There is an edge between a and b if and only if $a \prec b$. In other words, the diagram is an upward planar drawing of the *covering digraph* of a poset which contains the poset elements as vertex set and the covering pairs as directed edges from the lower to the upper covers. See Figure 3.15 for an example of a diagram. Posets and their diagrams are closely related to each other.

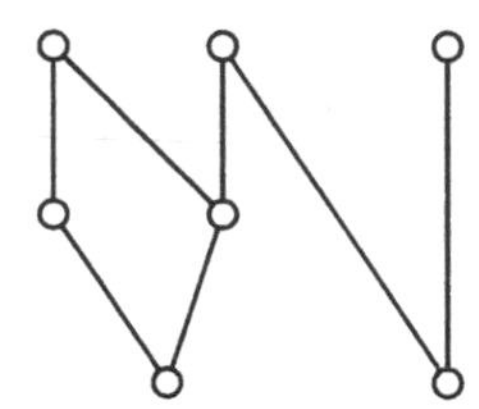

Fig. 3.15. Example of a Hasse diagram.

Lemma 3.13 (Uniqueness of the diagram). *A finite partially ordered set is determined up to isomorphism by its diagram.*

When $Q \subset P$, the set $\{p \in P : q \leq p \text{ in } P \text{ for every } q \in Q\}$ is the set of *upper bounds* for Q. Dually, the set $\{p \in P : q \geq p \text{ in } P \text{ for every } q \in Q\}$ is the set of *lower bounds* for Q. If it exists, the unique smallest upper bound of $a, b \in P$ is called the *join* $a \vee b$, and similarly, the unique greatest lower bound of $a, b \in P$ is the *meet* $a \wedge b$. A poset is a *lattice* if for every two elements $a, b \in P$ the join $a \vee b$ and the meet $a \wedge b$ exist.

Figure 3.16 shows all lattices with five elements.

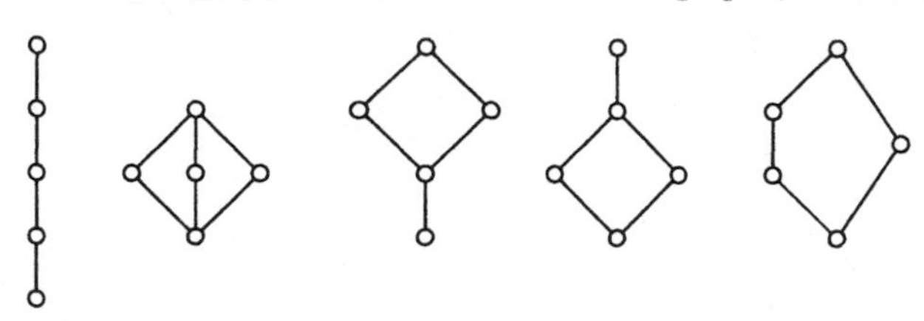

Fig. 3.16. Diagrams of lattices with 5 elements.

Criteria for Good Diagrams. To emphasize the ordering relation, edges in a diagram are usually drawn as "steep" as possible. An obvious goal is also to draw the diagram without crossings, if possible. In addition, whatever symmetries exist should be made "apparent."

Apart from that, general criteria for good drawings of lattices are hard to find. To see that such criteria depend on its intended use, consider the example of the lattice 2^4 (2^n denotes the lattice of all subsets of an n-element set ordered by inclusion). Figure 3.17, taken from Rival (1985), shows four different drawings of this lattice, the first is a drawing as the direct product $2^1 \times 2^3$, the second as the direct product $2^2 \times 2^2$, and the other two drawings are "merely" symmetric drawings. Obviously, there may be a conflict between the goals of highlighting symmetry and putting emphasis on certain structural properties.

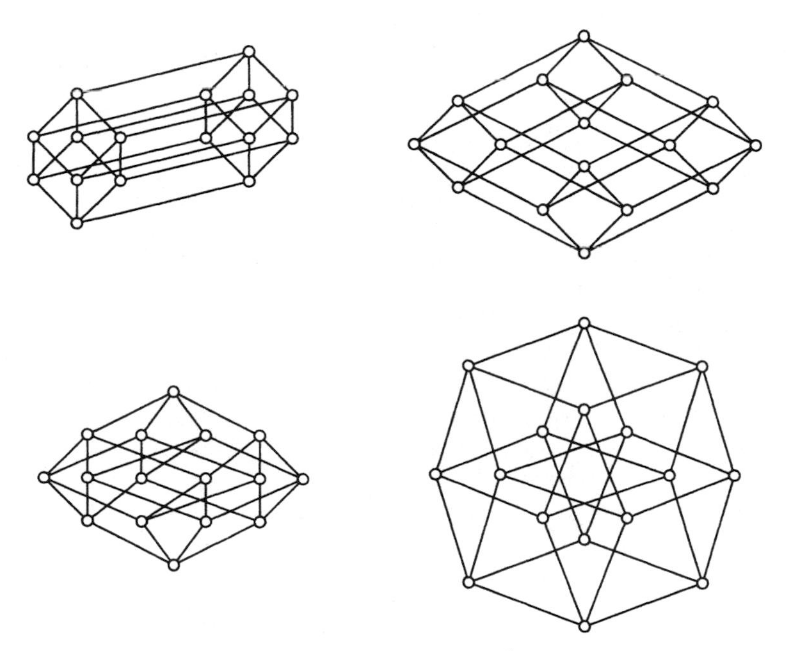

Fig. 3.17. Four different drawings of the lattice 2^4.

3.3.2 Planar Lattices

Special attention has been paid to planar lattices. A lattice is called *planar* if it has a planar diagram.

A first characterization of planar lattices needs some more definitions. A poset is a *chain* (also a *totally ordered set* or a *linearly ordered set*) if each pair of elements of the ground set is comparable. A *linear extension* of a poset $(P, \leq)$ is a chain $(P, \leq')$ defined on the same ground set which respects all comparabilities of the relation $\leq$. The *(order) dimension* of a poset $(P, \leq)$ is the least t for which there exists a family $\{L_1, L_2, \ldots, L_t\}$ of linear extensions of P such that $P = L_1 \cap L_2 \cap \cdots \cap L_t$. Baker et al. (1971) gave a characterization in terms of the dimension of a lattice.

Theorem 3.14 (Baker et al. 1971). *A lattice is planar if and only if it has order dimension at most two.*

In contrast, Kelly and Rival (1975) characterized finite lattices in terms of forbidden configurations, so-called *obstructions*. They showed that a certain family $\mathcal{L}$ of non-planar lattices is a minimal obstruction set for planar diagrams.

Theorem 3.15 (Kelly and Rival 1975). *A finite lattice is planar if and only if it contains no subset isomorphic to a member of the family $\mathcal{L}$.*

An necessary and sufficient condition for planar lattices that can be tested efficiently has been given by Platt (1976).

Theorem 3.16 (Platt 1976). *A finite lattice is planar if and only if the undirected covering graph plus an additional edge from the maximum to the minimum element is planar.*

In addition to planarity, one has studied the criterion to use as few slopes as possible for the drawing of the covering edges. For a while, it was thought that the minimum number of slopes needed to draw a lattice depends only on the maximum number of upper covers (the maximum *up-degree*) and the maximum number of lower covers (the maximum *down-degree*) among the elements of the lattice. Of course, for any ordered set, the maximum of the up-degrees and the down-degrees is a lower bound on the number of slopes needed. However, the conjecture that this is the actual number of slopes needed has been disproved for lattices in general (Czyzowicz et al., 1990; Czyzowicz, 1991). In contrast, nice positive results have been found for planar lattices.

Theorem 3.17 (Czyzowicz et al. 1990). *Every finite planar lattice with maximum up-degree and down-degree two has a planar, two-slope diagram.*

3.3.3 Concept Lattices

We conclude our section on lattices with an application of graph drawing for the visualization of formal concepts. The description is based on Wille (1997), Wille (1989), Vogt and Wille (1995), Ganter and Wille (1999), and Vogt (1996).

A *formal context* is a triple (G, M, R) where G is a set of *objects*, M a set of *attributes*, and R a relation between objects and attributes.

Such a context is often described as a cross table, see Table 3.2. We take the example of the context Living Beings and Water from Wille (1997) and Ganter and Wille (1999).

For formalizing concepts within a context we define the following derivation operators:

$$A \mapsto A' := \{m \in M \mid gRm \text{ for all } g \in A\} \text{ for } A \subseteq G,$$

$$B \mapsto B' := \{g \in G \mid gRm \text{ for all } m \in B\} \text{ for } B \subseteq M.$$

By means of the derivation operators, we define the *formal concept* of a context (G, M, R) as a pair (A, B) with $A \subseteq G, B \subseteq M$ and $A' = B, B' = A$. A is called the *extent*, B the *intent* of the concept. Note that for each $A \subseteq G$, the set A' is the intent of some concept, because (A'', A') is always a concept. The set A'' is the smallest extent of a concept which contains A. In the given example (Table 3.2), we have 19 formal concepts in total, among them the concepts $(\{1,2,3\}, \{a,b,g\})$ and $(\{2,3\}, \{a,b,g,h\})$.

The *subconcept-superconcept-relationship* describes the order relation between concepts:

$$(A_1, B_1) \leq (A_2, B_2) :\Leftrightarrow A_1 \subseteq A_2 (\Leftrightarrow B_1 \supseteq B_2).$$

The set of all concepts of a context (G, M, R) together with the subconcept-superconcept-relation is the *concept lattice* of (G, M, R).

Construction of the Concept Lattice. To build the concept lattice from a given context one has to determine all concepts first. An efficient approach computes the extents of all concepts in a certain order.

For simplicity we assume that $G = \{1, 2, \ldots, n\}$. A subset $A \subseteq G$ is smaller in the so-called *lectic order* as a set $B \neq A$, if the smallest element for which A and B differ from each other belong to B. More formally,

$$A < B :\Leftrightarrow \exists_{i \in B \setminus A} \; A \cap \{1, 2, \ldots, i-1\} = B \cap \{1, 2, \ldots, i-1\}.$$

Suppose that we are able to compute for any given set $A \subset G$ the smallest extent of a context which is larger than A with respect to the lectic order. Then there is an obvious algorithm to compute all extents. The smallest extent is $\emptyset''$. We get all other extents if we determine successively from the last found extent the next one in lectic order. The process terminates with the largest extent, namely G.

To do that, we define for $A, B \subseteq G, i \in G$,

$$A <_i B :\Leftrightarrow i \in B \setminus A \text{ and } A \cap \{1, 2, \ldots, i-1\} = B \cap \{1, 2, \ldots, i-1\},$$

$$A \oplus i := ((A \cap \{1, 2, \ldots, i-1\}) \cup \{i\})''.$$

Table 3.2. Cross table of the context 'Living Beings and Water' from Wille (1997).

		a	b	c	d	e	f	g	h	i
1	leech	X	X					X		
2	bream	X	X					X	X	
3	frog	X	X	X				X	X	
4	dog	X		X				X	X	X
5	spike-weed	X	X		X		X			
6	reed	X	X	X	X		X			
7	bean	X		X	X	X				
8	maize	X		X	X		X			

a = needs water to live
b = lives in water
c = lives on land
d = needs chlorophyll
e = two germ-layers
f = one germ-layer
g = can move about
h = has limbs
i = suckles its offspring

Theorem 3.18. *Let $A \subset G$ be a set, and let i be the largest element of G with $A <_i A \oplus i$. Then $A \oplus i$ is the smallest extent that is larger than A with respect to the lectic order.*

Hence, for a given set $A \subset G$ we can find the next extent in lectic order in the following way. We successively test for all elements $i \in G \setminus A$, starting with the largest element and continuing in decreasing order, whether $A <_i A \oplus i$. As soon as this condition becomes true, we get $A \oplus i$ as the next extent.

Labeled Line Diagrams. The concept lattice of the context in Table 3.2 is represented in Figure 3.18 as a *labeled line diagram*. An element labeled by an object g represents the concept with the smallest extent containing g, whereas an element labeled by an attribute m represents the concept with the smallest intent containing m. The extent of each concept can be obtained by collecting all objects which can be reached by descending paths, and conversely, the intent can be obtained dually by collecting all attributes which can be reached by ascending paths.

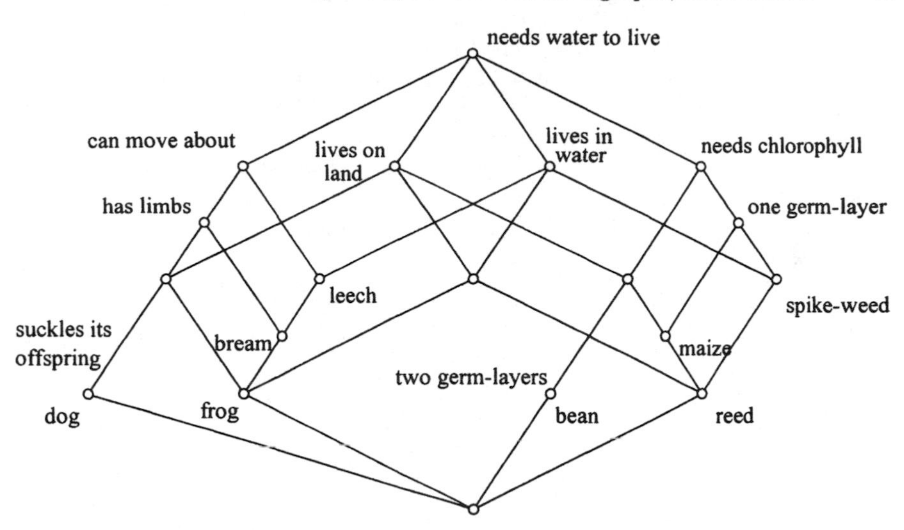

Fig. 3.18. Diagram of the concept 'Living Beings and Water'.

The automatic drawing of concept lattices is seemingly still a great challenge, as formal criteria like edge crossing minimization usually do not exhibit the lattice structure sufficiently. According to Stumme and Wille (1995), best results have been achieved by an interactive drawing method using geometrical heuristics and a lot of experience. Empirically it has been proved useful to develop the diagram from the top downwards (or vice versa) using lists of upper (lower) covers for the elements of the lattice. They also proposed two heuristics, the parallelogram rule and the straight line rule. The *parallelogram rule* recommends to place a new element such that, together with the introduced lines, it completes a parallelogram with three previously drawn elements. The *straight line rule* simply tries to place a new element such that the induced line segments extend pre-existing ones.

Nested Diagrams. Even a context of moderate size may have a concept lattice with many covering pairs which makes the diagram hard to read. Nested diagrams have been introduced to reduce the number of lines to be drawn, see Figure 3.19. The idea of *nested diagrams* is to factor out parts of the lattice into "blocks" (enclosed by rectangular boxes) with the understanding that a *single line* connecting two such blocks corresponds to parallel lines between identical pairs of the two blocks, whereas a *double line* means that all maximal elements of the lower block are pairwise covered by the minimal elements of the upper block.

We can obtain a nested line diagram from a formal context by first splitting the set of attributes into two parts, that is $M = M_1 \cup M_2$. The subsets M_1, M_2 are not necessarily disjoint but they should bear a certain meaning to allow an insightful interpretation. In a second step, the line diagrams are drawn for the two contexts $C_i = (G, M_i, R \cap (G \times M_i))$, for $i = 1, 2$. Finally, one takes one of the two line diagrams as the outer structure, enlarges the representation of each of its elements to a rectangular box, and inserts the other line diagram into each of the enlarged boxes. This way, one gets a representation of the direct product of the concept lattices of the contexts C_i, and the concept lattice of the original context can be embedded into this direct product.

The drawing tool TOSCANA of Vogt (1996), for example, makes extensive use of nested diagrams.

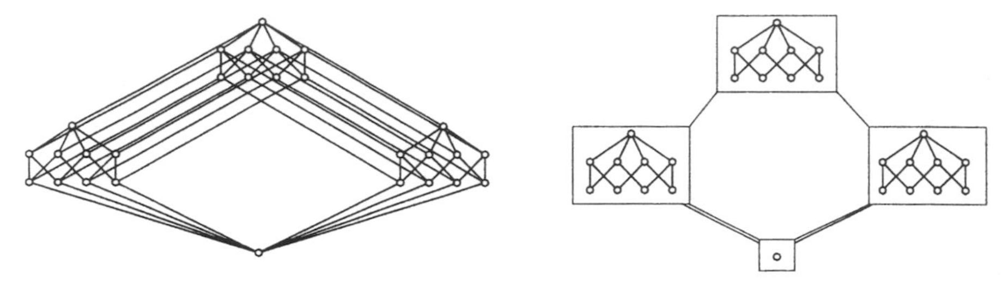

Fig. 3.19. Nested diagram of a lattice.

4. Drawing on Physical Analogies

Ulrik Brandes

Graph layout methods described in previous chapters were based on structural characteristics of the graph, or a preprocessed version of the graph. Often, such knowledge is not provided. In this chapter, we take a look at a class of methods applicable to general graphs, without prior knowledge of any structural properties. Their common denominator is that they liken the graph to a system of interacting physical objects, the underlying assumption being that relaxed (energy-minimal) states of suitably defined systems correspond to readable and informative layouts.

Methods based on physical analogies are quite popular, mainly for three reasons. First of all, they are very intuitive, because layout is related to the everyday experience of the surrounding physical world. Secondly, their basic instances are comparatively easy to understand and to program. The threshold to get started is thus very low. And finally, they often yield fairly satisfactory results on medium-sized graphs up to around 50 vertices. In general, these methods consist of two components,

- a *model* consisting of physical objects (representing the elements of the graph) and interactions between these objects, and
- an *algorithm* that (approximately) computes an equilibrium configuration of the system.

The specifications of a model fully represent the intuition behind what is considered a good layout, ideally depending on the specific context of the graph. Its associated algorithm merely serves as an optimization routine for the objective function explicated in the model.

First we introduce the fundamental concept behind physical modeling, that is the basis for all methods presented in this chapter. Then we describe a number of models and associated algorithms in Sections 4.2 and 4.3. An asset of physical modeling that is often overlooked is its inherent flexibility. For this reason, we conclude this chapter by listing examples of model specifications tailored to specific layout objectives.

4.1 The Springs

Given a connected undirected graph with no particular background information, the following two criteria of readable layout seem to be generally agreed upon for the conventional two-dimensional straight-line representation.

1. Vertices should spread well on the page.
2. Adjacent vertices should be close.

M. Kaufmann and D. Wagner (Eds.): Drawing Graphs, LNCS 2025, pp. 71-86, 2001.

Only intuitive explanations can be offered. While uniform vertex distribution reduces clutter, the implied uniform edge lengths leave an undistorted impression of the graph. Since "clutter" and "distortion" already have physical connotations, it seems fairly natural to start thinking of a more specific physical analogy.

We are used to observing even spacing between repelling objects. This makes it natural to imagine vertices behaving like charged balls to satisfy the first criterion. A physical analogy for the second criterion is also easy to find, since it states that we should not allow adjacent balls to drift too far apart. Springs replacing edges will do the job. Springs are better suited than, for example, sticks or ropes, because they can be both extended and compressed to allow moderate distortion, but exert increasingly strong forces when deviating further from their natural length. Moderate distortion may be inevitable, since it is impossible to represent every graph with straight edges of equal length. Note that it is $\mathcal{NP}$-hard to decide whether an arbitrary graph has a straight-line embedding with equal edge lengths in any number of dimensions (Johnson, 1982) or just any planar straight-line embedding (Eades and Wormald, 1990).

Figure 4.1 illustrates the imaginary substitution of vertices and edges with charged balls and connecting springs, respectively. If the system is let go, it attains an equilibrium state in which all forces cancel each other, and the substitution can be reversed to obtain a straight-line drawing that satifies the criteria from above at least approximately.

Formally, such a model can be expressed either in terms of forces acting on the physical objects, or in terms of a potential energy reflecting the internal stress of the system and thus describing how well a configuration matches the design goals modeled in the system. Algorithms to simulate a system's relaxation typically try to move the objects iteratively into stable states in which all forces cancel each other, or to minimize the energy directly. Prominent examples from each kind of formalization are described in the following two sections. In Section 4.4, we give examples for creative use of these analogies in more advanced layout models that include various criteria for good layout.

4.2 Force-Directed Placement

The seminal paper for physical modeling in graph drawing is a short text of Eades (1984), though closely related methods had already been described in the context of VLSI design (Fisk et al., 1967; Quinn and Breuer, 1979). Given a connected undirected graph $G = (V, E)$, let $p = (p_v)_{v \in V}$ be a vector of vertex positions $p_v = (x_v, y_v)$ in the plane. We denote by $\|p_v - p_u\|$ the length of the difference vector $p_v - p_u$, which is the Euclidean distance between positions p_u and p_v. Furthermore, we denote by $\overrightarrow{p_u p_v}$ the unit length vector $\frac{p_v - p_u}{\|p_v - p_u\|}$ pointing from p_u to p_v. The model of Eades (1984), now known as the

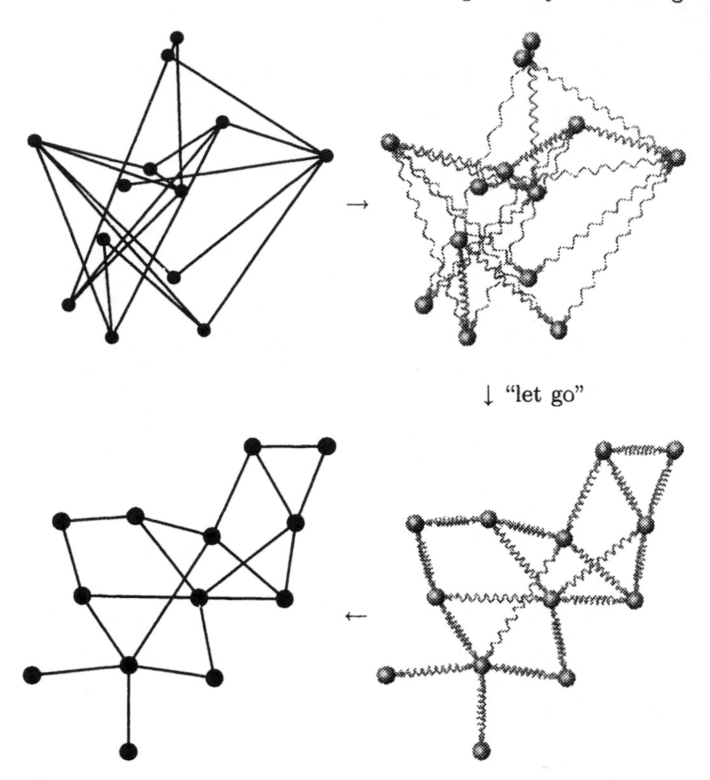

Fig. 4.1. The spring analogy.

spring embedder, implements the analogy described in the previous section. It is defined using repelling forces

$$f_{\text{rep}}(p_u, p_v) = \frac{c_\varrho}{\|p_v - p_u\|^2} \cdot \overrightarrow{p_u p_v}$$

between every pair of non-adjacent vertices $u, v \in V$, where c_ϱ is a repulsion constant. Complementary spring forces between adjacent vertices $u, v \in V$ shall keep these sufficiently apart, yet close to each other. However, instead of more realistic forces according to Hooke's law, (imaginary) logarithmic springs which exert weaker forces on far apart vertices are employed. They yield forces

$$f_{\text{spring}}(p_u, p_v) = c_\sigma \cdot \log \frac{\|p_u - p_v\|}{l} \cdot \overrightarrow{p_v p_u},$$

so that the direction depends on whether the actual distance is less or greater than a natural length l of the spring. Constant c_σ is a parameter controlling the strength of the spring. Figure 4.2 gives a qualitative impression of the forces a vertex u exerts on vertex v, depending on the distance between the two. The solid line shows the force in case u is adjacent to v (f_{spring}), while the dotted line indicates the force in case u is not adjacent to v (f_{rep}). Positive values signify a force dragging v towards u, whereas negative values signify a force repelling v from u.

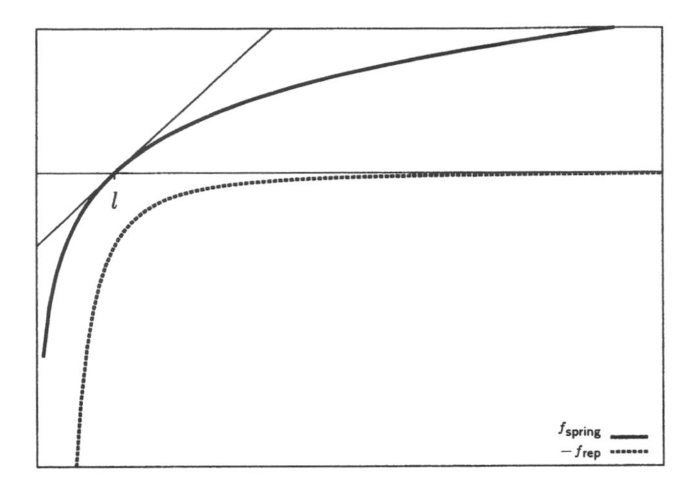

Fig. 4.2. Magnitude of spring embedder forces.

Next we address the question of how to obtain an equilibrium configuration. Vertex positions not corresponding to a system at equilibrium imply positive internal stress. To relax a stressed system, vertices are iteratively moved, at time t, according to a *net force vector* $F_v(t)$, which is the sum of all repulsion and spring forces acting on v. After computing $F_v(t)$ for all $v \in V$, each vertex is moved a constant δ times this vector. This constant is used to prevent excessive movement due to synchronous update. By iteratively computing the forces on all vertices and updating positions accordingly, the system approaches a stable state, in which no local improvement is possible. See Algorithm 6 for a concise description.

Algorithm 6: Spring embedder

Input: connected undirected graph $G = (V, E)$
initial placement $p = (p_v)_{v \in V}$

Output: placement p with low internal stress

for $t \leftarrow 1$ ***to*** ITERATIONS **do**
  for $v \in V$ **do**
    $F_v(t) \leftarrow \sum_{u:\{u,v\} \notin E} f_{\mathsf{rep}}(p_u, p_v) + \sum_{u:\{u,v\} \in E} f_{\mathsf{spring}}(p_u, p_v)$
  for $v \in V$ **do** $p_v \leftarrow p_v + \delta \cdot F_v(t)$

Despite its simplicity, the spring embedder produces satisfactory output in many cases. To even out some shortcomings of the method, several refinements have been developed. These refinements mainly aim at faster computation, but sometimes also at improved quality of the layout.

A number of heuristics is used by Fruchterman and Reingold (1991) to speed up many aspects of layout computation. Firstly, the forces are modified to allow faster evaluation. Repelling forces

$$f_{\mathsf{rep}}(p_u, p_v) = \frac{l^2}{\|p_u - p_v\|} \cdot \overrightarrow{p_u p_v}$$

are used between every pair of vertices, and additional attracting forces

$$f_{\mathsf{attr}}(p_u, p_v) = \frac{\|p_u - p_v\|^2}{l} \cdot \overrightarrow{p_v p_u}$$

are used between adjacent vertices. The combination of attraction and repulsion between adjacent vertices yields a spring-like force $f_{\mathsf{spring}}(p_u, p_v) = f_{\mathsf{attr}}(p_u, p_v) + f_{\mathsf{rep}}(p_u, p_v)$, similar in effect to the force used by Eades (1984). Since its magnitude increases more than proportionally with the distance (see Figure 4.3 for a comparison), one may also hope for faster convergence.

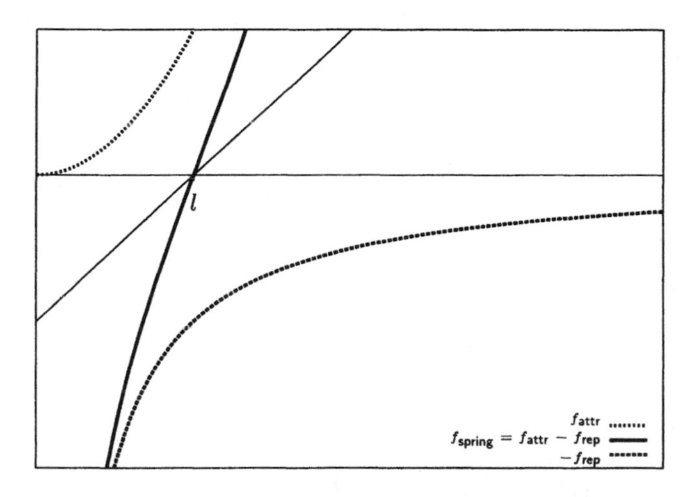

Fig. 4.3. Modified forces by Fruchterman and Reingold (1991).

A second heuristic to speed up computation does not change the objective function, but the precision of evaluation. Since repulsion from far away vertices does not contribute much to the displacement vector, such irrelevant vertices are omitted in the sum of repulsive forces using a grid technique. Only vertices lying in grid cells close to the cell of v are considered, and only if their distance is below a fixed threshold, a repulsive force is calculated and included in the sum of forces. See Figure 4.4(a).

Two other modifications with respect to Algorithm 6 are concerned with the displacement vector. Instead of applying a constant damping factor δ to the net force vector, the net force vector is clipped at a time-dependent maximum displacement $\delta(t)$ to prevent excessive changes, especially in later stages of the iteration when the placement is close to a stable state. The second modification to the displacement ensures that the graph is laid out inside of a given rectangular area, like a screen or a sheet of paper. If the displacement would position a vertex beyond a fixed boundary, the corresponding coordinate of the displacement vector is clipped.

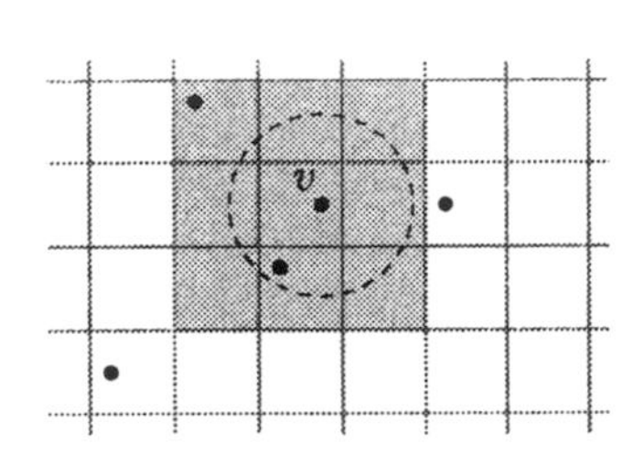

(a) neglecting weak repulsive forces

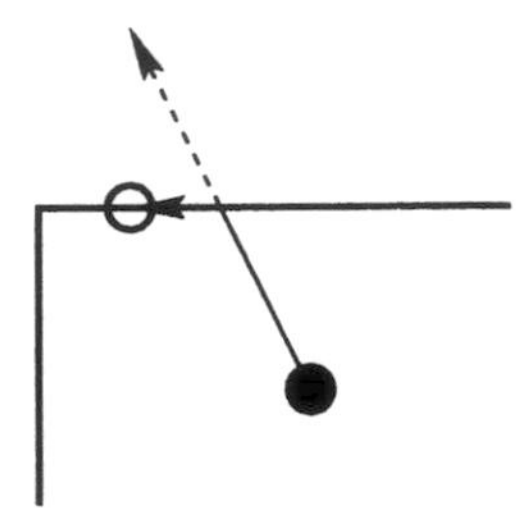

(b) coordinate clipping

Fig. 4.4. Spring embedder modifications of Fruchterman and Reingold (1991).

Another notable refinement of the basic spring embedder is described in Frick et al. (1995). Again, both forces and iteration scheme are modified to speed up the algorithm and to improve layout quality (under the same criteria). Repulsive and attractive forces are defined so that no square root has to be taken,

$$f_{\mathsf{rep}}(p_u, p_v) = \frac{l^2}{\|p_u - p_v\|^2} \cdot (p_u - p_v),$$

$$f_{\mathsf{attr}}(p_u, p_v) = \frac{\|p_u - p_v\|^2}{l^2 \cdot \Phi(v)} \cdot (p_v - p_u),$$

and all computations are performed using integer arithmetic. The denominator in the attractive force is defined as $\Phi(v) = 1 + \frac{d_G(v)}{2}$ and effectively slows down high-degree vertices. However, a new gravitational force is introduced, dragging each vertex towards the barycenter $\zeta = \sum_{w \in V} p_w$ of all vertices by

$$f_{\mathsf{grav}}(p_u, p_v) = \Phi(v) \cdot \gamma \cdot \left(\frac{\zeta}{|V|} - p_v \right),$$

which is stronger for high degree vertices. The contribution of this force is controlled via a gravitational constant γ. Similar to layout area restriction, gravitational forces prevent components of a disconnected graph from drifting arbitrarily far apart. In addition to the above forces, a small random force is added to the net force to make the algorithm more robust against poor equilibrium states.

To reduce the number of iterations, the net force vector of each vertex during the previous iteration, $F_v(t-1)$, is stored and compared to the current one. Each vertex has its own adaptive absolute displacement distance $\delta_v(t)$ that is modified according to the angular difference $\alpha = \angle(F_v(t-1), F_v(t))$ of the current and the previous net force vector. If a vertex is to be moved into roughly the same direction as before ($\sin \alpha \approx 1$), $\delta_v(t)$ is chosen larger, if it is to be moved in the opposite direction (*oscillation*: $\sin \alpha \approx -1$), $\delta_v(t)$ is chosen smaller. Like oscillation, *rotation* is an indicator of ineffective movement of a vertex. A *skew gauge* is updated, if the current and the previous displacement vector are almost perpendicular ($\cos \alpha \approx 1$), and $\delta_v(t)$ is lowered, if a large skew suggests that a vertex rotates around some position. During each iteration, vertices are visited in random order, and each position is updated immediately by $p_v \leftarrow p_v + \max\{\delta_{\mathsf{max}}, \delta_v(t)\} \cdot \frac{F_v(t)}{\|F_v(t)\|}$, where δ_{max} is a fixed maximum displacement. Our own experience confirms that these heuristics substantially reduce the number of iterations needed to reach a stable state.

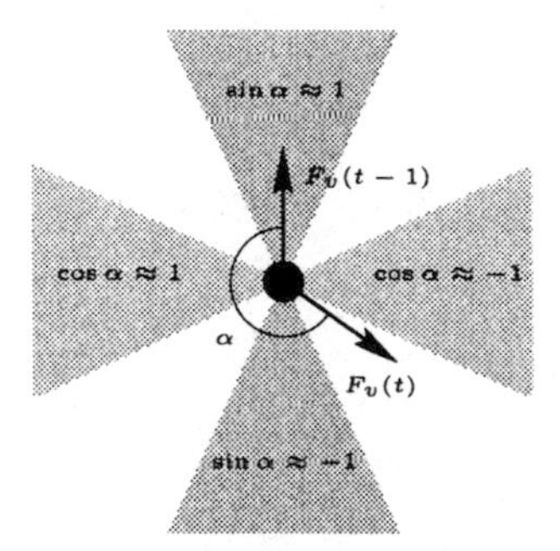

Fig. 4.5. Detection of oscillation and rotation (Frick et al., 1995).

The modifications of Fruchterman and Reingold (1991) and Frick et al. (1995) demonstrate that the spring embedder can be varied in many ways without changing its principle behavior. Clearly, many other heuristics are conceivable, but the two examples presented seem to cover sufficiently many aspects relevant to faster computation, faster convergence, and robust results.

In the next section, we turn to methods that try to satisfy the same criteria, but take a formally different approach by defining an explicit objective function.

4.3 Energy-Based Placement

Forces defined in the spring embedder variants described above indicate in which direction a vertex can be moved to reduce the forces acting on it, and thus an implicit internal energy of the physical system. Instead of displacing vertices according to these forces, one might as well attempt to minimize this energy directly. A spring of natural length l and of strength c_σ with actual length d (assumed to be within reasonable limits) has a *potential energy* of

$$U_{\text{spring}}(d) = c_\sigma \cdot (d - l)^2,$$

Kamada and Kawai (1989) avoid a second potential for repulsion by using springs of different length and strength between every pair of vertices. Their specific choice of springs is governed by the assumption that the ideal distance between two vertices is the length of a shortest path between them, multiplied by the ideal length of a single edge, i.e. every path in the graph is best represented by a straight line. The natural length of the spring connecting vertices $u, v \in V$ is therefore chosen proportional to $d_G(u, v)$, which denotes the length of a shortest path between them. Clearly, perfect relaxation of all springs is impossible for most graphs, so local distances are rendered more important by using springs of strength inverse to their length. The resulting objective function is the sum over the potential energies of all $n \cdot (n-1)/2$ springs,

$$U_{\mathsf{KK}}(p) = \sum_{u,v \in V} \frac{c}{d_G(p_u, p_v)^2} \cdot \left(\|p_u - p_v\| - l \cdot d_G(u, v)\right)^2,$$

where c is a scaling constant, and l is the ideal length of a single edge.

To obtain a local minimum of this objective function, a modified Newton-Raphson method is applied. In a local minimum, all partial derivatives of U_{KK} are zero. This condition can be expressed in a system of dependent non-linear equations. Similar to Quinn and Breuer (1979), the Newton-Raphson method is modified in that the coordinates of a single vertex are updated while all others are fixed. In each iteration, the vertex with the longest gradient is picked and moved several times until its gradient falls below a given threshold.

It is interesting to note that the physically inspired objective function U_{KK} is closely related to the objective function

$$U_{\mathsf{MDS}}(p) = \frac{1}{\sum\limits_{u,v \in V} d_G(u, v)^2} \cdot \sum_{u,v \in V} \left(\|p_u - p_v\| - l \cdot d_G(u, v)\right)^2$$

of multidimensional scaling defined in (Kruskal and Wish, 1978). The family

$$S_k(p) = \frac{1}{\sum\limits_{u,v \in V} l_{u,v}^{2-k}} \cdot \sum_{u,v \in V} \frac{1}{l_{u,v}^k} \cdot \left(\|p_u - p_v\| - l_{u,v}\right)^2$$

of objective functions, where $l_{u,v}$ is the desired distance between vertices u and v, and $k \in \{0, 1, 2\}$, is discussed by Cohen (1997). While S_0 corresponds to multidimensional scaling, S_2 corresponds to the above layout objective function. As early as in the 1960s, multidimensional scaling was the first technique for automatic layout of social networks,[1] and it is still in use (Krackhardt et al., 1994). It seems, though, that the bias of U_{KK} towards exact representation of short distances, due to stronger short springs, results in layouts that display less clutter and have fewer small angles.

While the optimization method used by Kamada and Kawai (1989) does not differ notably from the ones in the previous section, the following simplification of the objective function admits exact optimization in time that is polynomial in the number of vertices. Instead of springs of some varying length, Tutte (1963) uses springs of ideal length zero. Setting the partial derivatives of the resulting objective function

$$U_{\mathsf{center}}(p) = \sum_{\{u,v\} \in E} \|p_u - p_v\|^2$$

equal to zero yields two independent systems of linear equations (one for each coordinate), as opposed to the one non-linear system obtained from U_{KK}. These linear systems of equations can conveniently be written in the form

$$(D - A) \cdot x = \mathbf{0},$$
$$(D - A) \cdot y = \mathbf{0},$$

where A is the adjacency matrix of G, D is the diagonal matrix of vertex degrees, and x and y are the vectors of x- and y-coordinates. The famous Matrix Tree Theorem (Kirchhoff, 1847) then states that the determinant of any submatrix obtained from $D - A$ by deleting a positive number of rows and their corresponding columns equals the number of spanning trees of the graph obtained from G by contracting the vertices corresponding to these rows into a new one. See Chaiken and Kleitman (1978) for several variants of this theorem. Since this number clearly is positive for connected graphs, so is the determinant, which implies that there is a unique layout minimizing U_{center}, if only the position of at least one vertex in each connected component of the input graph is fixed. This optimal layout can be computed by solving the smaller system of linear equations obtained by adjusting the right hand sides using the coordinates of fixed vertices.

Optimal layouts with respect to U_{center} are called *barycentric*, because the optimality conditions imply that every vertex not fixed in advance is placed in the barycenter of its neighbors. The main theorem of Tutte (1963) assures even more: barycentric layouts of 3-connected planar graphs are planar with

[1] Charles Kadushin, personal communication (1999).

all internal faces convex, if the vertices of a single face in the unique planar embedding are fixed to lie on a convex polygon (in appropriate order). Since the adjacency matrices of planar graphs are sparse, such layouts can be obtained in time $\mathcal{O}(n \log n)$ (Lipton et al., 1979). See Figure 4.6 for an example.

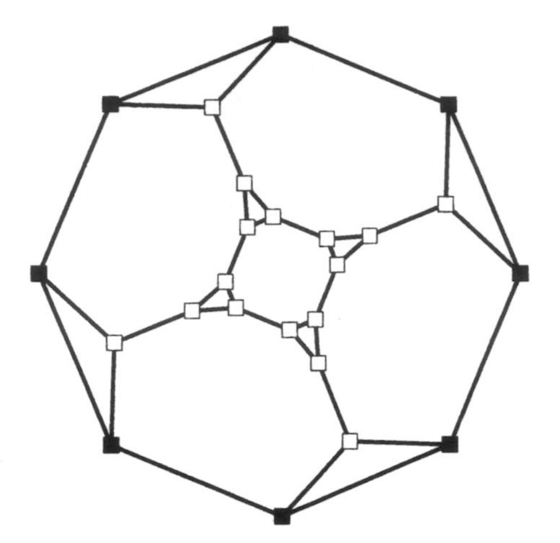

Fig. 4.6. Barycentric layout (darker vertices have been fixed in advance).

Davidson and Harel (1996) deviate even further from models of physical reality. They define attraction and repulsion potentials

$$U_{\mathsf{attr}}(p_u, p_v) = c_{\mathsf{attr}} \cdot \|p_u - p_v\|^2$$
$$U_{\mathsf{rep}}(p_u, p_v) = \frac{c_{\mathsf{rep}}}{\|p_u - p_v\|^2}$$

that are similar to the respective forces of Fruchterman and Reingold (1991). Note that the attraction potential is a scaled version of the zero-length spring potential of Tutte (1963). Combining attraction and repulsion yields a spring potential $U_{\mathsf{spring}}(p_u, p_v) = U_{\mathsf{attr}}(p_u, p_v) + U_{\mathsf{rep}}(p_u, p_v)$ that is again computed for pairs of adjacent vertices. Note that this definition results in an ideal edge length of $\sqrt[4]{c_{\mathsf{rep}}/c_{\mathsf{attr}}}$. Generalizing the repulsion analogy, they also define potentials penalizing vertices that lie close to the boundary of the layout area, and potentials penalizing short distances between a vertex and an edge.

A distinct ingredient of the objective function used in Davidson and Harel (1996) is a potential weighting of the number of crossings in the current layout. Crossings tend to be displeasing and hindering to a viewer (Purchase et al., 1997), and their minization is thus a reasonable criterion for good layout. However, counting crossings leads to a discrete objective function that can no longer be treated by an algorithm based on gradient methods. Moreover, minimizing the number of crossings is also an $\mathcal{NP}$-hard problem (Garey and Johnson, 1983). To obtain at least a local minimum of U_{DH}, *simulated annealing* (see Reeves 1995 for a more recent textbook), a general method for minimizing objective functions of combinatorial problems, is used. It is only by coincidence that this method also has a physical analogy. Given a candidate solution, a new solution is proposed by slightly modifying the current one. If the new solution reduces the value of the objective function, it becomes the new candidate solution. Otherwise, it becomes the new candidate solution only with probability $e^{\frac{-\Delta U}{T}}$, where ΔU is the increase of the objective function, and $T > 0$ is the *temperature* parameter controlling the algorithms ability to climb up hills in the energy landscape. Convergence is enforced by slowly lowering T to zero. A more precise description of a typical simulated annealing algorithm used for graph layout is given in Algorithm 7.

Algorithm 7: Simulated annealing

Input: graph $G = (V, E)$
initial placement $p = (p_v)_{v \in V}$

Output: placement p with locally optimal value $U(p)$

```
while T > THRESHOLD do
    for v ∈ V do
        p^old ← p
        p_v ← p_v + Δ_random
        if U(p^old) < U(p) then
            with probability 1 − e^((U(p^old) − U(p))/T) reset p ← p^old
    reduce T
```

A number of heuristics to speed up layout computation with the notoriously slow simulated annealing is introduced by Tunkelang (1994). Most notably, there is no initial placement, but vertices are introduced in a breadth-first-search order starting in the graph-theoretic center, and positions are restricted to a coarse grid with only few types of displacements allowed. A parallel implementation is described in Monien et al. (1996).

An experimental comparison (Brandenburg et al., 1996) of (sequential) implementations revealed that the approaches presented in these last two sections (excluding the barycentric model) yield comparable layouts. As a very general rule-of-thumb, energy-based placement approaches tend to produce better results for small to medium-sized graphs (around 30 vertices), while force-directed placement approaches are considerably faster (experience suggest that they need only one tenth of the number of iterations).

Besides gradient methods and simulated annealing, genetic algorithms have been applied to physically motivated objective functions (Kosak et al., 1994; Masui, 1992). Branke et al. (1997) use force-directed methods as a local fine-tuning step of their genetic algorithm.

4.4 Modeling with Forces and Energies

Forces and potential energies have been found to model elementary criteria for readable layouts of straight-line representations of graphs. While the results are usually satisfactory with respect to the two criteria mentioned in the introductory section of this chapter (uniform vertex distribution, uniform edge lengths, and, as a consequence, symmetry), such drawings may not be useful for graphs representing specific structural information.

An important aspect that has not been covered yet is the immense expressive power of force-directed or energy-based placement for formulating layout design goals. This section contains a number of examples, showing how forces can be used to formulate criteria for good layout far beyond vertex distribution and edge length.

The following paragraphs sketch modeling ideas from the literature, comprising a toolbox for the adjustment of force-directed or energy-based layout methods to account for a fairly broad range of requirements.

3D Layouts. For some purposes it may be necessary or desirable to represent a graph in three-dimensional space. In combination with interactive (preferably fly-through) browsers, these appear to be particularly useful for exploring large graphs. The methods outlined so far make no particular assumptions on the number of dimensions and are easily modified to produce three-dimensional (or one-dimensional, for that matter) layouts. For example, a straightforward generalization of the method by Frick et al. (1995) is given in Bruß and Frick (1996). The approach of Davidson and Harel (1996) is adjusted by Cruz and Twarog (1996), where the crossing count, superfluous in three dimensions, is substituted with edge-edge repulsion. See Chapter 7 for more on three-dimensional graph layout.

Clustering. Important information may be represented by intrinsic or extrinsic clusterings of vertices. Since experimental work suggests that users tend to group geometrically close vertices (McGrath et al., 1996), it seems desirable to keep vertices of the same cluster close to each other, while those in different clusters should be further apart.

By their very nature, spring-type methods geometrically cluster dense subgraphs. While this is not at all true for other definitions of clusters, these can be reduced to dense subgraphs by introducing dummy vertices that represent the clusters, and connecting them appropriately (Eades et al., 1997a; Huang and Eades, 1998a). For each cluster, a new vertex is introduced, attracting its contained vertices, while repelling vertices of other clusters. To achieve larger distances between clusters, additional repelling forces are introduced between dummy vertices and all other vertices not contained in their respective cluster. Figure 4.8(a) gives a small example.

Instead of introducing dummy vertices to represent cluster centroids, a *meta-graph* representing connections between clusters is introduced by Wang and Miyamoto (1996). It consists of a vertex for each cluster, and has an

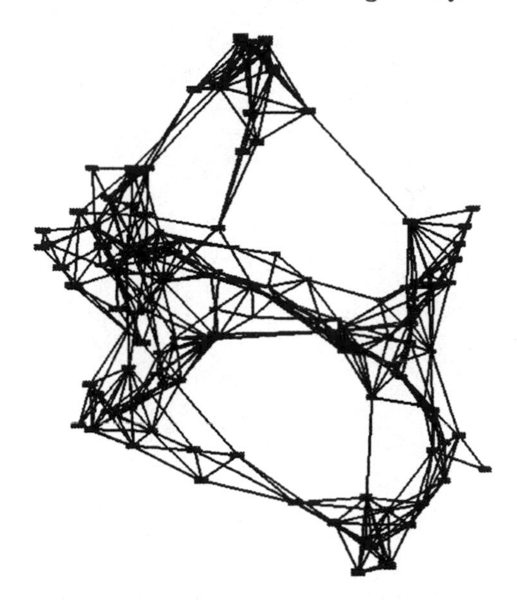

Fig. 4.7. 3D layout of travel distances between 31 U.S. cities (Bruß and Frick, 1996).

edge between each pair of distinct meta-vertices containing adjacent vertices of the original graph. To provide the necessary area for clusters, meta-vertices are not assumed to have point size, but to occupy rectangular areas. Consequently, first a technique is introduced that accounts for the fact that actual edge lengths depend on the relative positions of rectangularly shaped vertices. In the placement iteration, contribution weights are used to gradually shift from an emphasis of forces in the meta-graph (for cluster spacing) to forces in the input graph. More detail on force-directed methods to represent clusters is provided in Chapter 8.

Directed Edges. For directed graphs such as data-flow or time-dependency graphs, it is often desirable to have the directed edges point into roughly the same direction. Using forces exerted on edges by external fields of a chosen orientation, edges can be pushed to point in any given direction (Sugiyama and Misue, 1995). Let θ be the angle between an edge's prescribed direction and its current direction, and let $\overrightarrow{p_u p_v}^{\perp}$ be the unit length vector perpendicular to $\overrightarrow{p_u p_v}$ and pointing towards a decrease of θ, then rotative forces

$$f_{\text{rot}}(p_u, p_v) = b \cdot \|p_u - p_v\|^{c_1} \cdot \theta^{c_2} \cdot \overrightarrow{p_u p_v}^{\perp}$$

can be added to the net force vector, thus pressing the edge to reduce θ (see Figure 4.9). While b is a constant that controls the strength of the magnetic field acting on an edge between u and v with respect to the other forces, c_1 and c_2 are parameters that control the relative dependency of rotative forces on vertex distance and angle deviation, respectively.

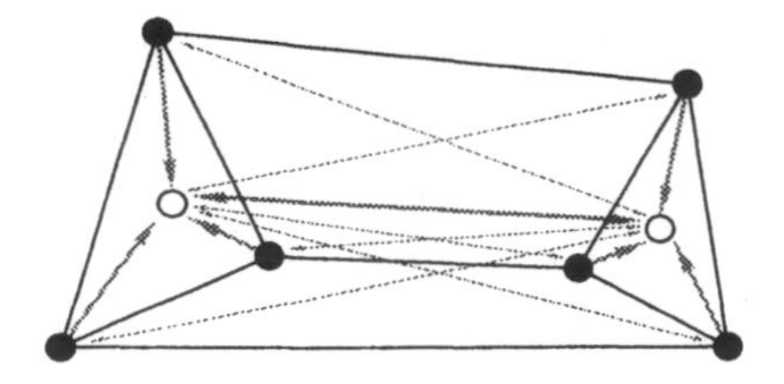

(a) Dummy vertices represent the centroids of clusters (Eades et al., 1997a; Huang and Eades, 1998a).

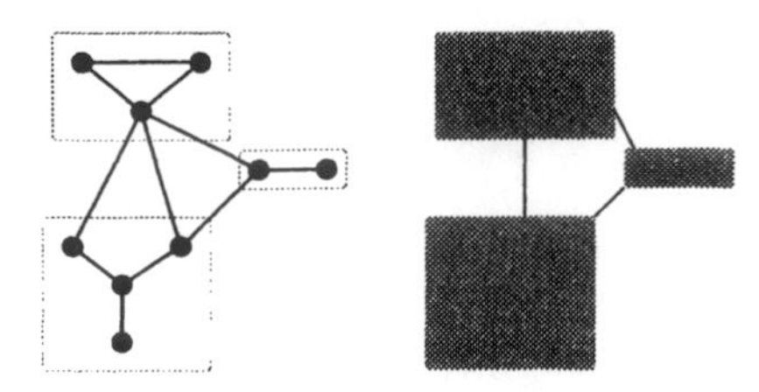

(b) A clustering with its corresponding meta-graph (Wang and Miyamoto, 1996).

Fig. 4.8. Force-directed models for clusters.

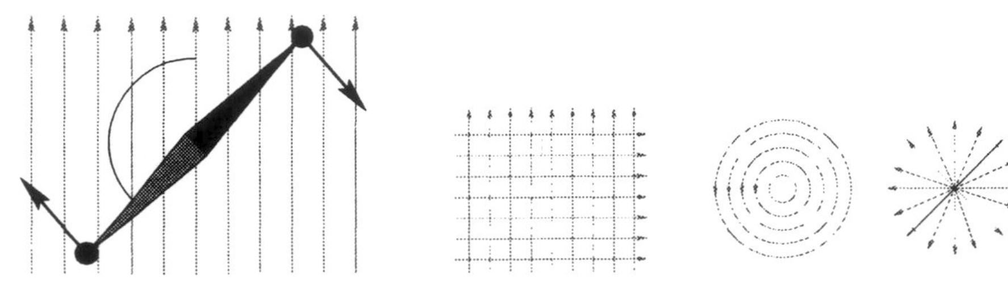

Fig. 4.9. Magnetic springs represent directed edges subject to an arbitrary force field (Sugiyama and Misue, 1995).

Curved Edges. So far, all representations considered here use straight lines to depict edges. It is shown in Brandes and Wagner (1998b) how the layout of curved line representations can be reduced to the straight-line case by placing control points instead of vertices. More precisely, edges that are to be depicted by Bézier curves are replaced with vertices representing their control points. Then an auxiliary graph is constructed by introducing a dummy edge of appropriate desired length for each control segment, and additional attractive and repulsive forces to ensure certain properties of the resulting layout. Figure 4.10 shows a graph of train connections with curved representation of edges that are classified as "transitive" (top left), the auxiliary graph constructed for this instance (bottom left), and a larger example (right). The same technique obviously works for polyline representations.

Dynamic Layout. When users look at the drawing of a graph, they familiarize themselves with this drawing by building a cognitive representation called the *mental map*. Dynamic graph layout is concerned with the trade-off between static layout quality and the user's mental map.

A framework to extend static graph layout methods to dynamic ones by incorporating a difference metric into the objective function is advocated

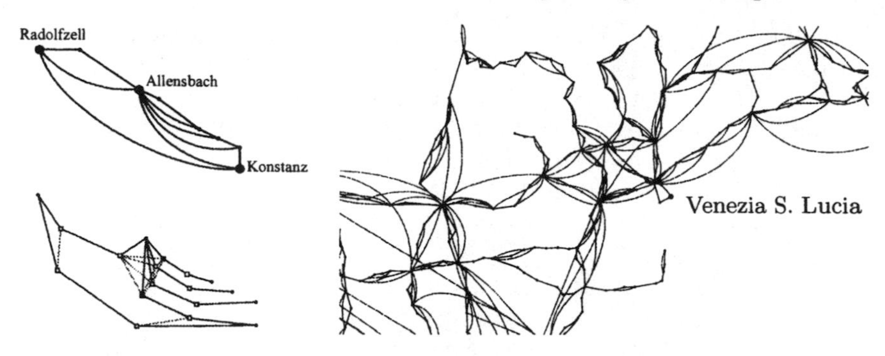

Fig. 4.10. Curved edge layout using an auxiliary graph (Brandes and Wagner, 1998b).

in Brandes and Wagner (1997). For energy-based methods, the difference metric summing over the Euclidean distances of vertex positions translates nicely into an attraction potential for imaginary springs of natural length zero, which can be used in force-directed approaches as well. An overview of approaches for drawing dynamic and interactive graphs is provided in Chapter 9.

Constraints. Constraints can be integrated into force-directed or energy-based placement methods in several ways, depending on the type of constraint. Unary constraints, i.e., restrictions on the admissible positions of single vertices, are easily satisfied if feasible positions form a connected region in the layout area or space (see Figure 4.11). In this case, every configuration remains reachable, and position updates can be modified to ensure that each vertex is at a feasible position at all times.

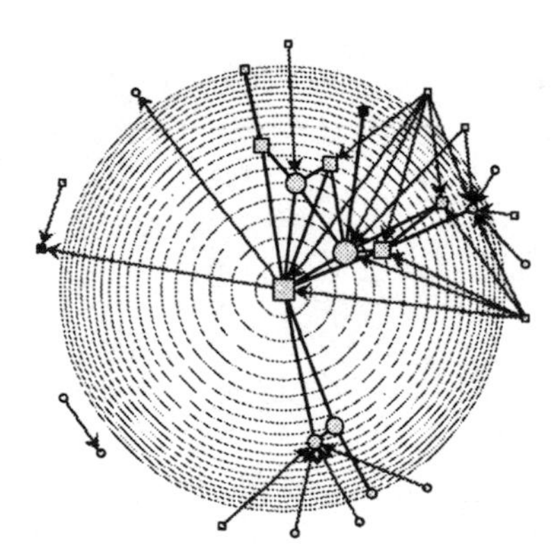

Fig. 4.11. Unary constraints. In this drawing of a social network, each vertex is restricted to lie on the circumference of a circle, signifying its centrality (Brandes, 1999).

A second way to satisfy unary constraints that also handles disconnected feasible regions is to allow vertices to temporarily occupy infeasible positions, but to use additional forces tht drag vertices to feasible positions or penalties pushing them away from infeasible positions, respectively. The penalty approach is easily extended to arbitrary constraints, but rarely guarantees satisfaction of all constraints in practice.

To satisfy pairwise linear (in)equalities, the placement iteration is interleaved with conflict detection in Wang and Miyamoto (1996). Every pair of vertices that would not satisfy a constraint if a displacement is carried out, is joined by a "rigid stick" that forces them to move together while maintaining the constraint. Note that criteria corresponding to weak linear pairwise constraints can be modeled using the magnetic springs of Sugiyama and Misue (1995).

Arbitrary constraints can be incorporated in methods based on physical analogies by applying more elaborate constraint satisfaction techniques, as in He and Marriott (1998). A survey of the use of constraints in graph layout is given by Tamassia (1998)

From a more general point of view, the above modeling techniques can be seen as instances of a generic framework for layout models, in which arbitrary layout variables are assigned values subject to some constraints and criteria that describe feasible and desirable configurations. Graph layout is thus cast as a general constraint optimization problem, for which objective functions can be devised by combining criteria for small configurations like pairs of adjacent or non-adjacent vertices, triples of connected vertices, pairs of edges, and so on. Force-directed and energy-based methods fit well into this framework, because they are intuitive and incorporate an implicit notion of compromise between the various criteria.

5. Layered Drawings of Digraphs

Oliver Bastert* and Christian Matuszewski**

5.1 Introduction

In this chapter, we present the standard criterea and techniques for drawing directed graphs[1]. The resulting drawings are called *layered drawings* of the given graphs.

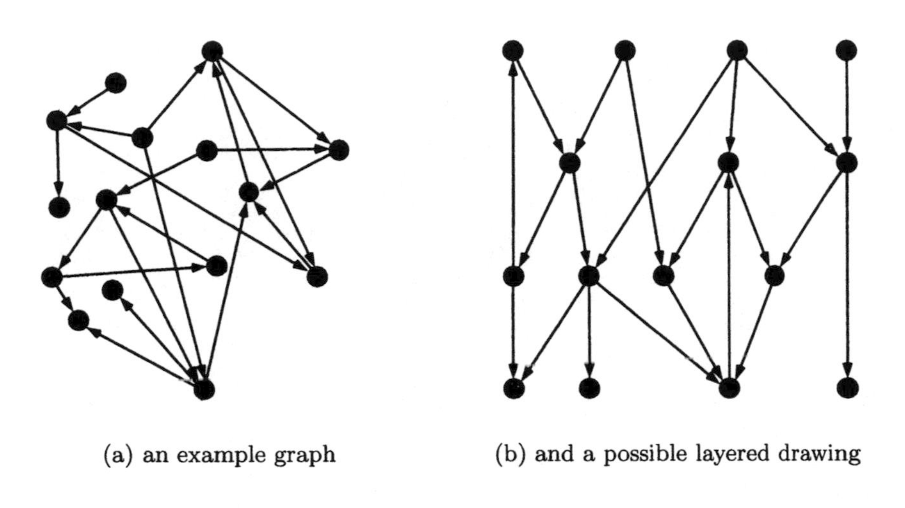

Fig. 5.1. Two different drawings of a graph.

The development of algorithms for computing layered drawings has started in the seventies. The first ideas were given in articles by Warfield (1977) and Carpano (1980b). The most popular method was introduced in 1981 by Sugiyama et al. and extended in the survey article by Eades and Sugiyama (1990). The method has attracted a lot of attention and an implementation of this approach can be found in many graph drawing tools.

Throughout this chapter, it is assumed that the graphs have an overall flow or direction. This will be emphasized by drawing most of the edges in

* Supported by the Deutsche Forschungsgemeinschaft (DFG), graduate program Angewandte Algorithmische Mathematik, Technische Universität München.
** Supported by the Deutsche Forschungsgemeinschaft (DFG), grant Mo645/5, Martin-Luther-Universität Halle-Wittenberg

[1] In this chapter, we only consider loopless directed graphs.

M. Kaufmann and D. Wagner (Eds.): Drawing Graphs, LNCS 2025, pp. 87-120, 2001.

one specific direction. We will assume that the preferred direction is top to bottom. Moreover, the drawing should meet some aesthetic and readability criteria which can be summarized as follows.

1. Edges pointing upward should be avoided.
2a. Nodes should be evenly distributed.
2b. Long edges should be avoided.
3. There should be as few edge crossings as possible.
4. Edges should be as straight/vertical as possible.

Figure 5.1(b) depicts a layered drawing of an example graph. Throughout this section, we show drawings of the same graph produced by several graph drawing tools.

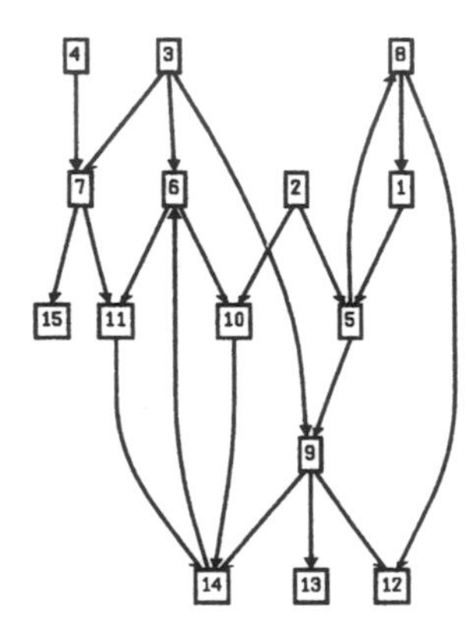

Fig. 5.2. The example graph drawn with the `VCG tool`.

Optimizing all of the above objectives is in general impossible, since some of them contradict each other, e.g., see Figure 5.20 on page 120. Furthermore, it is even hard, in the complexity theoretical sense, to compute drawings satisfying only some of the aesthetic criteria. To handle these problems, the approach described in this chapter is usually divided into the following four steps, each addressing one of the above optimization criteria:

1. Cycle Removal As few edges as possible are *reversed* to make the graph *acyclic*. This allows to draw all edges in one direction which is important for the next step. At the end of the algorithm, the reversed edges are reversed agian to obtain their initial orientation.
2. Layer Assignment A *layering* will be computed, i.e., an assignment of the vertices to layers such that all edges point downward. As we will see, for many algorithms which solve the succeeding step, a *proper layering* is needed. A layering is called *proper* if edges occur between adjacent layers only. To achieve the latter, *dummy vertices* are introduced along the edges.

3. Crossing Reduction For each layer, an *ordering* of the vertices is computed. The ordering should be computed in such a way that the number of *edge crossings* is kept small. This is usually done by examining adjacent layers and the edges between them.
4. X-coordinate assignment of vertices and placing the edges
   The requirements on the horizontal positions of the vertices are that the vertices do not overlap and that, preferably, no vertices lie on the straight lines between two adjacent vertices. Finally, the edges have to be placed either as polylines or as curves.

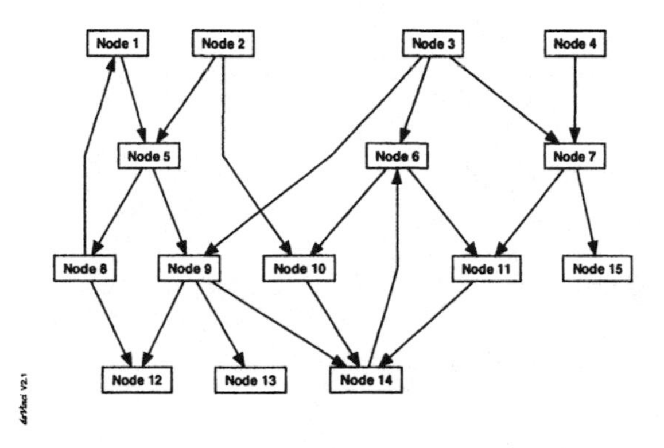

Fig. 5.3. The example graph drawn with **daVinci**.

Besides the mentioned articles, our presentation is mainly based on works of Gansner et al. (1993), Sander (1996b) and Di Battista et al. (1999). Furthermore, we describe the work of Berger and Shor (1990) in some detail. For further reference, see the articles of Messinger et al. (1991), Paulish (1993), Di Battista et al. (1994), or the manuals of software packages discussed in the appendix.

The organization of this chapter follows the above steps. Our aim is to present the state of the art of both, exact algorithms and heuristic approaches, to the single steps. Finally, we briefly present some related approaches.

It is not always necessary to perform all steps of the algorithm. In some cases, a layering is given together with the graph, e.g., the graph represents events on a timeline like a pedigree. Then only the last two steps have to be performed.

5.2 Cycle Removal

In this section, we will address solution methods for the *maximum acyclic subgraph problem*: find a maximum set $E_a \subset E$ such that the graph (V, E_a)

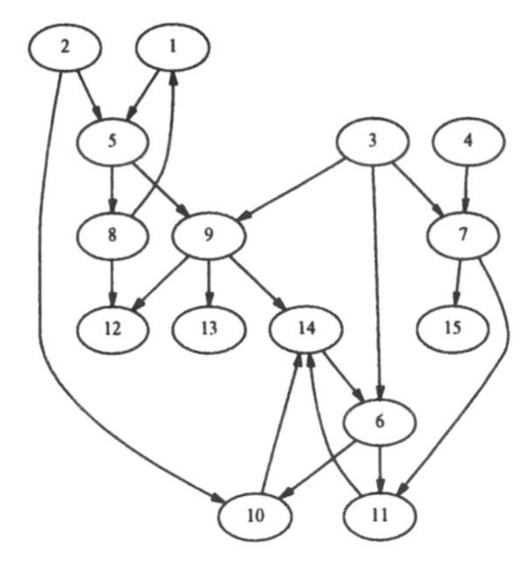

Fig. 5.4. The example graph drawn with **dot** (part of **GraViz**).

contains no cycles. The problem is often stated as the *feedback arc set problem*: find a minimum set $E_f \subset E$ such that the graph $(V, E \setminus E_f)$ contains no cycles.

Since we do not want to loose the information whether two vertices are adjacent or not, the edges in $E \setminus E_a$ will be reversed. It is an easy exercise to show that the resulting graph is acyclic.

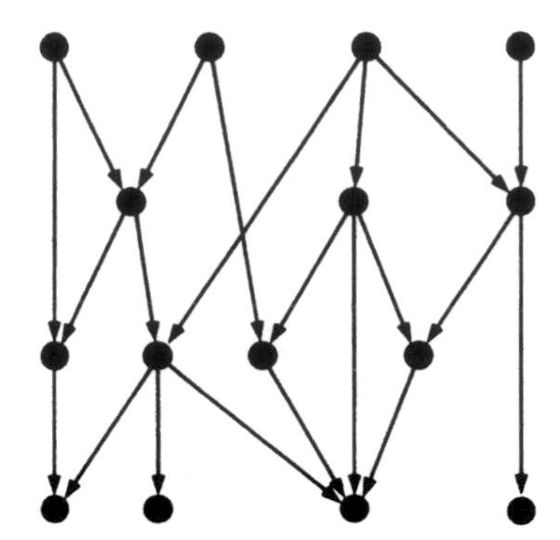

Fig. 5.5. The example graph made acyclic by reversing two edges.

Unfortunately, the maximum acyclic subgraph problem is NP-hard (Karp, 1972; Garey and Johnson, 1991).

To simplify the analysis of the forthcoming heuristics, we assume that the graph does not contain two-cycles. A *two-cycle* consists of two antipodal edges (u, v) and (v, u). Otherwise, we delete both edges of the two-cycle, apply an algorithm or heuristic for finding a maximum acyclic subgraph and insert two edges pointing in the same direction into the graph. The direction should be chosen in such a way that no cycles are generated by the insertion. Obviously, such a direction always exists (Berger and Shor, 1990).

We first present some fast heuristics in Section 5.2.1. Afterwards in Section 5.2.2, we give an overview over several variants of a frequently used greedy heuristic. Although these algorithms give very good results in practice, the best performance guarantee can be proven for another algorithm which will be presented in Section 5.2.3. Finally, we discuss an exact approach to the maximum acyclic subgraph problem.

5.2.1 Fast Heuristics

Observe that the maximum acyclic subgraph problem is equivalent to the *unweighted linear ordering problem*: find an ordering of the vertices of G, i.e., find a mapping $o : V \rightarrow \{1, 2, \ldots, |V|\}$ such that the number of edges $(u, v) \in E : o(u) > o(v)$ is minimized.

Thus, the easiest heuristic for the maximum acyclic subgraph problem is to take an arbitrary ordering of the graph and delete the edges (u, v) with $o(u) > o(v)$. We might use a given ordering or, e.g., use an ordering computed by applying breadth first search or depth first search (see Section 5.3.1) to the graph. These heuristics are fast but do not allow to give any quality guarantees.

Next, we present a heuristic which guarantees an acyclic set of size at least $\frac{1}{2}|E|$. The idea is to delete for every vertex either the incoming or outgoing edges. We define $\delta^+(v) = \{(v, u) \mid (v, u) \in E\}$, the set of the outgoing edges of v, $\delta^-(v) = \{(u, v) \mid (u, v) \in E\}$, the set of the ingoing edges into v, and $\delta(v) = \delta^+(v) \cup \delta^-(v)$, the set of edges incident to v, $v \in V$. $|\delta^+(v)|$ ($|\delta^-(v)|$) is called the *outdegree* (*indegree*) of v.

Algorithm 8: A Greedy Algorithm

```
E_a = ∅;
foreach v ∈ V do
    if |δ+(v)| ≥ |δ−(v)| then
        append δ+(v) to E_a;
    else
        append δ−(v) to E_a;
    delete δ(v) from G;
```

Trivially, Algorithm 8 computes an acyclic set E_a with size $|E_a| \geq \frac{1}{2}|E|$ and runs in linear time[2] (Berger and Shor, 1990).

[2] As usual, we define the input size of a graph by $n + m$ and hence, linear time means $O(n + m)$ time.

5.2.2 An Enhanced Greedy Heuristic

A closer look at the problem shows that sources and sinks (which may arise during the algorithm) play a special role: edges incident to sources or sinks cannot be part of a cycle. This observation is used in the following algorithm (Eades et al., 1993):

Algorithm 9: An Enhanced Greedy Heuristic

```
    E_a = ∅;
    while G is not empty do
1       while G contains a sink v do
            add δ⁻(v) to E_a and delete v and δ⁻(v) from G;
2       delete all isolated vertices from G;
3       while G contains a source v do
            add δ⁺(v) to E_a and delete v and δ⁺(v) from G;
4       if G is not empty then
            let v be a vertex in G with maximum value |δ⁺(v)| − |δ⁻(v)|;
            add δ⁺(v) to E_a and delete v and δ(v) from G;
```

The only difference between Algorithm 8 and Algorithm 9 is that the latter one processes the vertices in a special order. Hence, the output of Algorithm 9 is acyclic as well.

Theorem 5.1 (Eades et al. 1993). *Let $G = (V, E)$ be a connected digraph with no two-cycles. Then Algorithm 9 computes an acyclic edge set E_a with*

$$|E_a| \geq \frac{|E|}{2} + \frac{|V|}{6}.$$

Proof. The vertex set V can be partitioned into five sets V_{sink}, V_{iso}, V_{source}, $V_=$ and $V_<$. V_{sink} consists of the non-isolated sink vertices removed from G in Step 1, V_{iso} consists of the isolated vertices removed from G in Step 2, V_{source} consists of the non-isolated source vertices removed from G in Step 3, $V_=$ consists of the vertices whose indegree equals its outdegree, removed from G in Step 4 and $V_<$ consists of the vertices whose indegree is less than its outdegree, removed from G in Step 4. Note that these sets form a partition of V.

Denote by m_i the number of edges removed from G as the result of the removal of the vertices in V_i, $i \in \{sink, iso, source, =, <\} =: \mathcal{I}$, and by n_i the cardinality of V_i. Clearly,

$$|V| = \sum_{i \in \mathcal{I}} n_i \ , \ |E| = \sum_{i \in \mathcal{I}} m_i, \text{ and } m_{iso} = 0$$

holds.

Since the input graph is connected, isolated vertices can only be created in Step 1 and hence, $n_{iso} \leq m_{sink}$.

It is not hard to see that after the removal of a vertex from $V_=$, at least one vertex whose indegree is not equal to its outdegree exists. Since the resulting graph contains no isolated vertices, the next deleted vertex will be in $V_{sink} \cup V_{source} \cup V_<$. Hence, we get $n_= \leq n_{sink} + n_{source} + n_<$. This can be used to find an estimation of n by substituting $n_=$:

$$n \leq 2n_{sink} + n_{iso} + 2n_{source} + 2n_<.$$

This can be relaxed to

$$n \leq 2n_{sink} + n_{iso} + 3n_{source} + 3n_<.$$

Using the facts $n_{iso} \leq m_{sink}$ and $n_{sink} \leq m_{sink}$ we get

$$n \leq 3(m_{sink} + n_{source} + n_<). \tag{5.1}$$

Observe that the only step where edges from E are thrown away and not inserted in E_a is Step 4. Suppose $v \in V_=$, then the number of thrown away edges is exactly $\frac{|\delta(v)|}{2}$. Otherwise, if $v \in V_<$, this number is bounded from above by $\frac{|\delta(v)|-1}{2}$. Thus, the number of thrown away edges is at most

$$\begin{aligned} |E| - |E_a| &\leq \frac{m_=}{2} + \frac{m_< - n_<}{2} \\ &= \frac{m}{2} - \frac{m_{sink} + m_{source} + n_<}{2} \\ &\leq \frac{m}{2} - \frac{m_{sink} + n_{source} + n_<}{2}, \end{aligned}$$

where the last inequality is true since $n_{source} \leq m_{source}$.

By applying (5.1), we obtain

$$|E| - |E_a| \leq \frac{m}{2} - \frac{n}{6}.$$

This completes the proof.

The algorithm can be implemented in linear time and space (Eades et al., 1993). In addition, it can easily be shown that Algorithm 9 computes a set E_a with size at least $\frac{2}{3}|E|$ on graphs with $\Delta(G) \leq 3$. $\Delta(G)$ denotes the maximum degree of a vertex in G.

Sander (1996b) suggests a more elaborated version of Step 4. A graph is called strongly-connected if

Algorithm 10: A Variant of Step 4

...
4 **if** *G is not empty* **then**
compute the strongly-connected components (scc) of G;
add all edges not contained in a scc to E_a and delete them from G;
foreach *scc* G_{scc} *of* G *with* $|G_{scc}| > 1$ **do**
$W = \{v \in G_{scc} \mid \forall u \in V : |\delta^-(v)| \leq \delta^-(u)|\}$;
choose $v \in W$ that maximizes $|\delta^+(v)| + \sum_{u:N^+(u)=\{v\}} |\delta^-(u)|$;
add $\delta^+(v)$ and $\delta^-(v) \setminus \delta^-(v)$ to E_a and delete v and $\delta(v)$ from G;
foreach u *with* $N^+(u) = \{v\}$ **do**
add $\delta^-(u)$ to E_a and delete u and $\delta(u)$ from G;
...

$$\forall u, v \in V \quad \exists \text{ paths from } u \text{ to } v \text{ and from } v \text{ to } u.$$

A strongly-connected component of a graph is a maximal strongly-connected subgraph of the graph. The set of strongly-connected components forms a node partition. Let $N^+(u) = \{v \in V | \exists (u, v) \in E\}$.

Sander reports very promising practical results but no better theoretical bounds are known so far. Observe that the strongly connected components in each iteration of the algorithm are subdivisions of the components of the previous iteration and thus, the sizes of the components decrease quickly. Since computing the strongly-connected components of a graph takes linear time (Mehlhorn, 1984) and by using suitable data structures, Algorithm 10 can be implemented in $O(mn)$ time. The idea of the choice of v in the above algorithm is that the edges in $\delta^+(v) \cup \bigcup_{u:N^+(u)=\{v\}} \delta^-(u)$ are not contained in any cycle after the removal of $\delta^-(v)$.

Eades and Lin (1995) give a generalization of the ideas of Sander. Their idea is based on the observation that on induced paths at most one edge has to be deleted. They shrink long induced paths in the graph, which enables them to guarantee an acyclic subgraph with $|E_a| \geq \frac{3}{4}|E|$ for cubic graphs.

5.2.3 A Randomized Algorithm

The results presented in this section are due to Berger and Shor (1990).

To achieve a better performance guarantee as given in the previous sections, consider the following algorithm.

Algorithm 8': A Randomized Variant of Algorithm 8

order the vertices randomly at the beginning;
process them in this order in Algorithm 8;

The following bound can be proven:

Theorem 5.2 (Berger and Shor 1990). *The expected size $\mathcal{E}_{8'}(|E_a|)$ of E_a computed by Algorithm 8' is bounded from below by*

$$(\frac{1}{2} + \Omega(\frac{1}{\sqrt{\Delta(G)}}))|E|.$$

Berger and Shor also give a deterministic algorithm and are able to prove that the expectation value from Theorem 5.2 is valid as worst case bound on $|E_a|$ for their deterministic algorithm. Let *processed*$(v_1, v_2, \ldots, v_i)$ mean that the vertices in $v_1, v_2, \ldots, v_i$ have been processed in the order in which they are listed.

Algorithm 11: A Deterministic Variant of Algorithm 8'

```
for i = 1 to |V| do
    choose v ∈ V \ {v_1, v_2, ..., v_{i-1}} for which
            E_1(|E_a| | processed(v_1, v_2, ..., v_{i-1}, v)) is maximized;
    define v_i = v;
```

Theorem 5.3. *Algorithm 11 returns an E_a with size greater or equal to*

$$(\frac{1}{2} + \Omega(\frac{1}{\sqrt{\Delta(G)}}))|E|.$$

Proof. The expectation value for the size of E_a for the randomized algorithm in iteration i is $\mathcal{E}_{8'}(|E_a| \mid \textit{processed}(v_1, v_2, \ldots, v_{i-1}))$ which equals

$$\frac{1}{|V| - (i-1)} \sum_{v \in V \setminus \{v_1, v_2, \ldots, v_{i-1}\}} \mathcal{E}_{8'}(|E_a| \mid \textit{processed}(v_1, v_2, \ldots, v_{i-1}, v)).$$

Since Algorithm 11 chooses

$$\max_{v \in V \setminus \{v_1, v_2, \ldots, v_{i-1}\}} \mathcal{E}_8(|E_a| \mid \textit{processed}(v_1, v_2, \ldots, v_{i-1}, v)),$$

in each iteration $\mathcal{E}_{11}(|E_a|)$ of Algorithm 11 is at least as large as $\mathcal{E}_{8'}(|E_a|)$. Thus, by Theorem 5.2, the algorithm has the claimed performance guarantee.

For graphs with small maximum degree, Berger and Shor give more precise bounds on $|E_a|/|E|$.

$\Delta(G)$	$\lvert E_a\rvert/\lvert E\rvert \geq$
2	$\frac{2}{3}$
3	$\frac{13}{18}$
4, 5	$\frac{19}{30}$

It is also shown that the above algorithm can be implemented in time $O(mn)$. The crucial point is that the expectation values can be computed quickly. Furthermore, they prove that the bound given in Theorem 5.3 is tight.

5.2.4 An Exact Algorithm

Let $S(G)$ be the set of all incidence vectors of acyclic subgraphs of $G = (V, \{e_1, e_2, \ldots, e_m\})$, i.e.,

$$S(G) = \{x \in \{0,1\}^{|E|} \mid \exists E_a \subseteq E, E_a \text{ acyclic} : x_i = 1 \text{ iff } e_i \in E_a\}.$$

Denote by $conv(S)$ the convex hull of a set S and by 1 the $|E|-dimensional$ all-one vector. Now, $ASP(G) = conv(S(G))$ defines a polytope in $\mathrm{R}^{|E|}$, and maximizing $1^t x$ over $ASP(G)$ yields a maximum acyclic set.

An extensive study of the facial structure of the acyclic subgraph polytope and the more general linear ordering polytope can be found in (Grötschel et al., 1985; Reinelt, 1985). The results given there can be used to solve the maximum acyclic subgraph problem by a branch and cut approach.

5.3 Layer Assignment

After introducing the necessary definitions, we discuss the objectives a layering should reach. Afterwards, we present algorithms for the different objectives.

Let $\mathcal{L}$ be a *partition* of V, i.e., $\mathcal{L} = \{L_1, L_2, \ldots, L_h\}$, $\cup_{i=1}^{h} L_i = V$. $y : V \to \{1, 2, \ldots, h\}$ denotes the *characteristic function* of this partition, i.e., $y(u) = i$ iff $u \in L_i$. $\mathcal{L}$ is called a *layering* if $\forall (u,v) \in E : y(u) > y(v)$ holds.

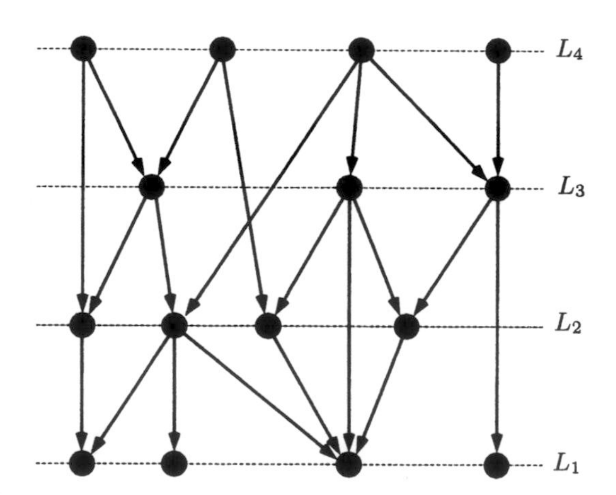

Fig. 5.6. A layering of the example graph.

The *height* of a layering is the number of layers h, the *width* of a layering is the number of vertices in the largest layer and the *span* of an edge (u, v) is defined as $y(u) - y(v)$. A layering is called *proper* if no edge has a span greater than one.

Sometimes it is not important to stress the general direction of the edges and in some cases such a direction does not exist. In these cases the algorithms presented in Section 5.3.1 can be applied without making the graph acyclic at first. In the other sections of this chapter, we assume that the graphs are acyclic. This guarantees the existance of a layering. Besides, there are some objectives a layering should fulfill. It should be compact. This means that the height and the width of the layering should be small. A simple algorithm to compute a layering with minimum height is given in Section 5.3.2. Unfortunately, minimizing the height with respect to a given width is NP-hard. In Section 5.3.3, we will present a heuristic to tackle this problem.
Most of the algorithms used in the subsequent steps need to have proper layerings. This can be easily achieved by introducing *dummy vertices* along edges (u, v) with span $k > 1$ into the layering. We replace (u, v) with a path $(u = v_1, v_2, \ldots, v_k = v)$ of length k. In each layer between $y(u)$ and $y(v)$, one dummy vertex will be placed (see Figure 5.7)

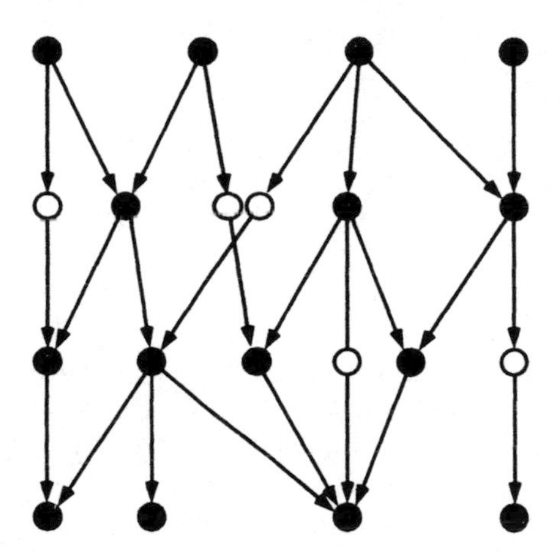

Fig. 5.7. Introducing dummy vertices into the example graph.

In addition, the number of dummy vertices should be small. There are three aspects why this should be. The running time of the following steps depends on the sum of the number of vertices and the number of dummy vertices. Bends in the drawing will only occur at dummy vertices and thus a small number of dummy vertices will increase the readability of the drawing. Furthermore, the edges will become long if many dummy vertices occur. We will present an algorithm for minimizing the number of dummy vertices in Section 5.3.4.

5.3.1 Layerings for General Graphs

Sander (1996b) summarizes some algorithms for general graphs, i.e., graphs do not need to be acyclic before applying these layering heuristics: edges which point upward are just reversed.

1. Calculate y by a depth first search or breadth first search. This generates an arbitrary partitioning in linear time.
2. Calculate the minimum cost spanning trees on the undirected instance of the graph. This is useful if edges e have weights $\omega(e)$. The cost will be $\frac{1}{\omega(e)}$. The edges with high priority will have small spans.
3. Apply a spring embedder. It is sufficient to take only f_{rep} and f_{att} (see Section 4.2) into account. Calculate only the one-dimensional coordinate y. This computes a layering where edges tend to have the same length.

5.3.2 Minimizing the Height

From now on, we assume that the graphs are acyclic. The following algorithm computes layerings of minimum height. Each sink is placed in layer L_1. For the remaining vertices the layer will be recursively defined by

$$y(u) := \max\{i \mid v \in N^+(u) \text{ and } y(v) = i\} + 1$$

and $N^+(u) := \{v \in V \mid \exists (u, v) \in E\}$.

This produces a layering where many vertices will stay close to the bottom. The algorithm can be implemented in linear time by using a topological ordering of the vertices (Mehlhorn, 1984).

5.3.3 Layerings with Given Width

Given a fixed width greater or equal to three, the problem of finding a layering with minimum height is NP-complete. The *precedence-constrained multiprocessor scheduling problem* can easily be reduced to it (Karp, 1972; Garey and Johnson, 1991).

In the following, we will present the *Coffman-Graham-Algorithm* (Coffman and Graham, 1972). It computes a layering with width at most w and tries to minimize the height. The Coffman-Graham-Algorithm takes as input a reduced digraph, i.e., no transitive edges are present in the graph, and a width w. An edge (u, v) is called *transitive* if a path $(u = v_1, v_2, \ldots, v_k = v)$ exists in the graph. Observe that the absence of transitive edges does not affect the width of the layering significantly and that transitive edges can be found in linear time (Mehlhorn, 1984).

The weakness of a simple greedy heuristic is illustrated in Figure 5.8. w is assumed to be 2. The example graph consists of $\frac{n}{2}, n \mod 4 = 0$, isolated vertices and a directed path of $\frac{n}{2}$ vertices (Figure 5.8(a)). The greedy heuristic would probably assign the isolated vertices to the $\frac{n}{4}$ first layers. This would result in a layering of height $\frac{3n}{4}$ (Figure 5.8(b)). An optimal solution is depicted in Figure 5.8(c).

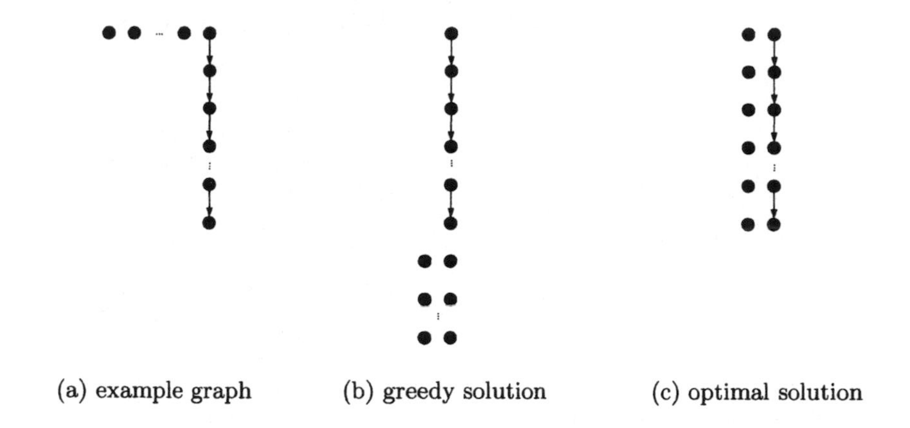

(a) example graph (b) greedy solution (c) optimal solution

Fig. 5.8. The worst and the best layering for the given graph.

The greedy solution is far from optimal since it does not consider the long path in the graph. This is exactly what the Coffman-Graham-Algorithm tries to avoid. It proceeds in two phases. The first orders the vertices mainly by their distance from the source vertices of the graph, the second assigns the vertices to the layers. Vertices with large distances from the sources will be assigned to layers as close to the bottom as possible.

We need a special lexicographical ordering on finite integer sets to describe the algorithm in more detail:
$S \prec T$ if either

1. $S = \emptyset$ and $T \neq \emptyset$, or
2. $S \neq \emptyset$, $T \neq \emptyset$ and $\max(S) < \max(T)$, or
3. $S \neq \emptyset$, $T \neq \emptyset$, $\max(S) = \max(T)$ and $S \setminus \{\max(S)\} \prec T \setminus \{\max(T)\}$.

We are now able to state the algorithm (Algorithm 12).

Lam and Sethi (1977) have shown that the height h of the computed layering with width w is bounded with respect to height of an optimal layering h_{opt}:

$$h \leq (2 - \frac{2}{w})h_{opt}.$$

Thus, the Coffman-Graham-Algorithm is an exact algorithm for $w \leq 2$.

5.3.4 Minimizing the Total Edge Span

The objective to minimize the total edge span (or edge length) is equivalent to minimizing the number of dummy vertices. As seen before this is a reasonable

Algorithm 12: Coffman-Graham-Algorithm

```
foreach v ∈ V do π(v) := n + 1;
for i = 1 to |V| do
    choose a vertex v with π(v) = n + 1
            and minimum set {π(u) | (u,v) ∈ E} with respect to ≺;
    π(v) := i;
k := 1; L_1 := ∅; U := V;
while U ≠ ∅ do
    choose u ∈ U such that every vertex in {v | (u,v) ∈ E} is in V \ U
            and π(u) is maximized;
    if |L_k| < w and N^+(u) ⊆ L_1 ∪ L_2 ∪ ... L_{k-1} then
        add u to L_k;
    else
        k := k + 1; L_k := {u};
    delete u from U;
```

objective. It can be shown that minimizing the number of dummy vertices guarantees minimum height (Eades and Sugiyama, 1990).

To solve this problem we formulate it as an integer program. The properties of a layering can be stated as follows.

$$y(u) - y(v) \geq 1 \text{ for all } (u,v) \in E \tag{5.2}$$

$$y(v) \in \mathrm{Z}^+ \text{ for all } v \in V \tag{5.3}$$

Thus minimizing over $\sum_{(u,v)\in E} (y(u) - y(v))$ minimizes the total span of the edges and thus the number of dummy vertices. Frick (1997) presents a detailed study on the number of dummy vertices.

It is sometimes useful to introduce weights or priorities ω on the edges to keep certain edges short. Furthermore, sometimes it is intended that an edge has a length of at least λ. Plugging this into the above linear program we obtain

$$\min \sum_{(u,v)\in E} \omega(u,v)(y(u) - y(v)) \tag{5.4}$$

$$y(u) - y(v) \geq \lambda(u,v) \text{ for all } (u,v) \in E \tag{5.5}$$

$$y(v) \in \mathrm{Z}^+ \text{ for all } v \in V \tag{5.6}$$

This problem can be easily solved by linear programming since the constraint matrix is totally unimodular and so standard linear programming will find an optimal integer solution. As an alternative, the network simplex algorithm might be used as described in (Gansner et al., 1993). For further reference on total unimodularity see (Schrijver, 1986), and for details concerning the network simplex algorithm see (Cunningham, 1976; Chvátal, 1983a). Observe that although the network simplex method like the simplex method itself does not guarantee a polynomial running time, it proves to be very efficient in practice.

5.4 Crossing Reduction

The aim of this part of the algorithm is to reduce the crossings between edges to improve the readability of the drawing. We assume that the layering was made proper in the previous layer assignment step.

A first observation to make is that the number of edge crossings does not depend on the exact positions of the vertices but only on the relative positions, i.e., the ordering of the vertices. This makes the problem somewhat easier to understand because we do not have to deal with the exact x-coordinates of the vertices. But, unfortunately, the problem of finding vertex orderings which minimize the crossings in a layered graph is $\mathcal{NP}$-hard even if we restrict the problem to bipartite (two-layered) graphs (Garey and Johnson, 1983). The problem remains $\mathcal{NP}$-hard if the ordering of the vertices in one layer of the bipartite graph is fixed (Eades and Whitesides, 1994; Eades and Wormald, 1994).

Many methods have been developed to reduce edge crossings but only a few work globally, i.e., minimize the crossings in the whole graph at once. Most algorithms use the layer-by-layer sweep described in the next section.

5.4.1 The Layer-by-Layer Sweep

This technique works as follows: First, a vertex ordering of the layers is chosen. An initial ordering of the layers that avoids crossings if the graph is a tree is given in the paper of Gansner et al. (1993): Do a depth first search or a breadth first search starting with the vertices in the layer with the minimum rank. The vertices get their positions in left-to-right order as the search progresses.

In the next step of the layer-by-layer sweep, a layer with a precomputed ordering, e.g., layer V_1, is chosen and for $i = 2, 3, \ldots, h$, the vertex ordering of layer V_{i-1} is held fixed while the vertices in V_i are reordered to reduce the crossings between layer V_{i-1} and V_i. After that, we can sweep back from layer V_h to layer V_1 and repeat these two steps until no further reduction of crossings can be achieved. Other ways of sweeping are possible, for instance we can hold a layer in the middle fixed and sweep from here to the bottom- and to the top-layer. However, the key problem of the layer-by-layer sweep is to reduce the crossings between two layers with the permutation of one side fixed. This problem is called the one sided crossing minimization problem and will be deeply treated in the next section.

Experiments by Jünger and Mutzel (1997) for the 2-layer crossing minimization problem show that the results of the layer-by-layer sweep are far from optimum. One can expect better results from considering all layers simultaneously, but k-layer crossing minimization is a very hard problem. A quick help is to start the layer-by-layer sweep several times with randomly permuted layers. This approach can tremendously improve the results (Jünger and Mutzel, 1997).

5.4.2 One Sided Crossing Minimization

To state the problem of the one sided crossing minimization precisely, we need some definitions and notations.

A *bipartite graph* is an undirected graph $G = (V, E)$ in which V can be partitioned into two sets V_1 and V_2 such that $\{u, v\} \in E$ implies either $u \in V_1$ and $v \in V_2$ or $u \in V_2$ and $v \in V_1$.

An ordering of layer V_i is specified by a permutation π_i of V_i. We express the ordering of V_1 by the permutation π_1 and the ordering of V_2 by π_2. Note, that we do not distinguish between permutations of vertices and permutations of positions since every vertex is clearly associated with its position and it will be always clear form the context what is meant. Let $\text{cross}(G, \pi_1, \pi_2)$ be the number of edge crossings in a straight-line drawing of G given by π_1 and π_2. If we fix the permutation of V_1, the minimum number of edge crossings we can achieve by reordering the vertices in V_2 is denoted by $\text{opt}(G, \pi_1)$. Thus

$$\text{opt}(G, \pi_1) = \min_{\pi_2} \text{cross}(G, \pi_1, \pi_2).$$

We can now formulate the *one sided crossing minimization problem* as follows.

> Given a bipartite Graph $G = (V_1, V_2, E)$ and a permutation π_1 of V_1. Find a permutation π_2 of V_2 that minimizes the edge crossings in the drawing of G, i.e., $\text{cross}(G, \pi_1, \pi_2) = \text{opt}(G, \pi_1)$.

The notion of the *crossing number* (Eades and Kelly, 1986) is important for many heuristics. Assume the permutation π_1 of V_1 is fixed. We define for each pair of vertices $u, v \in V_2$ the crossing number c_{uv} as the number of crossings between edges incident on u and edges incident on v, when $\pi_2(u) < \pi_2(v)$. Furthermore, we define $c_{uu} = 0$ for all $u \in V_2$. Observe, that the number of crossings between edges incident on u and edges incident on v depends only on the relative positions of u and v but not on the positions of the other vertices. To give an example, Figure 5.9 shows a drawing of a 2-layered graph. The corresponding crossing number matrix is depicted in Table 5.1.

We can use the crossing numbers to compute $\text{cross}(G, \pi_1, \pi_2)$:

$$\text{cross}(G, \pi_1, \pi_2) = \sum_{\pi_2(u) < \pi_2(v)} c_{uv} = \sum_{i=1}^{n_2 - 1} \sum_{j=i+1}^{n_2} c_{ij}$$

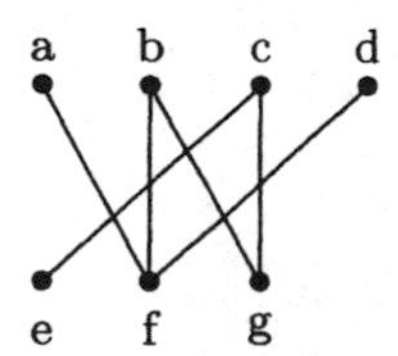

Fig. 5.9. A 2-layered graph.

Table 5.1. Crossing number matrix for the graph in Figure 5.9.

C	e	f	g
e	0	2	1
f	1	0	2
g	0	3	0

Sander (1994) presents a sweep line algorithm which computes the crossing numbers in time $O(|V_1| + |V_2| + |E| + c)$ where c is the number of crossings.

The crossing numbers are helpful to give a lower bound on the number of crossings:

$$L = \sum_{\pi_2(u)<\pi_2(v)} \min\{c_{uv}, c_{vu}\}$$

As experiments show (Jünger and Mutzel, 1997), this simple lower bound is very tight to the optimum.

We will now consider the most interesting heuristics in more detail and shortly describe the others.

Barycenter Heuristic. The *barycenter* heuristic (Sugiyama et al., 1981), which is also called *averaging*, is based on the intuition that in a drawing with few crossings, each node should be close to its adjacent nodes. The barycenter method is very popular because it is easy to implement, runs fast, and gives good results.

In this heuristic, we choose the position of a vertex as the barycenter (average) of the x-coordinates of its neighbours $N(u)$, where $N(u) := \{v : \{u, v\} \in E\}$. In order to do this, we compute

$$\text{bary}(u) = \frac{1}{\deg(u)} \sum_{v \in N(u)} \pi_1(v)$$

for every $u \in V_2$. If two values are equal we separate them arbitrarily by a small amount. Then we sort the vertices by their values. Because the vertices are likely to be presorted after some previous steps in the layer-by-layer sweep, Sander (1994) suggests to use a randomized quicksort (see for instance (Cormen et al., 1990)).

It is interesting, that the barycenter method gives a drawing without crossings if one is possible. Since the running time for computing bary(u) is proportional to the degree of u, the barycenters of all vertices can be found in linear time. Thus, with the subsequent sorting, the time complexity is $O(|V_2| \log |V_2|)$. Note, that the crossing number matrix has not to be precomputed as in several other heuristics.

Median Heuristic. In the *median* heuristic (Eades and Wormald, 1994), the x-coordinate of each vertex u is given by the median of the x-coordinates of the neighbours of u. Here, the median is defined as follows: Suppose the neighbours of u are $v_1, v_2, \ldots, v_j$ with $\pi_1(v_1) < \pi_1(v_2) < \ldots < \pi_1(v_j)$, then $\text{med}(u) = \pi_1(v_{\lceil j/2 \rceil})$. This definition differs from the classical notion of the median since there are actually two medians at $\frac{j}{2}$ and $\frac{j}{2}+1$ if j is even. Here, we take always the left median. Furthermore, if u has no neighbours, we set $\text{med}(u) = 0$.

As with the barycenter heuristic we have to sort V_2 according to $\text{med}(u)$. If two vertices have the same median they are separated by a small amount, with the restriction that if one vertex has odd degree and the other vertex has even degree, then the odd degree vertex is placed on the left of the vertex with even degree. If the degrees of the vertices have same parity, we can choose their order arbitrarily. This tie breaking method is necessary to prove the following performance guarantee, where the number of crossings in the output of the median heuristic is denoted by $\text{med}(G, \pi_1)$.

Theorem 5.4. *For all bipartite graphs $G = (V_1, V_2, E)$ and all permutations π_1 of V_1,* $\text{med}(G, x_1) \leq 3\,\text{opt}(G, x_1)$.

The proof of this theorem can be found in (Eades and Wormald, 1994) and (Di Battista et al., 1999). Like barycenter, the median method produces an output with zero crossings if possible. For each u, the median can be determined in $O(|N(u)|)$ (see for instance (Cormen et al., 1990)), so we get the same running time as for the barycenter heuristic.

The following two variants of the median method were defined by Mäkinen (1990). The *average median* assigns the arithmetic mean of the two medians if u has even degree, whereas in the *semi median* heuristic, we set $\text{smed}(u) = \text{bary}(u)$ if the degree of u is even. Both variants return the value of $\text{med}(u)$ if the degree of u is odd. As tests show, both heuristics improve the results of the original algorithm.

Gansner et al. (1993) refine the average median heuristic even further. If a vertex u has odd degree, the *weighted median* is defined by $\text{wmed}(u) = \text{med}(u)$. If $\deg(u) = 2$ then $\text{wmed}(u)$ is the arithmetic mean of the positions of the two neighbours. The difference to the average method occurs when $\deg(u)$ is even and $\deg(u) > 2$. In this case, the weighted median is defined as

$$\text{wmed}(u) = \frac{\pi_1(v_{j/2}) \cdot right + \pi_1(v_{j/2+1}) \cdot left}{left + right}$$

where $left = \pi_1(v_{j/2}) - \pi_1(v_1)$ and $right = \pi_1(v_j) - \pi_1(v_{j/2+1})$. This strategy puts the vertex toward the side where the neighbours are more closely packed.

Greedy Switch Heuristic. The *greedy switching* heuristic (Algorithm 13), also called *adjacent-exchange*, works in a way similar to bubble-sort. If u and v are two consecutive vertices in V_2, then switching their positions changes the total number of crossings by exactly $c_{vu} - c_{uv}$. The algorithm scans all consecutive pairs and switches them if this reduces the number of crossings. This process is repeated until no further switching occurs, i.e., for all consecutive pairs (u, v) the inequality $c_{uv} \leq c_{vu}$ holds. Such a vertex ordering is called *stable.*

Algorithm 13: greedy_switch

```
repeat
    for u := 1 to |V_2| - 1 do
        if c_{u(u+1)} > c_{(u+1)u} then
            switch vertices at positions u and u + 1;
until the number of crossings was not reduced;
```

Since one scan of the vertices can be implemented in $O(|V_2|)$ and there are at most $|V_2|$ scans, the time complexity of the greedy switching heuristic is $O(|V_2|^2)$.

As suggested by Mäkinen (1990) and Gansner et al. (1993), the greedy switch heuristic is preferable as a post processing step in combination with other heuristics such as barycenter or median. This is because greedy switching does not recompute the ordering completely but makes changes only when it improves the result.

Split Heuristic. The *split* heuristic (Eades and Kelly, 1986) gives better results than the median or barycenter heuristics at the expense of longer running times. The algorithm is comparable with quicksort. First, a "pivot"-vertex p is chosen. This can be done by arbitrarily selecting the leftmost vertex. A more sophisticated method would choose the pivot vertex randomly. In the next step of the algorithm, every other vertex v is placed to the left or to the right of p according to whether $c_{vp} < c_{pv}$ or $c_{pv} \leq c_{vp}$. After this partition, the algorithm is applied recursively to the left set and to the right set until both sets are ordered and can be concatenated. We start the Algorithm 14 by calling $split(1, |V_2|)$.

The split heuristic has a worst case running time of $O(|V_2|^2)$ but in practice it runs in time $O(|V_2| \log |V_2|)$ if we do not consider the computation of the crossing number matrix.

Sifting. The *sifting* algorithm was introduced by Rudell (1993) to reduce the number of nodes in reduced ordered binary decision diagrams (ROBDDS). An ROBDD is a graph which represents a boolean function and is primarily used in logic synthesis and verification. It is possible to adapt the sifting algorithm for the crossing minimization problem (Matuszewski et al., 1999).

Algorithm 14: split $(i, j : 1, \ldots, |V_2|)$

```
if j > i then
    pivot := low := i; high := j;
    for k := i + 1 to j do
        if c_{k pivot} < c_{pivot k} then
            π(k) := low; low := low + 1;
        else
            π(k) := high; high := high − 1;
    /* low == high */
    π(pivot) := low;
    copy π(i...j) into π_2(i...j);
    split (i, low − 1);
    split (high + 1, j);
```

Sifting yields very good results especially for sparse graphs but again this is paid with longer running times.

The algorithm determines the optimal position for every vertex u under the condition that the positions of the other vertices remain fixed (Algorithm 15).

Algorithm 15: sifting

```
foreach u ∈ V_2 do
    move u to the leftmost position;
    crossings := Σ_{π_2(u)<π_2(v)} c_{uv};
    min_crossings := crossings;
    for p := 1 to |V_2| − 1 do
        crossings := crossings − c_{p(p+1)} + c_{(p+1)p};
        switch vertices at positions p and p + 1;
        if crossings < min_crossings then
            min_crossings := crossings;
            best_position := p;
    move u to position best_position;
```

Since every vertex has to be set on every position, the time complexity is $O(|V_2|^2)$.

Other Heuristics. The *greedy insertion* algorithm (Eades and Kelly, 1986) proceeds by successively choosing the next vertex u to be the one which minimizes the number of crossings that edges adjacent to u make with edges adjacent to vertices to the right of u (if we start from left). The running time is in $O(|V_2|^2)$.

The *stochastic* heuristic (Dresbach, 1995) was originally designed to change the ordering of both layers but can be adapted to compute a solution if one layer is fixed. The algorithm greedily puts the vertices of V_1 and

V_2 to open positions according to *assessment numbers*. These numbers are derived from the *frequency numbers* that estimate the probability for an edge to cause a crossing in the drawing of the graph. The frequency number of an edge is the number of edges that cross that edge if the complete bipartite graph is drawn.

The *assignment* heuristic (Catarci, 1995) finds the linear assignment (see for instance (Lengauer, 1990)) for a simplified crossing minimization problem. We define d_{ij} as an upper bound on the number of edge crossings that are caused by edges incident on vertex i if we place i to position j. Such a number is easy to get. First we place vertex i on position j. Then we make the rest of the graph complete, i.e., every vertex in V_2 except vertex i is connected to every vertex in V_1. At last we count the number of edge crossings caused by edges incident on i. If all d_{ij} are computed we solve the assignment problem for matrix $D = ((d_{ij}))$, i.e., we choose n_2 elements from D such that each row and each column is covered and the sum of the elements is minimized.

Several meta heuristics were used to solve the one sided crossing minimization problem. A *genetic algorithm* is given by Mäkinen and Sieranta (1994) and the results are compared with the barycenter heuristic. The genetic algorithm is better with the expense of long computation times. Similarly, the *tabu search* by Laguna et al. (1997) gives high quality results but is usable only when fast computation is not necessary. The *GRASP* (greedy randomized adaptive search procedure) by Laguna and Martí (1999) gives good results especially for sparse graphs and has moderate computation times.

Optimal Crossing Minimization. The computation of exact solutions is very desirable to estimate the quality of the results produced by the heuristics. The branch-and-cut algorithm by Jünger and Mutzel (1997) does the computation of exact solutions in surprisingly short time. To name it, for instances with up to 60 vertices in the permutable layer the best permutation can be found faster or as fast as every other heuristic, except the barycenter and the median heuristic.

Let us state the one sided crossing minimization problem as an integer program. For each layer i we define $\delta^i_{kl} = 1$ if $\pi_i(k) < \pi_i(l)$, otherwise $\delta^i_{kl} = 0$. Thus, we can characterize π_i by the vector $\delta^i \in \{0,1\}^{\binom{n_i}{2}}$ and compute the number of crossings with

$$\mathrm{cross}(\pi_2) = \mathrm{cross}(\delta^2) = \sum_{i=1}^{n_2-1} \sum_{j=i+1}^{n_2} \sum_{k \in N(i)} \sum_{l \in N(j)} \delta^1_{kl} \cdot \delta^2_{ji} + \delta^1_{lk} \cdot \delta^2_{ij}.$$

The crossing numbers can be computed with

$$c_{ij} = \sum_{k \in N(i)} \sum_{l \in N(j)} \delta^1_{lk}.$$

Then

$$\begin{aligned}\text{cross}(\delta^2) &= \sum_{i=1}^{n_2-1}\sum_{j=i+1}^{n_2} c_{ij}\delta_{ij}^2 + c_{ji}(1-\delta_{ij}^2)\\ &= \sum_{i=1}^{n_2-1}\sum_{j=i+1}^{n_2} (c_{ij}-c_{ji})\delta_{ij}^2 + \sum_{i=1}^{n_2-1}\sum_{j=i+1}^{n_2} c_{ji}\end{aligned}$$

With $n = n_2$, $x_{ij} = \delta_{ij}^2$ and $a_{ij} = c_{ij} - c_{ji}$ we have to solve the linear ordering problem

$$\begin{aligned}\text{(LO) minimize } & \textstyle\sum_{i=1}^{n-1}\sum_{j=i+1}^{n} a_{ij}x_{ij} && && (5.7)\\ \text{subject to } & 0 \le x_{ij} + x_{jk} - x_{ik} \le 1 && \text{for } 1 \le i < j < k \le n && (5.8)\\ & 0 \le x_{ij} \le 1 && \text{for } 1 \le i < j \le n && (5.9)\\ & x_{ij} \in \mathbb{Z} && \text{for } 1 \le i < j \le n. && (5.10)\end{aligned}$$

To get the minimum number of crossings we have to add $\sum_{i=1}^{n-1}\sum_{j=i+1}^{n} c_{ji}$ to the optimum value of (LO). The "3-cycle-constraints" (inequalities (5.8)) ensure that the vector x indeed defines a permutation π_2 of V_2. To solve the linear ordering problem, a *branch-and-cut* approach is used.

First, a relaxation of (LO) is defined by dropping the integrality conditions. Since the space required for writing down all 3-cycle inequalities is in $O(|V_2|^3)$, solving the corresponding linear program is impractical. Therefore a cutting plane approach is used. The algorithm starts using only the hypercube inequalities (constraints (5.9)) and iteratively adds violated 3-cycle constraints and deletes non binding 3-cycle constraints until the relaxation is solved. If an integral solution is found, the algorithm stops. Otherwise, a fractional x_{ij} is chosen and the algorithm is applied recursively to two subproblems, one with $x_{ij} = 0$ and one with $x_{ij} = 1$.

Planarization. An alternative method to crossing reduction is given by Mutzel (1997). The idea is to remove a minimal set of edges such that the remaining k-layer graph can be drawn without edge crossings. In the final drawing, the removed edges are reinserted which, however, may produce much crossings. Let us for example consider the drawing of a graph from (Mutzel, 1997) obtained by 2-layer planarization in Figure 5.10 which has 34 crossings. The same graph with the minimum of 24 edge crossings is shown in Figure 5.11.

It is not quite clear whether the drawing with fewer crossings is more readable. On the contrary, in Figure 5.10 the four removed and reinserted edges stand out clearly and all edges can be easily followed by the eye. Still another motivation for studying the k-layer planarization problem is that it might be easier to attack than the k-layer crossing minimization problem.

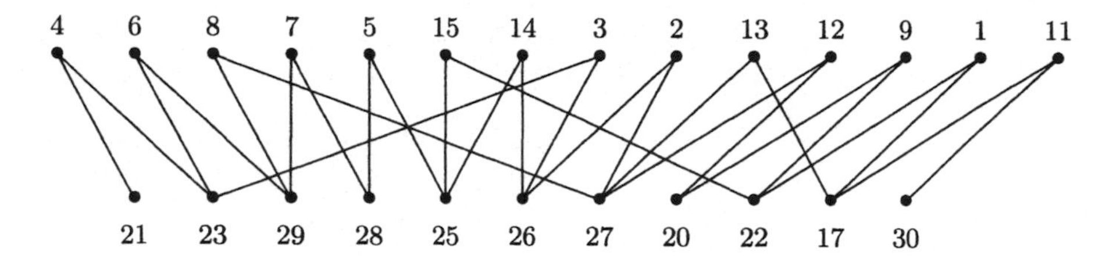

Fig. 5.10. Graph drawn using planarization.

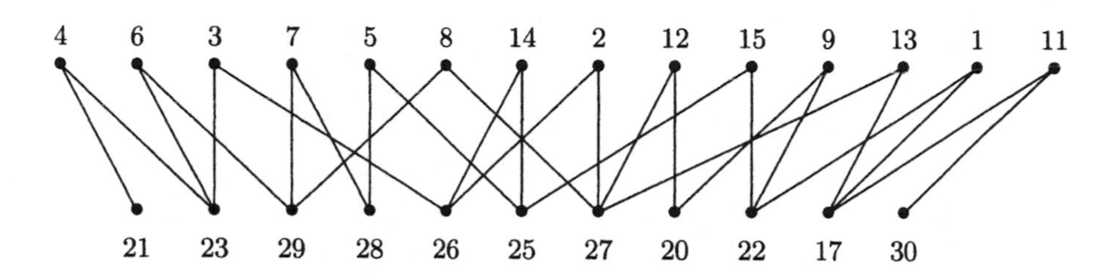

Fig. 5.11. Graph drawn with the minimal number of crossings.

However, even the 2-layer planarization problem is $\mathcal{NP}$-hard. To solve the problem, a formulation as an integer program and studies on the associated polytope are given by Mutzel (1997). The results are used in an efficient branch-and-cut algorithm and a heuristic is derived by setting a time limit of five minutes for the program. The results are close to the upper bound determined by the optimal solution of the linear programming relaxation.

5.4.3 K-layer Crossing Minimization

Tutte's Algorithm. The results of *Tutte's Algorithm* (Eades and Sugiyama, 1990) are similar to barycenter. First, the positions of the vertices in the first and in the last layer are fixed. In each other layer the x-coordinate of a vertex u is chosen as a weighted average of the x-coordinates of its neighbours:

$$x(u) = \frac{1}{2\,\text{outdeg}(u)} \sum_{v \in N^+(u)} x(v) + \frac{1}{2\,\text{indeg}(u)} \sum_{w \in N^-in(u)} x(w)$$

Now we have to solve a system of sparse linear equations to compute the value $x(u)$ for each vertex u. In the last step the vertices of each layer are sorted by x.

Optimum Crossing Minimization. Jünger and Mutzel (1997) and Lino et al. (1996) give branch-and-bound algorithms for the (two sided) 2-layer crossing minimization problem. In the approach of Jünger and Mutzel (1997), all permutations of the smaller layer are enumerated and the lower bound L is employed to make the search tree smaller. The branch-and-cut algorithm for the one sided crossing minimization is applied to obtain complete permutations of the larger layer.

To solve the 2-layer crossing minimization problem directly, an integer linear program and studies on the associated polytope are given by Jünger et al. (1997). Furthermore, they generalize the formulation to solve the k-layer problem. Note, that all cutting planes arising from the bipartite graphs are valid cutting planes for the combined k-layer graph. For both, 2-layer and k-layer problems, branch-and-cut algorithms were implemented and tested on 2-layer and 3-layer instances. The results indicate that the branch-and-cut approach may only be practicable, if deeper polyhedral studies are conducted.

5.4.4 Dense Graphs and Edge Concentration

The density of a graph is defined as the ratio of the number of edges of the graph to the number of edges of the corresponding complete graph. Thus, every complete bipartite graph has density 1. It can be proven, that for dense graphs the number of crossings is close to the optimum for every vertex ordering (Di Battista et al., 1999). The basic intuition is that if two vertices u and v have many common neighbours, then c_{uv} and c_{vu} are both large. More precisely, if χ_{uv} is the number of common neighbours of u and v, then

$$c_{uv} \geq \binom{\chi_{uv}}{2} \text{ and } c_{vu} \geq \binom{\chi_{uv}}{2}.$$

One consequence is that any crossing minimization heuristic will perform "better" if graphs become denser. The other is that large and dense graphs are hardly readable, even after the best crossing reduction, because every edge is hidden in a confusing mass of crossings. Two examples for such graphs are call graphs and graphs depicting relations between include files and source files of a system.

One way out of this dilemma is *edge concentration* (Paulish, 1993). This technique identifies complete bipartite subgraphs of a bipartite graph and replaces them with an equivalent tripartite graph as described next.

If a complete bipartite graph $G = (V_1, V_2, E)$ is given, the equivalent tripartite graph $G^* = (V_1, EC, V_2, E^*)$ is constructed by inserting a level with a single node, the *edge concentration node*, between the two levels. The edges in E are replaced by $E^* = (V_1 \times EC) \cup (EC \times V_2)$. We call the set of edges in the complete bipartite graph an *edge concentration* and the resulting tripartite graph is said to be *concentrated.*

Consider the complete bipartite graph in Figure 5.12(a). All edges are concentrated to a single node as shown in Figure 5.12(b). The resulting tripartite graph has fewer edges and no edge crossing. Note that after edge concentration edge labels are lost and an additional level is inserted.

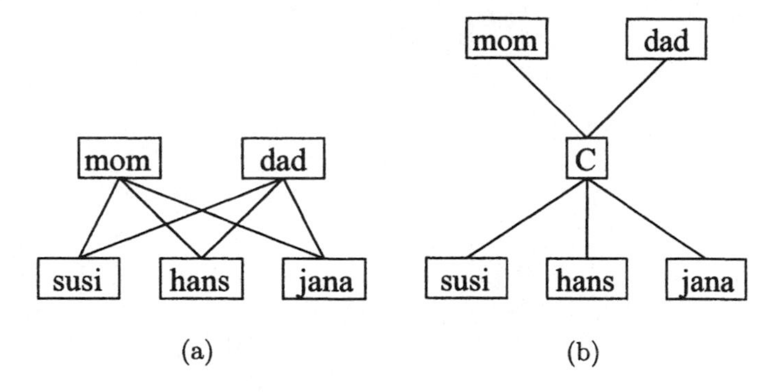

Fig. 5.12. A complete bipartite graph and the equivalent tripartite graph.

Since the concentration of a complete bipartite subgraph is easy, the main problem is to identify such subgraphs in an arbitrary bipartite graph. We state the *edge concentration problem* as follows.

> Given a bipartite graph $G = (V_1, V_2, E)$. Find a set of (possibly overlapping) complete bipartite subgraphs $G_1, G_2, \ldots, G_s$ of G that cover all edges of G and minimize the number of all edges in the equivalent, tripartite representations G_i^* of G_i.

It is an open question whether the edge concentration problem is $\mathcal{NP}$-hard or not. There is a similar and somewhat simpler problem, the *covering by complete bipartite subgraphs* problem, which is known to be $\mathcal{NP}$-hard (Garey and Johnson, 1991) and can be stated as follows:

> Given a bipartite graph $G = (V_1, V_2, E)$. Find the smallest set of (possibly overlapping) complete bipartite subgraphs $G_1, G_2, \ldots, G_s$ of G that cover all edges in G.

The edge concentration problem seems to be at least as hard as this problem, although no reduction has yet been found.

In the book of Paulish (1993) a heuristical algorithm for the edge concentration problem is given. First, potential edge concentrations are identified by considering the complete bipartite subgraph formed by each pair of source nodes and their common successors. Such a subgraph is called an *intersection.* For example, let us consider the graph in Figure 5.13. The intersection formed by the source nodes 1 and 2 and their commom successors has the node set $\{1, 2, A, B, C\}$. The node sets of the other two intersections are $\{1, 3, B, C\}$ and $\{2, 3, B, C, D\}$.

After determining all intersections, the algorithm tries to find good edge concentrations. If the target nodes of an intersection are a subset of the target nodes of a previously determined intersection, then the set of target nodes

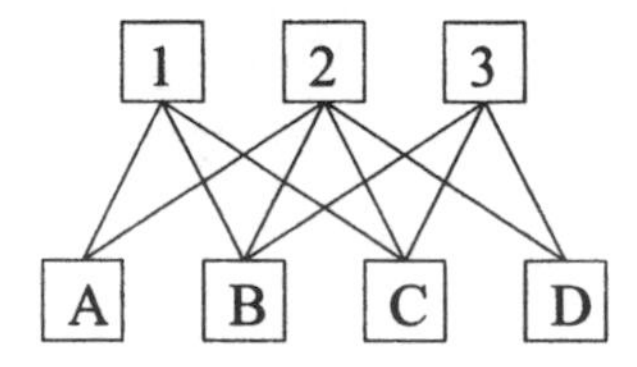

Fig. 5.13. A bipartite graph.

is partitioned into two sets. The first contains the target nodes which are in both intersections while the second contains the remaining target nodes. This partitioning of the target nodes helps to avoid overlap between concentrations which would lead to additional edges and crossings.

5.5 Horizontal Coordinates

The computation of the horizontal coordinates has mainly two different objectives. The layout should have as few bends as possible. As mentioned before, bends only occur at dummy vertices. This is true unless the expansion of the vertices is very large. For details concerning this particular problem, see for example (Sander, 1996b). In some applications not only straight edges but also vertical edges are preferred.

We will present exact approaches to this problem and afterwards introduce a heuristic.

5.5.1 Exact Algorithms

The problem of finding a layout with as straight edges as possible can be formulated as follows. Consider a directed path $p = (v_1, v_2, \ldots, v_k)$ where $v_2, v_3, \ldots, v_{k-1}$ are dummy vertices. We call this an *edge-path*. If the edge-path would be drawn straight, the dummy vertices would satisfy

$$x(v_i) - x(v_1) = \frac{i-1}{k-1}(x(v_k) - x(v_1))$$

for all $1 < i < k$. Observe that this formula is only valid for equidistant layers, but it is straight forward to adjust this formula for unequal layer distances.

To be able to state the objective function more compact, we introduce the term $\overline{x}(v_i) := \frac{i-1}{k-1}(x(v_k) - x(v_1)) + x(v_1)$ which would be the the x-coordinate of v_i if it would lie on the straight line between $x(v_1)$ and $x(v_k)$. We can now formulate a measure for the deviation of the path from a straight line

$$dev(p) = \sum_{i=2}^{k-1}(x(v_i) - \overline{x}(v_i))^2$$

To make the edges as straight as possible, we minimize the sum

$$\sum_{p \text{ is edge-path}} dev(p)$$

subject to the constraints

$$x(w) - x(v) \geq \rho(w, v)$$

for all pairs w, v of vertices in the same layer with w to the right of v.

The constraints ensure that the ordering within each layer computed by the crossing reduction step is preserved and that the horizontal distance $\rho(w, v)$ between the vertices is observed. The value $\rho(w, v)$ usually is calculated from the size of the vertices and the requested minimum horizontal distance between two succeeding vertices.

An optimal solution to this optimization problem may result in exponential width of the drawing and thus, if the width should be kept small further inequalities would have to be added. The main disadvantage is that since this problem has a quadratic objective function, it can only be solved to optimality for small instances.

Another objective is to draw the lines as close to vertical lines as possible. In this case the objective function can be stated as (Gansner et al., 1993)

$$\sum_{(u,v) \in E} \Omega(u, v)\omega(u, v)|x(u) - x(v)|,$$

where ω is a measure for the importance of an edge and Ω denotes an internal weight for straightening long edges. Therefore, the authors suggest higher priorities for edges between dummy vertices than between the other vertices ($\Omega(e) = 8$ if both endvertices are dummy vertices, $\Omega(e) = 2$ if exactly one endvertex is a dummy vertex and $\Omega(e) = 1$ otherwise). Of course, the introduction of a weight function may improve the layouts computed by the preceding model as well.

A new idea to solve this problem efficiently was introduced in (Gansner et al., 1993). Gansner et al. construct an auxiliary graph on which this problem transforms to a layering problem introduced in Section 5.3.4 which can be solved easily. The x-coordinates correspond to the layers and vice versa.

The auxiliary graph $G_a = (V_a, E_a)$ contains as vertices all vertices of G plus a vertex for each edge in G. Hence, $V_a = V \cup \{[uv] \mid (u, v) \in E\}$. We introduce two kinds of edges in G_a. The first class of edges encodes the original edges and is needed to eliminate the absolute values in the objective function. For every edge $(u, v) \in E$, we introduce two edges $([uv], u)$ and $([uv], v)$ in G_a. We define $\omega_a([uv], u) = \omega_a([uv], v) = \Omega(u, v)\omega(u, v)$ and $\lambda_a([uv], v) = \lambda_a([uv], v) = 0$. The second class of edges separates the vertices with the

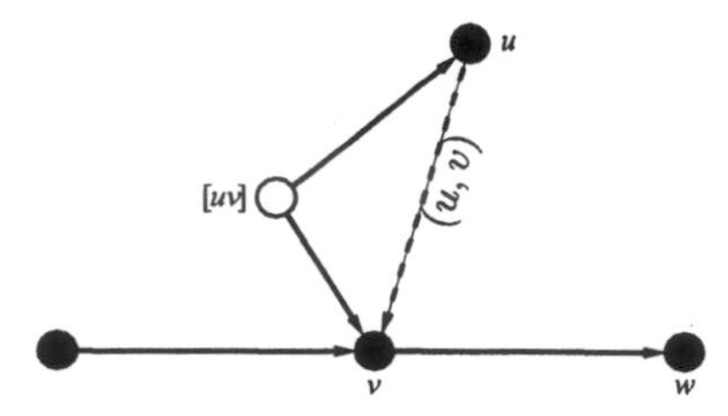

Fig. 5.14. Introducing auxiliary variables.

same rank. If v is a left neighbor of w in G, we insert an edge (v, w) in E_a and define $\omega_a(v, w) = 0$ and $\lambda_a(v, w) = \rho(v, w)$.

In the following, we will describe how a solution of the layering on G_a corresponds to a solution of the positioning problem on G and that both have the same cost. Let a solution of the positioning problem on G be given. Assign $[uv]$ to the layer $\min\{x(u), x(v)\}$. Conversely, in an optimal layer assignment in G_a, the vertex $[uv]$ lies in either the layer of u or the layer of v. Thus, one of the edges $([uv], u)$, $([uv], v)$ has length 0 and the other has length $|x(u) - x(v)|$. Hence, optimality in G_a implies optimality in G and a layering for G_a gives a solution for G.

5.5.2 A Heuristic

Another possibility is to obtain the x-coordinates by an improvement heuristic which can roughly be stated like the following:

---

initial coordinates;
while *some condition* **do**
    positioning;
    straightening;
    packing;

One possibility for computing an initial solution is to position the vertices with minimal distance from left to right in the order given by the crossing minimization.

In the positioning phase essentially the ideas used in the previous section for crossing reduction between two layers might be applied, like the median or barycenter heuristic. Another idea is to think of the vertices as balls and the edges as strings of a pendulum (Sander, 1996b).

Since these strategies compute layouts with many bends, in the straightening phase one tries to assign paths of dummy vertices to the same x-coordinate. The edges can be seen as rubber bends with vertices at both ends and the dummy vertices in between. This enlarges the drawing in x-direction of course. Hence, the drawing is compressed by moving the vertices closer together again without introducing new bends. These steps might be iterated to obtain a satisfying solution.

5.6 Positioning of Edges

Drawing of edges is easy if all nodes have the same size and shape. We simply draw arcs with span one as straight lines and longer arcs as polygons using the dummy nodes as intermediate points. If the gap between two nodes is wide enough, there will be no intersection between nodes and edges.

But if the nodes differ in size and shape, the problem to draw the edges such that no visible node is intersected is much more complex (see Figure 5.15(a)). A possible solution is to draw the edges orthogonal (in Manhattan layout) as shown in Figure 5.15(b). An algorithm for doing this was described by Sander (1996a).

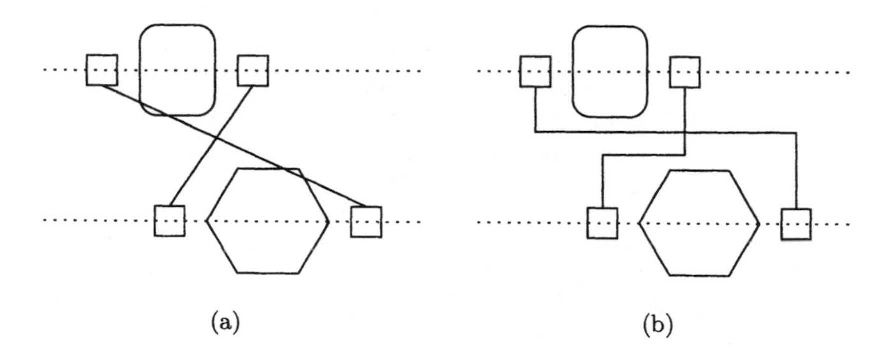

Fig. 5.15. Problem of node intersection and solution with orthogonal edges.

Another approach is to bend the edges at two points. An easy way to do this is to set the length of the vertical segments of all edges to the size of the largest node in the layer. Thus, all nonvertical segments between layer V_i and layer V_{i+1} start at the same y-coordinate and end at the same y-coordinate. Figure 5.16(a) shows the result.

Clearly, no edge can go through a node or can cross the vertical segment of another edge. The disadvantage is that too many bendings are produced. One can get better-looking graphs by bending edges only if necessary (see Figure 5.16(b)). But now, bent edges may cross neighboured edges. This can be avoided by also bending an edge which crosses a vertical segment of an already bent edge. The iterative Algorithm 16 implemented in the VCG-tool (Sander, 1994) starts with unbent edges and introduces vertical segments or enlarges them until no overlapping or additional edge crossing remains.

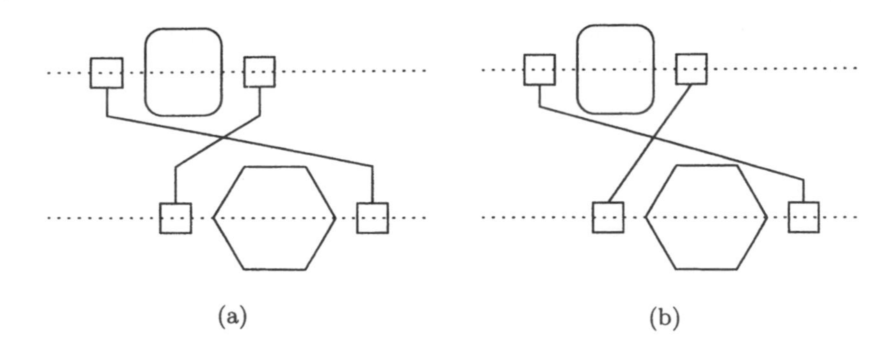

Fig. 5.16. Solutions with bending of edges.

Algorithm 16: bend_edges

while $\exists e \in E$ *that overlaps a node or crosses a vertical segment of an edge* e' **do**
 enlarge the vertical segments of e;

We can get even nicer drawings if we draw the edges as curves instead of straight lines and polygons. Among the different methods to interpolate and approximate points, Bézier curves have properties which make them suitable to represent edges between nodes. We will shortly describe this kind of curves next; for a deeper view see for instance (Foley et al., 1990).

Bézier curves are specified by the control points $b_0, \ldots, b_n$. The points b_0 and b_n are the interpolated end points of the curve, all other control points are approximated. The gradients in b_0 and b_n are given by the gradient of the straight line segment between b_0 and b_1 and between b_n and b_{n-1}, respectively. Figure 5.17 shows a Bézier curve with three control points and the associated polygon. Note that the entire curve is completely enclosed by the convex hull of the control points.

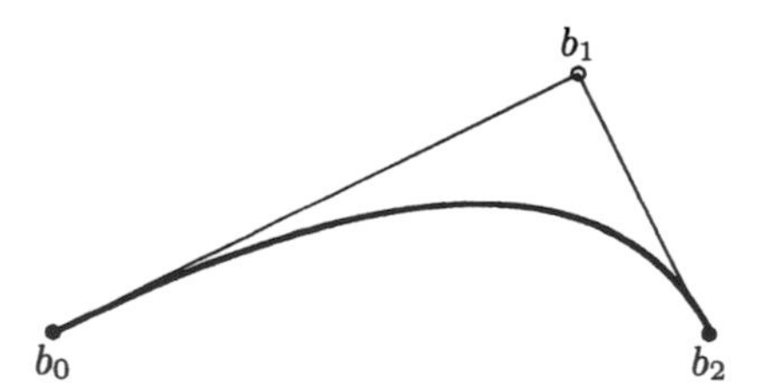

Fig. 5.17. A Bézier curve.

The Bézier curves used in the VCG-tool also have three control points which are determined by finding appropriate triangles at the bendpoints of the polygons. Thereby, the middle control point is set to the bendpoint itself and we try to find good positions for the two end points on the adjacent polygon segments. In order to do this, the end points are set on the adjacent segments such that an isoscele triangle is produced (see Figure 5.18(a)). Then the size of the triangle is reduced until all nodes and bendpoints are outside the channel (Figure 5.18(b)). Finally, the curve is drawn inside the triangle as shown in Figure 5.18(c).

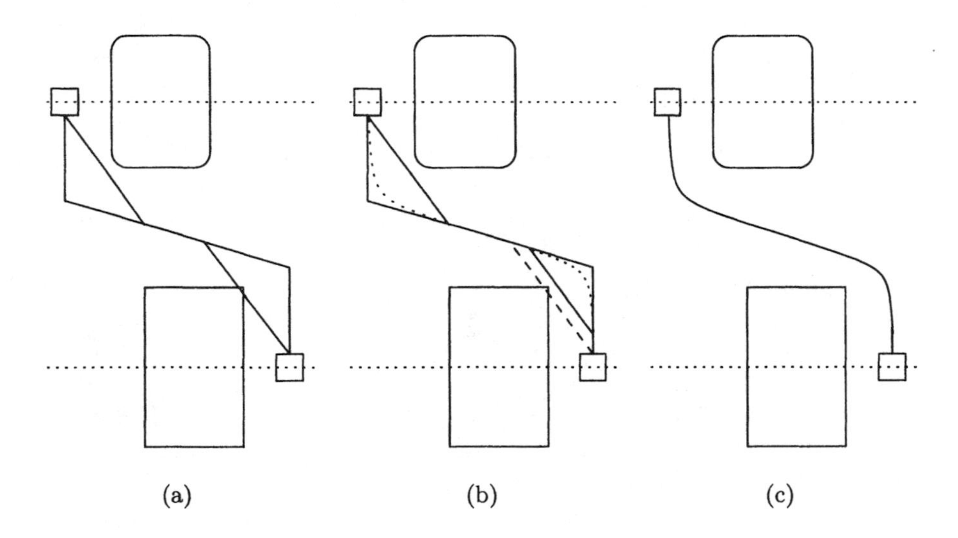

Fig. 5.18. Calculation of the curve.

A sophisticated method to find the smoothest curve between two points is given in the article of Gansner et al. (1993). First, a region where the curve may be drawn is determined. This space is represented by a set of boxes parallel to the coordinate axes. Then, the best curve within the region is drawn as a piecewise Bézier curve which is done in the following way. First, a polygon is generated which lies entirely inside the region. The endpoints and intermediate points are used as hints for the control points of the Bézier curves. The actual Bézier curve is determined in an iterative process that perturbs the control points until the curve fits in the region.

5.7 Related Approaches

5.7.1 Upward Planarity

A directed graph is *upward planar* if it can be drawn in a way such that no two edges intersect and every edge is monotonically nondecreasing in the vertical direction (see Chapter 2 for an introduction to planar graphs). So, two of the aesthetic criteria mentioned in Section 5.1 can be optimally satisfied. Unfortunately, although testing whether a digraph admits a planar drawing or an upward drawing can be done in linear time, upward planarity testing is NP-complete (Garg and Tamassia, 1995a). Nevertheless, there are many special classes of directed graphs for which upward planarity testing can be done in polynomial time (Garg and Tamassia (1995b) give a survey of testing algorithms).

Note, that ordered sets and upward planar digraphs are closely related. Ordered sets with a special treatment of lattices are covered in Section 3.3 in this book.

We will now mention some classes of digraphs for which efficient upward planarity testing and drawing methods exist.

(s,t)-digraphs. An *(s,t)-digraph* is an acyclic digraph with exactly one source s, exactly on sink t, and the edge (s,t). A planar (s,t)-graphs is always upward planar. More general, a digraph is upward planar if and only if it is a subgraph of a planar (s,t)-digraph (Garg and Tamassia, 1995b). Garg and Tamassia (1993) describe an efficient algorithm for drawing a planar (s,t)-digraph. Since every upward planar digraph can be easily extended to an planar (s,t)-digraph, we can draw every upward planar digraph efficiently.

Embedded digraphs. An *embedding* of a graph associates to each vertex v a circular clockwise ordering of the incidence list of v. An *embedded graph* is a graph with a given embedding. Bertolazzi et al. (1994b) give an polynomial-time algorithm for testing if an embedded digraph has a upward planar drawing. If such a drawing exist the algorithm allows to construct one easily.

Single source digraphs. Hutton and Lubiw (1991) give an $O(n^2)$ time algorithm for testing if a single source graph is upward planar where n is the number of vertices in the graph. Improvements made by Bertolazzi et al. (1993) yield an $O(n)$ time algorithm.

Embedded single source digraphs. Hutton and Lubiw (1991) and Bertolazzi et al. (1993) give an $O(n^2)$ time resp. $O(n)$ time algorithm for testing if whether an embedded single source digraph has an upward drawing which preserves the embedding.

Planar bipartite digraphs. Planar bipartite digraphs are always upward planar. We refer to the paper of Di Battista et al. (1990).

Series-Parallel digraphs. Series-parallel digraphs are always upward planar and covered in Section 3.2 in this book.

Trees and Forests. Directed graphs whose underlying undirected graph is a forest are always upward planar. See Section 3.1 for more information about trees.

Triconnected digraphs. A graph G is *triconnected* if it is biconnected and has no separation pair which is a pair of vertices whose removal increases the number of connected components. Since a triconnected digraph has a unique embedding we can use the algorithm of Bertolazzi et al. (1994b) for testing upward planarity.

Outerplanar digraphs. Outerplanar graphs are planar graphs which admit an embedding such that all the vertices are on the same face. Testing if an embedded outerplanar digraph with n vertices is upward planar can be done in time $O(n)$ while testing if an outerplanar digraph is upward planar can be done in time $O(n^2)$ (Papakostas, 1995).

5.7.2 Clustered Graphs and Hierarchical Graphs

Sometimes some grouping of the vertices (clustering) is given together with the graph. These graphs and more general graphs will be treated in Chapter 8. Another possibility is to compute a grouping, draw the groups and then combine the partial drawings to complete the drawing. This leads to a divide and conquer approach. Concerning the computation of layered drawings this strategy is discussed in (Messinger et al., 1991). See Chapter 8 as well.

5.7.3 Recurrent Hierarchy

Sugiyama et al. (1981) suggest another hierarchy, called recurrent hierarchy. Edges are allowed between consecutive layers and between the last and the first. This might be useful to illustrate graphs with large feedback arc sets. Unfortunately, this problem is still not well studied.

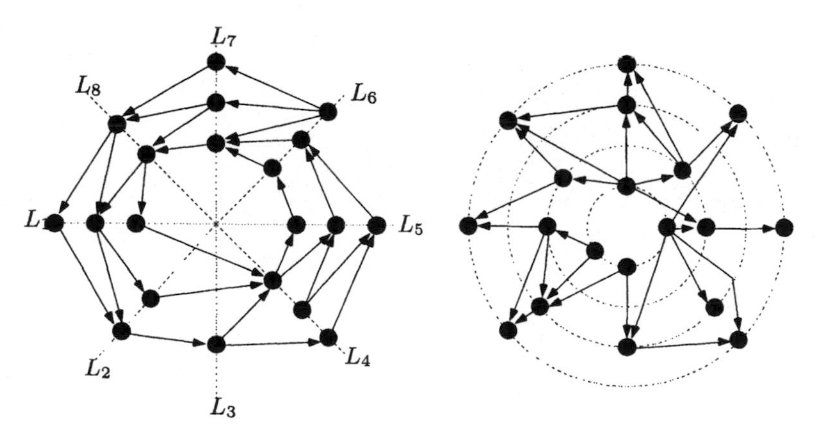

Fig. 5.19. A recurrent hierarchy and a ring diagram.

5.7.4 Ring Diagram

Another idea is to display the flow radial where the layers lie on concentric circles and the arcs are pointing outward (Reggiani and Marchetti, 1988). This is very similar to the three-dimensional case of layered drawings, see Section 7.3 for details on this.

5.7.5 Combining the Steps

Usually the single steps are performed independently. Specially the layering step and crossing minimization step are strongly related and it seems to be reasonable to solve them together in one step. A convincing example is depicted in Figure 5.20. A first attempt using an evolutionary algorithm is presented in Utech et al. (1998).

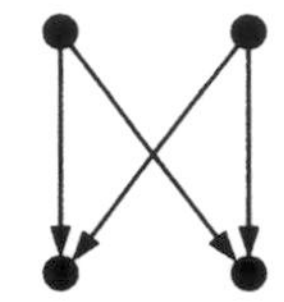
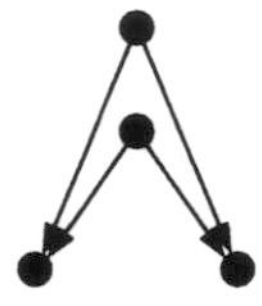

Fig. 5.20. A graph which be drawn either in two layers but with crossings or in at least three layers without crossings but with long edges.

6. Orthogonal Graph Drawing

Markus Eiglsperger, Sándor P. Fekete, and Gunnar W. Klau

6.1 Introduction

There are various criteria to judge the quality of a drawing of a graph. From a human point of view, one of the most important issues is the readability of a drawing: ideally, it should be easy to understand the structure of a graph with just a few glances, and the chance of confusion over connections between different vertices should be small. From an algorithmic point of view, it is necessary to capture this quality by means of an objective function. Various objective functions have been studied, with a great deal of effort put into their optimization by means of combinatorial algorithms.

An undesired property of a drawing that may impede its legibility is the presence of edges that are too close together. Keeping different edges apart may be particularly difficult in the vicinity of vertices, where several adjacent edges have to meet. Clearly, there is some correlation between the involved angles and the optical distinctiveness of the drawn edges. This motivates a particular objective function that is considered at the beginning of this chapter: find a drawing of a graph such that the minimum angle between adjacent edges is maximized. The first section discusses upper and lower bounds on angles in straight-line drawings.

There is a particularly nice way to guarantee maximal distinctiveness of adjacent edges in a drawing: when forcing all angles between adjacent edges to be multiples of $\frac{\pi}{2}$, edges will correspond to axis-parallel paths. The price we may have to pay for this type of clarity is to admit bends in the path representing an edge. In order to avoid confusion by too complicated paths, it is desirable to minimize the number of these bends. This setup has given rise to the area of orthogonal graph drawing – probably one of the most prolific in all of graph drawing, with scores of methods, heuristics, and sophisticated algorithms like KANDINSKY, others extending to mainly theoretical research areas like three-dimensional drawings. Over the years, orthogonal graph drawing has become far more important than the issues of angles in drawings, but both types of problems have their own motivation and have been studied independently.

It is the main objective of this chapter to combine a description of the key ideas (like the flow methods in the landmark paper of Tamassia 1987) with an overview of some of the main consequences and applications.

This chapter is organized as follows: after discussing angles in drawings in Section 6.2, Section 6.3 characterizes the correspondence between orthogonal drawings and combinatorial descriptions: how can we give a compact

M. Kaufmann and D. Wagner (Eds.): Drawing Graphs, LNCS 2025, pp. 121-171, 2001.

combinatorial encoding of the orthogonal shape of a drawing (called an "orthogonal representation" in the literature), and how can we realize a given combinatorial encoding as an orthogonal drawing?

Section 6.4 describes a number of heuristics that have been developed for finding a good orthogonal drawing without employing orthogonal representations.

When trying to find orthogonal drawings with few bends, we concentrate on the space of orthogonal representations. Optimizing over this space is the subject of Section 6.5, where we describe an efficient combinatorial algorithm for this task. Extensions to planar graphs of possibly high degree are sketched.

The final Sections 6.6 and 6.7 deal with improving orthogonal drawings. In many cases, the output of orthogonal drawing algorithms can be compacted further by assigning different – but still consistent – lengths to the edge segments. Section 6.6 gives an overview of compaction techniques, ranging from efficient heuristics to optimal techniques. The following Section 6.7 presents some efficient postprocessing methods that operate directly on the orthogonal drawings and try to improve aesthetic criteria like the number of bends or the number of crossings. We conclude with Section 6.8 and present some open problems in orthogonal graph drawing.

In many algorithms presented in this chapter, flow algorithms play the key role. Many problems can be reduced to a maximum or minimum cost flow problem. A good overview of network flow problems, modeling, and algorithms can be found in Ahuja et al. (1993) or in Bertsekas (1998).

6.2 Angles in Drawings

As described above, the following optimization problem comes up naturally when trying to create drawings with high resolution:

Problem 6.1 (Angular Resolution). Given a graph $G = (V, E)$ with n vertices and m edges, how can we draw the vertices as points in the plane, and the edges as straight lines between adjacent vertices, such that the angular resolution, i.e., the smallest angle between adjacent edges, is as large as possible?

Over the years, a number of researchers have given various kinds of answers to this question. One type of result is to establish the complexity of ANGULAR RESOLUTION. It was shown by Formann et al. (1990) that it is $\mathcal{NP}$-hard to check whether a planar graph with maximum degree 4 can be drawn with angular resolution at least $\frac{\pi}{2}$. Before discussing how to relax the requirements on drawings, such that a drawing with resolution $\frac{\pi}{2}$ is always possible, we describe some lower and upper bounds for straight-line drawings.

If d is the maximum degree of a vertex in G, it is clear that the angular resolution cannot exceed $\frac{2\pi}{d}$. It was shown in (Formann et al., 1990) that for planar graphs, a resolution of $\Omega(\frac{1}{d})$ can indeed be achieved. For general graphs, a resolution of $\Omega(\frac{1}{d^2})$ can be guaranteed. The key is to use a coloring of $G^2 = (V, E^2)$ of G. (Recall that two vertices in V are adjacent in G^2 if and only if they have distance at most 2 in G.) It can be shown that there is a coloring of G with $O(d)$ colors for planar graphs G and with $O(d^2)$ colors for general graphs G. Then points of a color class are drawn as a cluster of points on a unit circle, with different clusters distributed at equal distance around the circle. This guarantees that any angle between adjacent edges in G involves points from three different color classes of G^2, implying the claimed bounds.

The method by Formann et al. suffers from a very serious drawback for practical purposes. The number of crossings in the straight-line drawing may be much higher than necessary. In particular, the resulting drawing of a planar graph may not be free of edge crossings. Thus, it remained open whether the resolution of a planar drawing of a planar graph could be bounded from below in a satisfactory manner.

A first partial answer to this problem was given by Malitz and Papakostas (1992, 1994) who described a method to guarantee a lower bound of $\frac{1}{7^d}$ for planar straight-line drawings of planar graphs. Their approach relies on so-called "disc-packings" or "coin graph representations" of a planar graph $G = (V, E)$, where vertices $v \in V$ are represented by disjoint discs, and two vertices $v_1, v_2 \in V$ are adjacent if and only if the discs corresponding to v_1 and v_2 touch. The existence of this type of representation has been proved independently by a number of researchers, including Koebe (1936), Andreev (1970a,b), Colin de Verdière (1989), and Thurston (unpublished). The lower bound arises from the fact that in certain subsets of adjacent discs, the radii can vary by at most a factor of $\frac{1}{7}$. Malitz and Papakostas also conjectured that a lower bound of $\Omega(\frac{1}{d})$ for the angular resolution of planar straight-line drawings of planar graphs might be achievable. As partial evidence for this conjecture, they showed that a particular relaxation of the problem ANGULAR RESOLUTION has an optimum of $O(\frac{1}{d})$:

Any set of angles in a feasible drawing has to satisfy a set of linear equalities – see Figure 6.1. Around every vertex v, the sum of the angles $\Phi(v)$ at v must be 2π; for every interior cycle the sum of the angles $\Phi(f)$ must be $(|\Phi(f)| - 2)\pi$, and $(|\Phi(f_0)| + 2)\pi$ for the exterior cycle f_0:

$$\sum_{\phi_i \in \Phi(v)} \phi_i = 2\pi \text{ for all } v \in V. \tag{6.1}$$

$$\sum_{\phi_i \in \Phi(f)} \phi_i = \pi(|\Phi(f)| - 2) \text{ for all } f \in F \setminus \{f_0\}. \tag{6.2}$$

$$\sum_{\phi_i \in \Phi(f_0)} \phi_i = \pi(|\Phi(f_0)| + 2). \tag{6.3}$$

If we only impose these necessary conditions while maximizing the minimum angle, we get a linear program that can be solved efficiently. By using linear programming duality, it can be shown that there is always an optimum of value $\Omega(\frac{1}{d})$.

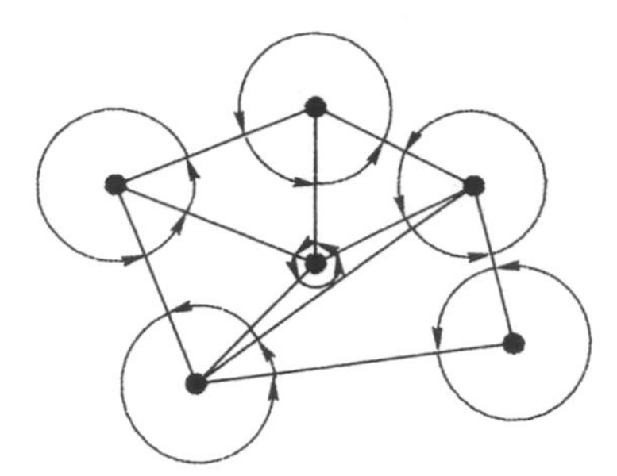

Fig. 6.1. Angles in a drawing must satisfy certain conditions.

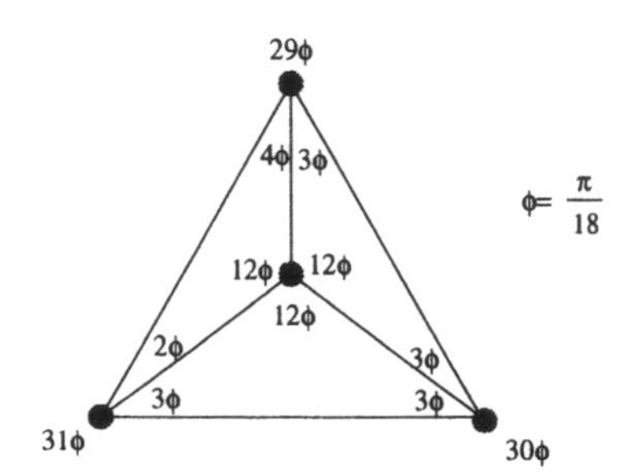

Fig. 6.2. A solution to the linear system that does not correspond to a feasible drawing.

It can be seen from Figure 6.2 that not every solution to the linear programming relaxation corresponds to a set of angles that allow a drawing. After drawing the bottom triangle with the given angles, the three edges incident to the top vertex cannot intersect in one point. It was shown by Di Battista and Vismara (1993, 1996) that for triangulated triconnected planar graphs with a designated external face f_0, the conditions on angle sums can be amended by the following requirement to get a set of necessary and sufficient conditions – see Figure 6.3:

Around each vertex v of degree $d(v)$, we have angles γ_i, $i = 1, \dots, \deg(v)$. Each γ_i lies in the same triangle as the angles α_i and β_i, as shown in the figure. In any feasible drawing, the angles α_i, β_i satisfy

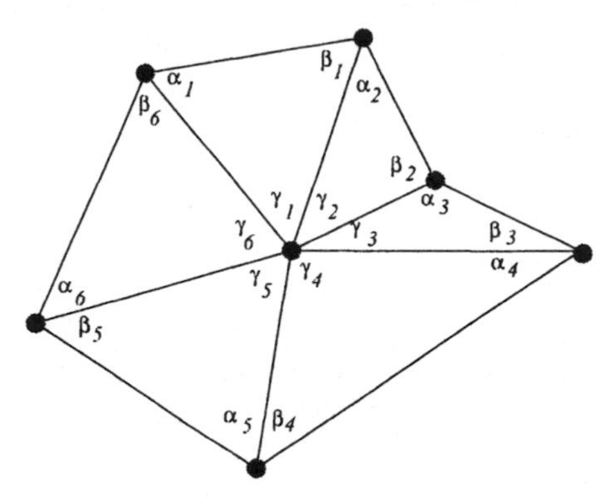

Fig. 6.3. Angles around a vertex in a drawing.

$$\prod_{i=1}^{d(v)} \frac{\sin \alpha_i}{\sin \beta_i} = 1.$$

The necessity of these conditions is a consequence of planar trigonometry; conversely, it can be shown that this additional condition suffices to get a feasible drawing for each "wheel", as shown in Figure 6.3, since any triple of edges that are incident to the same vertex will indeed meet at a single point. By induction, it follows that we get a feasible drawing for the full graph. Adding these conditions results in a nonlinear program for optimizing the angular resolution.

Garg and Tamassia (1994) managed to disprove the conjecture of Malitz and Papakostas (1994) by giving a family of graphs with an upper bound of $O(\sqrt{\log d/d^3})$. See Figure 6.4. In the same paper, they showed that good angular resolution may come at the expense of high numerical resolution, i.e., large angles require high precision in the coordinates. In other terms, they showed that there is a family of graphs, such that if all n vertices are drawn at integer coordinates, a drawing of angular resolution ρ requires area $\Omega(c^{\rho n})$. See Figure 6.5.

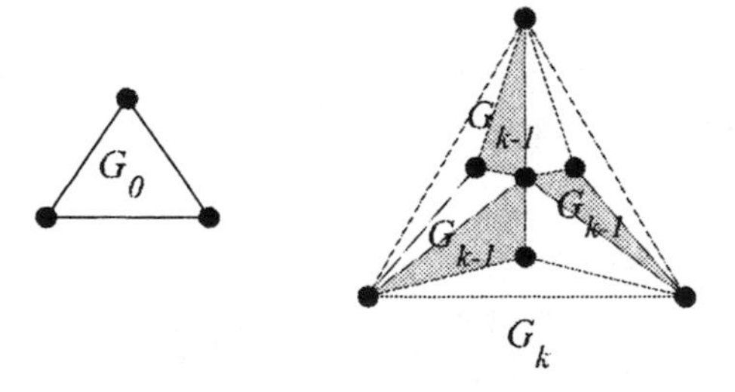

Fig. 6.4. A family of graphs with small angular resolution.

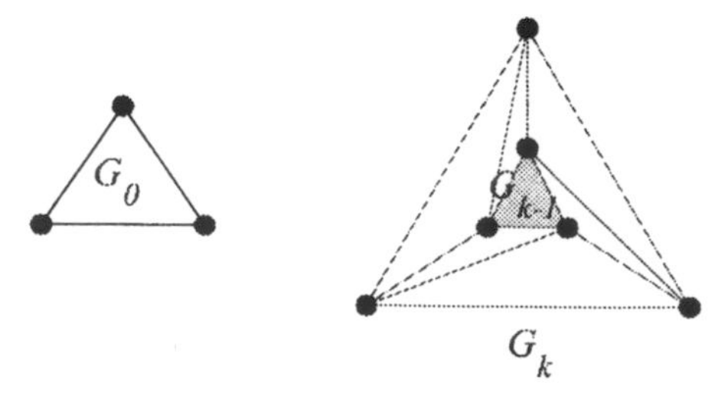

Fig. 6.5. A family of graphs with tradeoff between angular resolution and area of a drawing.

6.3 Orthogonal Drawings and Their Encoding

6.3.1 Why Orthogonal Drawings?

When trying to come up with a drawing of a graph that makes it easy to distinguish different edges, there is an alternative to using straight edges at arbitrary angles: if edges are allowed to be drawn as a path consisting of several line segments, it is possible to let all edges be represented by axis-parallel paths, called an *orthogonal drawing*. The price we may have to pay is the introduction of additional nodes where changes in direction occur in a path, i.e., a number of *bends* of the edges.

See Figure 6.6 for an example. A formal definition of an orthogonal drawing can be given as follows:

Definition 6.2. *An* orthogonal grid embedding *Γ of a graph $G = (V, E)$ is a mapping into the plane, which maps vertices $v \in V$ to integer grid points $\Gamma(v)$ and edges in $(v, w) \in E$ to non-overlapping paths in the grid such that the images of their endpoints $\Gamma(v)$ and $\Gamma(w)$ are connected. A grid embedding is* simple *if its number of bends is zero. A simple embedding induces a partition of the edge set E into a horizontal set E_h and a vertical set E_v. For simplicity, we assume that edges in E_h are directed from left to right and edges in E_v from bottom to top.*

Another issue is the numerical resolution, i.e., the difference of the involved coordinates for a drawing of given size. If we scale the final drawing such that all coordinates are integers, this translates to trying to find a drawing that uses small area.

The arrangement of edges in an orthogonal drawing is well-structured (there are only two classes of line segments, and the segments for each individual edge interchange between horizontal and vertical), so it is conceivable that drawings with good visual and structural properties are possible. Moreover, there is a close relationship to problems of VLSI layout. The components

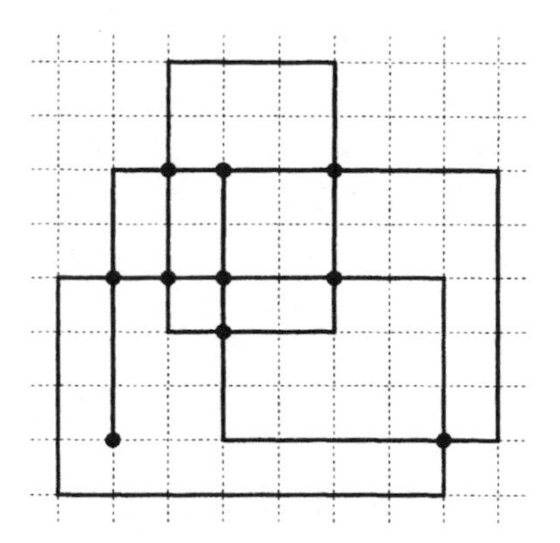

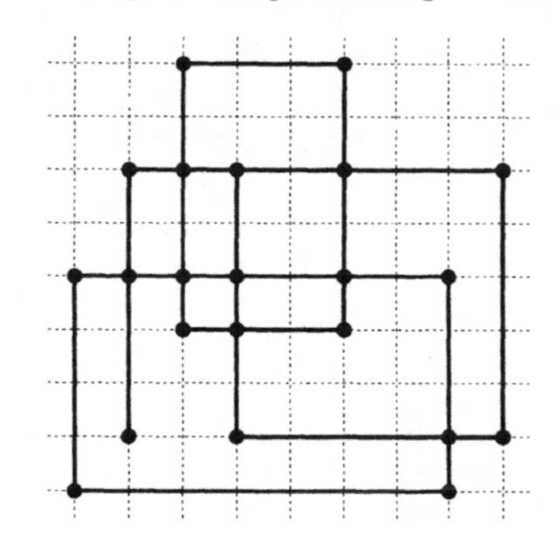

Fig. 6.6. An orthogonal grid embedding (left) and its simple counterpart (right).

in a chip layout correspond to the vertices, their connecting wires to the edges of a graph. This allows it to make use of existing methods; see Sections 6.4 and 6.6. In VLSI, however, research has concentrated on very efficient methods due to the large sizes of the instances. Often, superlinear running time is too slow for practical application, whereas in the area of graph drawing higher running time is tolerated to get better drawings[1]. However, the presence of crossing edges is highly undesirable in well-structured drawings of a graph as well as in routing wires of a chip; thus, planar orthogonal graph drawing deals almost exclusively with plane drawings of planar graphs. Since it is impossible to avoid overlap between edges if a graph has degree more than 4, we start by focusing on planar graphs with maximum degree 4 – so-called *4-planar graphs*. It will be discussed later how to deal with planar graphs of higher degree.

6.3.2 Encoding Planarity

Before dealing with heuristic and exact algorithms for orthogonal drawings and their optimization in the following sections, we now describe a way to encode a graph and a drawing of a graph, such that we can use these encodings for input and output of our algorithms. A graph $G = (V, E)$ can be simply described as a list of vertices V, and the edges E connecting them. A plane drawing of a planar graph contains additional topological information: if the edges of a graph are represented by a set of curves, the plane is subdivided into a number of open regions. These regions are called *faces*. For a given embedding, the structure of these faces is characterized by the cycles of edges surrounding them, so that it is also legitimate to speak of these cycles as faces.

[1] Another difference (at least compared to problems within the topology-shape-metrics paradigm that will be discussed in Section 6.3.4) is that the order of wires connected to a component is not necessarily fixed – this corresponds to a scenario with an arbitrary embedding.

It is not hard to see that faces and their adjacencies are determined by the circular order of edges around each vertex. Thus, we can describe a particular (topological) embedding of a planar graph by a list of its vertices V, a list of its edges E, a list of faces F, and a list $P(f)$ of edges for each face f. See Figure 6.7 for an example; the corresponding list of faces and list of edges around each face is as follows:

$$
\begin{aligned}
&F = \{f_1, \dots, f_5\} \\
&P(f_1) = (e_1, e_3, e_5, e_9, e_{10}, e_4, e_3, e_2) \\
&P(f_2) = (e_2, e_1) \\
&P(f_3) = (e_5, e_6, e_8) \\
&P(f_4) = (e_6, e_4, e_7) \\
&P(f_5) = (e_8, e_7, e_{10}, e_9).
\end{aligned}
$$

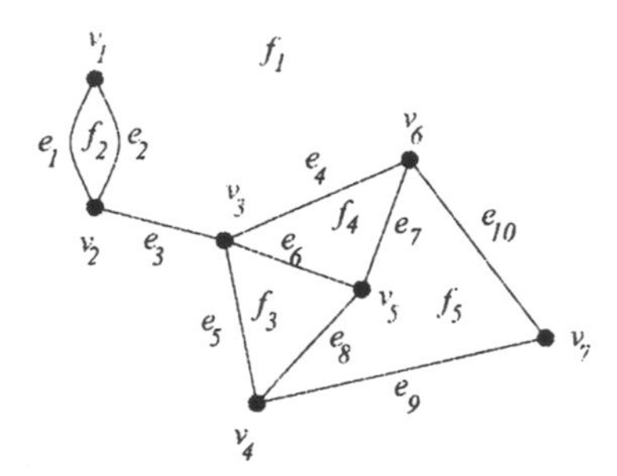

Fig. 6.7. A plane drawing of a graph G.

6.3.3 Encoding Orthogonality

When considering an orthogonal drawing of a planar graph, we need to provide even more information – in particular, we need to describe the bends along an edge, and the type of turn that an edge takes at a bend. This can be done as follows: for each edge in the edge list of a face, we describe the sequence of bends encountered while traveling along the edge by a 0-1 string. A "1" in the string indicates a left-hand turn taken at the bend, while a right-hand turn is indicated by a "0". If an edge has no bends, this is indicated by the empty string ε. Finally, each edge is assigned an angle (as a multiple of $\frac{\pi}{2}$) that is enclosed by its last line segment and the first line segment of the next edge. See Figure 6.8 for an example with the following encoding of the edges:

$$H(f_1) = ((e_1, \varepsilon, 3), (e_5, 11, 1), (e_4, \varepsilon, 3), (e_2, 1011, 1)),$$
$$H(f_2) = ((e_1, \varepsilon, 1), (e_6, \varepsilon, 2), (e_5, 00, 1)),$$
$$H(f_3) = ((e_2, 0010, 1), (e_4, \varepsilon, 1), (e_6, \varepsilon, 1), (e_3, 0, 4), (e_3, 1, 1)).$$

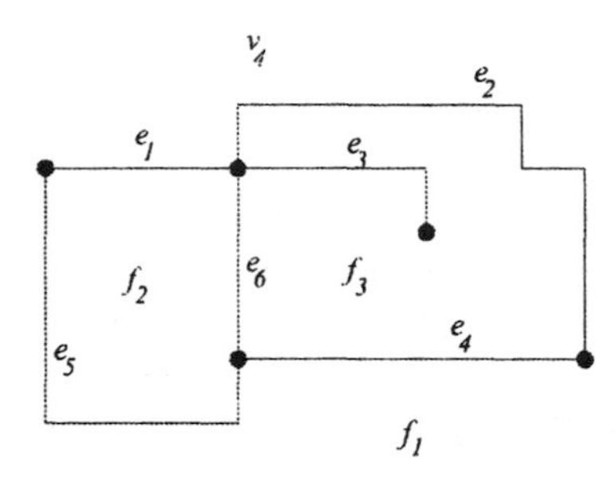

Fig. 6.8. An orthogonal drawing of a graph G.

This type of encoding of the orthogonal shape of a drawing has been called an *orthogonal representation* of the drawing; for historical reasons, we will use this term, even though it can be argued that in the true meaning of the word, the orthogonal drawing itself is a representation of the graph. (Strictly speaking, the code should be called an "encoding of the representation".)

The orthogonal representation carries only the combinatorial and some of the geometric information of the drawing; in particular, there is no information about the edge lengths. As described above, there is a well-defined orthogonal representation for each drawing of a graph, and it is not hard to see that there are a number of necessary conditions on an orthogonal representation:

1. There is a 4-planar graph corresponding to the lists for V and E.
2. Each edge is encoded twice, once for each of the two faces it bounds. Both these encodings must be consistent.
3. The sum of angles along the perimeter of a face f_i described by $H(f_i)$ must be consistent with the fact that f_i is a simple rectilinear polygon.
4. For each vertex v, the sum of angles between consecutive edges around v must sum to 4.

All four of these conditions are easy to check. As it turns out, they are also sufficient. We describe in the next section how to find an orthogonal drawing that realizes a given orthogonal representation.

6.3.4 Getting a First Drawing

We now discuss how to assign consistent integer lengths to the edge segments that are contained in an orthogonal representation. This task is also referred to as the third phase in the *topology-shape-metrics approach* as presented in (Di Battista et al., 1999). The first phase deals with fixing the topology of the eventual drawing by determining a combinatorial embedding and an outer face of the planar input graph. In the second phase the shape of the orthogonal drawing is fixed. Its output is an orthogonal representation H as described in Section 6.3.2. The best-known member of this class of algorithms is the bend-minimization algorithm by Tamassia (1987). Section 6.5 describes these flow-based algorithms. For now, we concentrate on the third task: fixing the metrics of the drawing resulting in an orthogonal grid embedding.

We present an algorithm which first finds an orthogonal grid embedding for H by describing a simplified variant of the approach presented in Tamassia (1987). This grid embedding can be further improved by applying compaction and postprocessing algorithms described in Section 6.6.

The idea of this method is to add artificial vertices and edges to H so that it is easy to find a drawing Γ' for the resulting orthogonal representation H'. Note that the insertion of artificial objects still happens on the level of the orthogonal representation and does not involve geometric operations. The removal of the artificial vertices and edges leads to an orthogonal grid embedding Γ for the original representation H. The algorithm presented in this section runs in time $O((n+b)^{7/4}\sqrt{\log(n+b)})$ and guarantees a drawing of $O((n+b)^2)$ area, where b is the number of bends.

In a first step, each bend in H is replaced by a virtual vertex, resulting in a simple orthogonal representation with $n+b$ vertices. Consider for every face f in H the circular bit string $S(f)$ that is built by traversing the edges of f in clockwise order as follows. Depending on the angle $a(e)$ (expressed in multiples of $\frac{\pi}{2}$) that an edge e forms with its succeeding edge, we add a "0" (if $a(e)=1$), a "1" (if $a(e)=3$) or a "11" (if $a(e)=4$) to the bit string. Angles of 2 do not contribute to $S(f)$, since they correspond to two collinear line segments that do not form a bend. The resulting string describes the shape of face f as in Figure 6.9 (a). Now the algorithm looks for occurrences of the substring "100" in $S(f)$, which corresponds to a rectangular "ear". Whenever such a string is found, the corresponding part of the face is removed and "100" is replaced by "0" (see Figure 6.9 (b)). Clearly, this operation affects only the shape of face f, other faces remain untouched. The procedure finds all such substrings in time $O(|S(f)|)$ for each face f by a circular traversal of $S(f)$ and stacking the encountered "1"s. The bit string of a face f in the resulting orthogonal representation H' is either $S(f)=$"0000" – in this case f is an internal face – or it does not contain two consecutive zeros – then we are dealing with the external face. (See Figures 6.9 (c) and (d). The example will be continued in Section 6.6.)

An orthogonal grid embedding for H' is found by assigning the same length to opposite sides in the rectangular faces. This can be done optimally by the following algorithm; its key step is the computation of two minimum cost flows. Build two networks N_h and N_v-one for each direction, as in Fig-

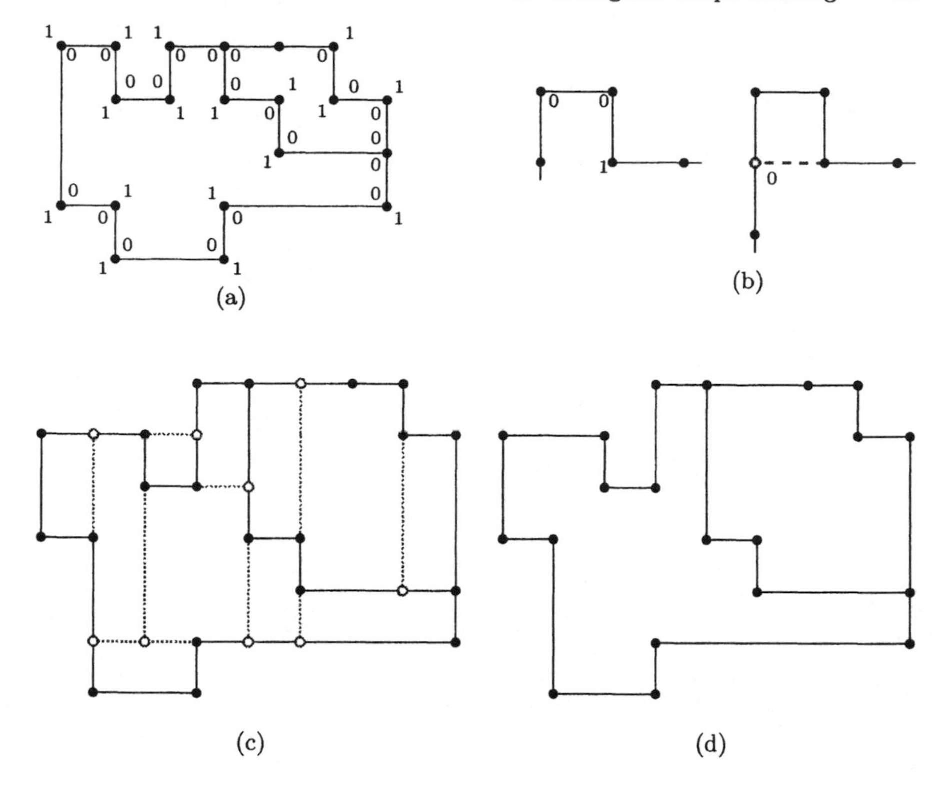

Fig. 6.9. (a) A sketch of an orthogonal representation H with bit strings $S(f)$ for each face f. (b) Substring "100" and the corresponding cut. (c) A sketch of the dissected representation H'. (d) Deleting the artificial objects in a drawing for H' yields a drawing for H.

ure 6.10. The union of N_h and N_v is the dual graph of H', the arcs in N_h are directed from bottom to top, those in N_v from left to right. Each arc a corresponds to an edge e in H', and the flow through a is interpreted as the length of e. Intuitively, this implies a lower bound of one and unit cost for the flow through a. Flow conservation ensures that opposite sides are of equal length. The cost of the flows add up to the total edge length, and – since they are minimized – lead to an optimal drawing for H'. Note however, that the artificially introduced vertices and edges have to be removed from the drawing in order to get an orthogonal grid embedding for H. As Figure 6.9(d) shows, the initial solution is not optimal for the original input.

An optimal flow can be computed in time $O(n^{7/4}\sqrt{\log n})$ with the algorithm described in Garg and Tamassia (1996b). This corresponds to a drawing for H' with minimum total edge length, width, height, and area. The networks are similar to the ones used for optimal one-dimensional compaction, which are introduced in Section 6.6.2. Furthermore, there is a linear-time method for

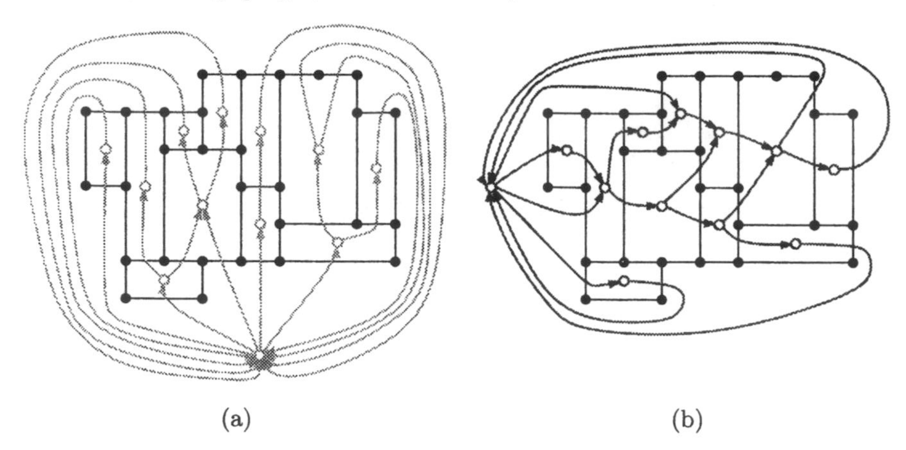

Fig. 6.10. The two networks N_h and N_v serve to construct a first drawing for an orthogonal representation.

computing a feasible flow in the networks. This method for finding a drawing for H' is optimal with respect to width, height and area, but not necessarily to total edge length. It is based on finding a topological numbering of special directed acyclic graphs and will be explained in Section 6.6.2, because it can also be used as a heuristic for compaction.

6.4 Heuristics

We have seen in the last two sections that there is a close connection between orthogonal drawings of planar graphs and a certain type of combinatorial description. Before we proceed to describe a combinatorial algorithm that uses a network flow approach based on this characterization to optimize the number of bends for a fixed embedding, we discuss a number of heuristic methods that construct good layouts. A main advantage of these methods is their fast running time: typically, it is linear. This makes them more suitable for large problems than the network flow approach, which produces almost quadratic running time. Furthermore, they provide easy procedures for local improvements of a drawing, yielding worst-case bounds for the number of bends and the area, depending on the size of the graph. A final advantage of the heuristic approaches is the fact that some of them work on non-planar graphs.

Typically, the methods work best on 2-connected graphs. A graph is called *2-connected* if removing any vertex and its incident edges leaves a connected graph. The *2-connected components* (or *blocks*) of a connected graph are (a) its maximal 2-connected subgraphs, and (b) its bridges together with their endpoints. If removing $\{v_1, v_2\}$ disconnects the graph we call $\{v_1, v_2\}$ a *cutting pair* of G.

The following result is folklore, see the textbook Sedgewick (1988) for details:

Lemma 6.3. *Testing 2-connectivity and finding all cutting pairs can be done in linear time.*

A natural approach to drawing a graph is to proceed by adding one vertex at a time to an existing drawing. To make sure that any new vertex has the necessary space, a special type of order is chosen:

Definition 6.4. *Let G be a graph. An* st-order *of G is an ordering $\{v_1, v_2, \ldots, v_n\}$ of the vertices of G such that every v_j $(2 \leq j \leq n-1)$ has at least one "predecessor" v_i and at least one "successor" v_k that are neighbors of v_k with $i < j < k$. The edges from v_i to its predecessors (successors) are called* incoming (outgoing) *edges of v_i. Their number is the in-degree indeg(v_i) (out-degree outdeg(v_i)) of v_i.*

If the graph is not 2-connected, there may not be an *st*-order; otherwise, the following theorem holds:

Theorem 6.5 (Lempel et al. 1967, Even and Tarjan 1976).
Let G be a 2-connected graph and $s, t \in V$. Then there exists an st-order such that s is the first and t is the last vertex. It can be computed in $O(m)$ time.

6.4.1 Visibility Representations

A *visibility representation* (Rosenstiehl and Tarjan, 1986) Γ for a graph G maps every vertex v of G to a horizontal segment $\Gamma(v)$ (called *vertex segment*), and each edge (u, v) of G to a vertical segment $\Gamma(u, v)$ (called *edge segment*), such that for each edge (u, v), the edge segment $\Gamma(u, v)$ has its endpoints on the vertex segments $\Gamma(u)$ and $\Gamma(v)$, and does not intersect any other vertex segment. A vertex segment $\Gamma(s)$ is called *source*, if all of its incident edges are above $\Gamma(s)$. A vertex segment $\Gamma(t)$ is called a *sink*, if all its incident edge segments are below $\Gamma(t)$. Figure 6.11 shows a visibility representation of a graph. Visibility representations were introduced by Otten and van Wijk (1978) and are often used as a starting point for drawing a planar graph. The following lemma was proven independently by Rosenstiehl and Tarjan (1986), and Tamassia and Tollis (1986):

Lemma 6.6. *Let G be a 2-connected planar graph with n vertices. Then for any two vertices s and t on the same face of G, there is a visibility representation Γ for G such that:*

(a) Γ has exactly one source, $\Gamma(s)$, and exactly one sink, $\Gamma(t)$.
(b) all the remaining vertex segments $\Gamma(v), v \neq s, t$, have two edge segments incident to $\Gamma(v)$ at its left endpoint (one from below and one from above).

A representation with these properties can be constructed in $O(n)$ time.

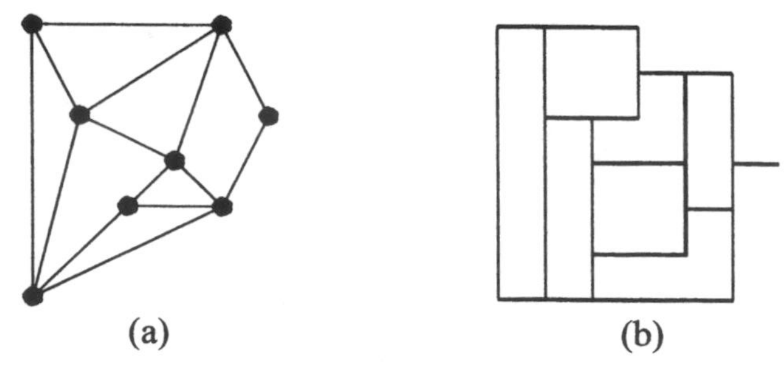

Fig. 6.11. A graph G and a visibility representation for G.

6.4.2 The Algorithm by Tamassia and Tollis

The following algorithm constructs an orthogonal grid embedding as described in Definition 6.2 for a 4-planar graph G. It is due to Tamassia and Tollis (1989) and has linear running time; it produces drawings with at most $2.4n + 2$ bends, and no edge has more than four bends. The length of each edge is $O(n)$, and the total area of the orthogonal grid embedding is $O(n^2)$. If G is 2-connected, we get even better bounds: the number of bends is bounded by $2n + 4$, and only two edges have more than two bends.

The algorithm has four phases. In the first phase, a visibility representation of the graph is constructed. In the second phase, this is transformed into an orthogonal grid embedding. In the third phase, a number of modifications are applied to the orthogonal representation of the grid embedding in order to reduce the number of bends. The last phase computes an orthogonal grid embedding for the orthogonal representation. We start by presenting the individual phases, then summarize the overall algorithm.

Visibility Representation. In Section 6.4.1 it was stated that for each planar 2-connected graph a visibility representation with one source and one sink can be computed in linear time. For planar graphs that are not 2-connected, it cannot be guaranteed that there exists a visibility representation with exactly one source and exactly one sink. However, the number and the degree of the sources and sinks of the visibility representation are crucial for the quality of the drawings produced by the algorithm.

Therefore, the algorithm constructs a visibility representation of a connected graph G with only one source and a low number of sinks. This is done by decomposing the graph into 2-connected blocks that are separated by cut vertices. For each of these blocks a visibility representation is computed according to Lemma 6.6. A cut vertex is chosen as the source, and if possible a cut vertex is chosen as a sink. The visibility representations of the distinct blocks can then be merged, such that one visibility representation for the entire graph is created. For details see Tamassia and Tollis (1989).

Lemma 6.7. *Let $G = (V, E)$ be a connected planar 4-graph. Then a visibility representation Γ for G can be constructed in $O(n)$ time, such that:*

(a) Γ has exactly one source
(b) all the non-source and non-sink vertex segments $\Gamma(v)$ have two edge segments incident to $\Gamma(v)$ at its left endpoint (one from below and one from above).

Creating an Orthogonal Embedding. In the second phase, the transformation of the visibility representation into an orthogonal embedding is performed. This is done by substituting every vertex segment $\Gamma(v)$ of Γ by a structure consisting of a single vertex v and some vertical and horizontal segments. Figure 6.12 shows the corresponding structure for every possible shape of v. Symmetric cases are omitted. It is not hard to see that this can be done in $O(n)$ time, and only a constant number of bends is inserted per edge, for a total of $O(n)$ bends.

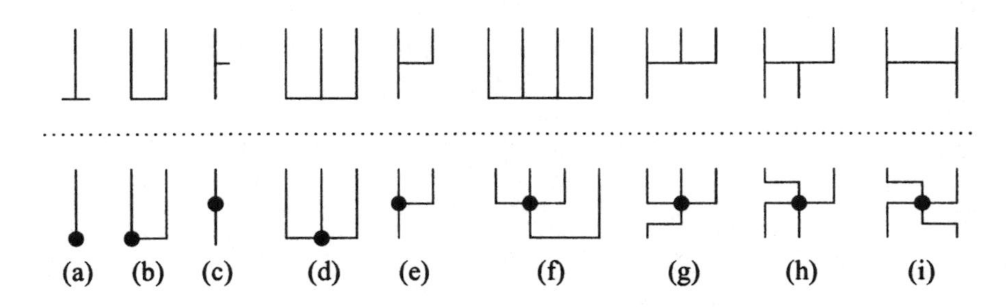

Fig. 6.12. Substitutions of vertex segments with structures.

Bend-Stretching Transformations. Because only one vertex at a time is treated during the substitution process, the embedding may contain a considerable number of artificial bends. During the third phase we try to remove these bends by performing a series of bend-stretching transformations, which are local optimization steps. There are three types of bend stretching transformations, each working on the orthogonal representation H of the embedding. H can be obtained from the orthogonal embedding that is computed in the second phase. A bend on an edge is called *convex* if it forms an angle of $\frac{\pi}{2}$ and *concave* if it forms an angle of $\frac{3\pi}{2}$. The transformations are as follows – see Figure 6.13:

Transformation T1. If an edge (u, v) has both convex and concave bends, remove one bend of each type until the edge has only bends of one type.
Transformation T2. If all edges around a vertex have bends of the same type, these bends can be removed.
Transformation T3. If two edges e_1, e_2, following each other in the clockwise order around a vertex v, form an angle of π, and e_2 has a convex bend with respect to v, then the bend can be removed.

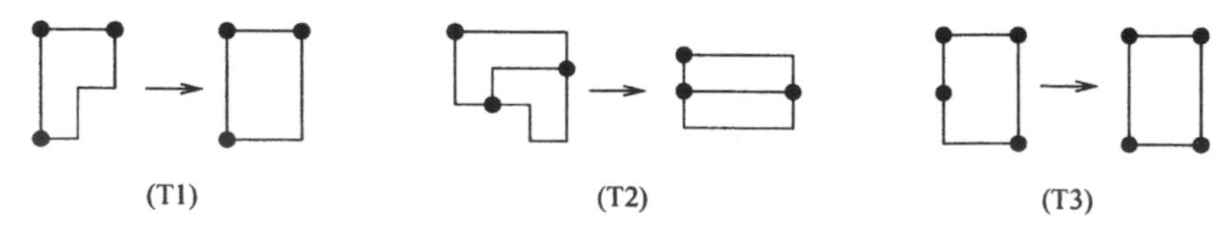

Fig. 6.13. Examples of bend-stretching transformations T1, T2, and T3.

The above transformations generate a new orthogonal representation H' with fewer bends than H. In particular, any transformation of type T1 reduces the number of bends by at least 2, while transformations of type T2 and T3 reduce it by at least 1. Each transformation can be executed in constant time. Since each bend-stretching transformation removes at least one bend, and in the previous step only $O(n)$ bends are introduced, performing these transformations takes only time $O(n)$.

Construction of a Grid Embedding. In the last phase, an orthogonal grid embedding is computed from the orthogonal representation. Details are as described in Section 6.3. Here we only note that for an orthogonal representation with $O(n)$ bends, the drawing can be computed in $O(n)$ time, and the area is $O(n^2)$.

Algorithm 17: Visibility-Grid-Embedding

Input: a 1-connected planar 4-graph G
Output: an orthogonal planar grid embedding of G

construct a visibility representation Γ of G using algorithm *Visibility*;
create an orthogonal embedding $\hat{G}$ of G by substituting each vertex-segment of Γ with the appropriate structure listed in Figure 6.12;
calculate the orthogonal shape H of $\hat{G}$;
apply T_1 on every edge of H (if possible);
apply T_2 on every vertex of H(if possible);
apply T_3 on every vertex of H with degree ≤ 3 while possible;
use H to construct an orthogonal planar grid embedding of G;

Analysis of the Algorithm.

Lemma 6.8. *If the visibility representation constructed in Step 1 of algorithm* Visibility Grid Embedding *has one source, t_i sinks, and n_i non-source vertex segments of degree i, $i = 2, 3, 4$, then the number of bends is at most $n_3 + 2n_4 + t_2 + t_3 + 2t_4 + \delta$, where $\delta = 0, 1, 2, 4$, depending on whether the source has degree 1, 2, 3, or 4.*

Proof. Let n'_4 be the number of vertices resulting from the substitutions of Figure 6.12 (g), (h), and let n''_4 (i) be the number of vertices resulting from substitutions of type (i). Notice that $n'_4 + n''_4 \leq n_4 + t_4$. From the substitutions of Figure 6.4.2, the number of bends after Step 2 is given by

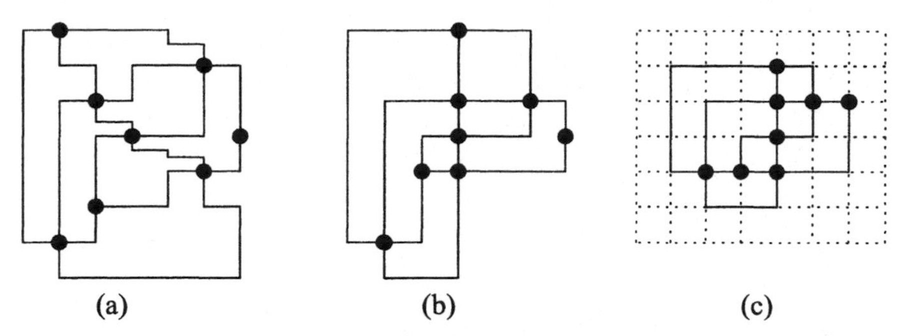

Fig. 6.14. (a) shows the graph of Figure 6.11 (b) after Step 2 of the Algorithm *Visibility Grid Embedding*, (b) after Step 6, and (c) after Step 7.

$n_3 + 4n_4' + 6n_4'' + t_2 + t_3 + 4t_4 + \delta$. In Step 4, transformation T_1 is applied at least once for each vertex in case of Figure 6.12 (g), (h), and at least twice for each vertex in the case of Figure 6.12 (i). This means that in Step 4 at least $2n_4' + 4n_4''$ bends are eliminated. Thus, the above equation yields a number of bends of $n_3 + 2n_4' + 2n_4'' + t_2 + t_3 + 4t_4 + \delta \leq n_3 + 2n_4 + t_2 + t_3 + 2t_4 + \delta$. During steps 5 and 6, this number may only improve.

Theorem 6.9. *Let G be a connected 4-graph with n vertices. Then algorithm* Visibility Grid Embedding *produces a grid embedding of G with at most* $2n+4$ *bends if G is 2-connected, and* $2.4n+2$ *bends if G is connected. The running time is $O(n)$.*

Sketch of Proof. For the case of a 2-connected graph, the visibility representation computed by algorithm *Visibility* has only one sink, and the bound follows immediately from Lemma 6.8. For other connected graphs, it can be shown that the bound $t_3 + 2t_4 \leq m/5$ holds, unless the graphs belong to a certain family of graphs. With the help of this bound, Lemma 6.8 implies the claim. For the exceptions to this bound it is possible to use their special structure for showing that algorithm *Visibility Grid Embedding* needs not more than $2.4n + 2$ bends to draw them. The linear running time follows directly from the above discussion. See Tamassia and Tollis (1989) for the complete proof.

6.4.3 The Algorithm by Biedl and Kant

The algorithm by Tamassia and Tollis relies on the planarity of the graph. It is natural to investigate heuristics that also work for non-planar graphs. Biedl and Kant (1994) have designed an algorithm for constructing an orthogonal grid-embedding of a 2-connected 4-graph G. It starts by computing an *st*-order of the input graph (as defined in Definition 6.4), and then embeds the vertices consecutively in the grid, according to their order in the *st*-order. For each vertex a new row is added to the layout. For each uncompleted edge, i.e., an edge with exactly one endpoint embedded in the grid, a column on the left or right boundary of the existing layout is added.

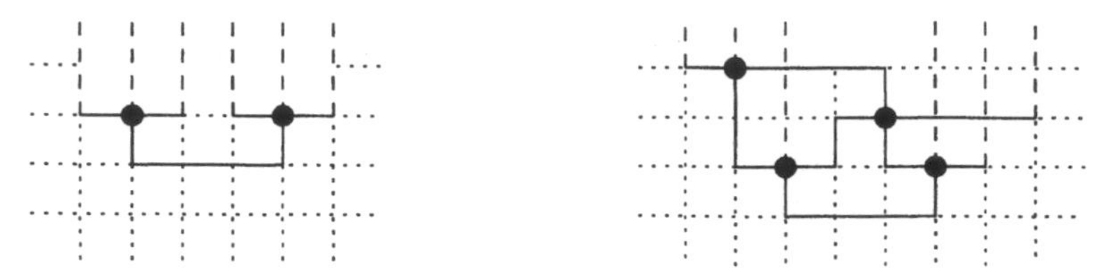

Fig. 6.15. Embedding of the first two vertices and the layout at a later stage.

Algorithm 18: Grid-Embedding

Input: a 2-connected 4-graph G
Output: an orthogonal grid-embedding of G
obtain an st-order $\{v_1, v_2, \dots, v_n\}$ for G;
place vertices v_1 and v_2 on the grid and connect them;
allocate one column in the grid for each edge of v_1 and v_2, except for the edge connecting v_1 and v_2;
for $3 \leq i \leq n$ **do**
 place v_i on a new row;
 place v_i on a column that is allocated to an incoming edge of v_i; if possible, do not take the leftmost or rightmost column;
 draw all its incoming edges using the columns allocated to it;
 allocate columns to the outgoing edges of v_i on the left or right boundary.

Lemma 6.10. *The grid size is at most* $(m - n + 1) \times n$.

Proof. Observe that the height of the constructed layout is one less than the number of rows, and the width is one less than the number of columns. In order to embed vertices v_1 and v_2, two rows are used as shown in Figure 6.15. Every following vertex increases the height by one, the last vertex by at most two. Thus, the height is bounded by n. When embedding v_1 and v_2, we use a width of $outdeg(v_1) + outdeg(v_2) - 2$. Every following vertex v increases the width by $outdeg(v) - 1$, except for the last vertex, which increases it by $0 = outdeg(v_n) - 1 + 1$. Thus, the width is $\sum_{v \in V}(outdeg(v) - 1) + 1 = m - n + 1$.

Lemma 6.11. *At most one edge has three bends, all other edges have at most two bends. Overall, there are at most* $2m - 2n + 4$ *bends in the drawing.*

Proof. Every edge (v_i, v_j), $i < j$, bends at most once when v_i is embedded. Completing the edge needs at most one additional bend if $v_j \neq v_n$.

Embedding v_n bends one edge twice, all others at most once, thus only this edge can have three bends. With the embedding of $v \notin \{v_1, v_2, v_n\}$, there are $indeg(v) - 1$ and $outdeg(v) - 1$ new bends, hence $deg(v) - 2$ new bends. Embedding v_1 and v_2 gives $outdeg(v_1) + outdeg(v_2) - 1$ bends, and v_n requires $indeg(v_n)$ bends if $indeg(v_n) = 4$, and $indeg(v_n) - 1$ bends otherwise. As $indeg(v_0) = 0$, $indeg(v_1) = 1$, and $outdeg(v_n) = 0$, we have $\sum_{v \in V}(deg(v) - 2) + 4 = 2m - 2n + 4$ bends if $deg(v_n) = 4$ and $2m - 2n + 3$ bends otherwise.

In (Biedl and Kant, 1994) it is shown that edges with three bends can be avoided unless G is the *octahedron*, i.e., the unique planar 4-regular graph with six vertices. The area bound proven in Lemma 6.10 can be improved even further. The authors show that if G has at most one vertex of degree two, one column and two bends can be saved. This is done by using a special *st*-order. For $m \geq 2n - 1$, there is at most one vertex of degree two. So we have a width of at most n if $m = 2n$, and a height of $n - 1$ if $m = 2n - 1$. If in addition G is not 4-regular, a row can be saved by choosing a vertex of degree less than four as v_n. This leads to the following theorem:

Theorem 6.12. *Let $G = (V, E)$ be a 2-connected 4-graph. Then G can be embedded in an $n \times n$ grid with at most $2n + 2$ bends. If G is not 4-regular, an $(n-1) \times (n-1)$ grid and $2n - 1$ bends suffice. Each edge is bent at most twice, unless G is the octahedron.*

There is also a variant of the algorithm that produces planar orthogonal drawings of planar graphs with the same bounds on area and number of bends. The above algorithm works only for 2-connected graphs, but it can be extended to connected graphs as follows. Break the graph into its blocks and compute the *block cut vertex tree* of G. The block cut vertex tree of a graph G has a *B-node* for each block of G and a *C-node* for each cut vertex of G. Edges in the block cut vertex tree connect each B-node to the C-nodes associated with the cut vertices of the block. Now the graph is drawn inductively: in the base case, execute the algorithm for a 2-connected graph. In the induction step, consider a subtree of the block cut vertex tree of G and split the subtree into a block G_0 (i.e., the root of the subtree) and the connected subgraphs $G_1, G_2, \ldots, G_q$. By induction hypothesis, each G_i already has a drawing. Hence, the process of drawing G reduces to drawing G_0 and merging each G_i appropriately. The merging process can be done such that the area bounds for the 2-connected case hold also for connected graphs. For details see Biedl and Kant (1994).

6.4.4 Pairing Technique

The algorithm in the previous section made generous use of new columns and rows for additional vertices and edges. The following algorithm tries to reuse as many rows and columns as possible for placing new vertices. In this way, a better area bound is achieved. The algorithm requires 2-connectivity of the input graph. Its authors Papakostas and Tollis (1997d) show that the algorithm can be extended to the simply connected case in a way similar to the algorithm by Biedl and Kant, though, unlike the latter, it does not necessarily produce planar drawings of planar graphs.

The central idea of the algorithm is to form *pairs* of vertices. There are two different kinds of pairs:

Row pairs. The two vertices of a pair are placed in a way that reuses a row in the final drawing of G, i.e., at least two vertices are placed in the same row.

Column pairs. The two vertices of a pair are placed in a way that reuses a column in the final drawing of G, i.e., at least two vertices are placed in the same column.

In order to obtain the pairs, an *st*-order of G is computed. As a next step, each vertex is assigned a type. A vertex with a incoming edges and b outgoing edges is called a vertex of type a-b, or an *a-b vertex* $(1 \leq a, b \leq 4)$. If there are 1-1 vertices whose outgoing edges are entering a 1-2 or a 1-3 vertex, we remove these 1-1 vertices and create a new edge between its predecessor and its successor. These removed vertices can be inserted in the drawing at the end of the algorithm without affecting the bounds for area and the number of bends. The graph obtained by removing these vertices is called the *reduced* graph G', and the number of its vertices is denoted by n'. For forming the pairs, the vertices are considered in reverse order of the *st*-order. If a vertex of type 1-2 or 1-3 is encountered, it is paired with its immediate predecessor in the *st*-order. If a vertex of type 2-2 is encountered, the vertex is paired with the next vertex in the *st*-order that is not a 1-1, 2-1, or 3-1 vertex, or a predecessor of the 2-2 vertex. After this step, all 1-2, 1-3, and 2-2 vertices v_i for $3 \leq i \leq n$ are paired. Paired vertices are called *assigned*, vertices not belonging to a pair are *unassigned*. After the pairs are computed, the vertices are embedded into the grid according to the *st*-order. Consider a vertex v that has not yet been embedded. If v is not paired, it is embedded in the grid like in the algorithm by Biedl and Kant. If v is paired, it is embedded together with the second vertex in the pair either as a column pair or as a row pair. The concrete embedding of the pairs is rather technical, for a description see Papakostas and Tollis (1997d). Algorithm 19 summarizes the above steps.

An analysis of the algorithm shows that only for unpaired vertices of type 4-0, 0-4, and 3-1 both a new column and a new row must be allocated when these vertices are embedded. For vertices that do not fulfill these conditions either a new row or a new column must be added to the drawing, but not both. By solving a system of linear equations it can be shown that there can be at most $\lfloor \frac{n+2}{2} \rfloor$ unpaired vertices of type 4-0, 0-4, and 3-1. This leads us to the following theorem:

Algorithm 19: Pair-Orthogonal

Input: a 2-connected 4-graph G
Output: an orthogonal grid drawing of G

```
compute an st-order of G;
construct the reduced graph G' from G;
obtain a pairing for G';
place v1 and v2 on the grid;
i := 3;
while i < n' do
    if vi has not already been placed then
        if vertex vi is unassigned then
            place vertex vi in a new row;
            connect vi with the incoming edges;
            allocate columns for the outgoing edges of vi;
        else
            place vertex vi along with the other vertex in the pair, following
            the placement rules described above for the specific type of pair;
    i := i + 1;
```

Theorem 6.13. *Let G be a 2-connected 4-graph with n vertices. Algorithm* Pair-Orthogonal *constructs an orthogonal grid drawing of G in $O(n)$ time with area $0.77n^2$. The total number of bends of the drawing is at most $2n+4$, and no edge has more than two bends.*

Proof. Let k_1 and k_2 denote the number of vertices that do not allocate a new row or column when they are embedded. It follows from our above discussion that there are at most $\lfloor \frac{n-2}{2} \rfloor$ vertices contributing neither to k_1 nor to k_2. Thus, $k_1 + k_2 \geq \lceil \frac{n-2}{4} \rceil$. The area is maximized when $k_1 = k_2 = \frac{n-2}{8}$. The claim is now verified by simple calculation. The analysis of bend costs is similar to Lemma 6.11; in addition, the construction allows it to avoid edges with three bends. Using the data structure by Dietz and Sleator (1987), the algorithm can be implemented in $O(n)$ time.

6.4.5 Algorithms for Drawing High-Degree Graphs

So far we have only considered graphs with maximum degree 4, i.e., *4-graphs.* When considering graphs of higher degree, we cannot avoid overlap of edges if we continue to draw vertices as points, so it makes sense to draw vertices as boxes with a sufficient number of grid lines for adjacent edges. In order to use the existing machinery, a straightforward approach is to split high-degree vertices into chains or cycles of vertices before applying an algorithm for 4-graphs like the algorithm by Tamassia (1987) or the algorithm by Biedl and Kant (1994). From this layout, the boxes for the vertices are created. The GIOTTO system (Tamassia et al., 1988) and the quasi-orthogonal drawing algorithm by Klau and Mutzel (1998) follow this approach. Unfortunately, this concept allows no control over the box size of the vertices, so the generated layouts may contain vertices of unrestricted size. Other examples are algorithms for visibility representations (Rosenstiehl and Tarjan, 1986; Tamassia and Tollis, 1986).

A different approach is the KANDINSKY framework, where each vertex is represented by a square. As shown in Figure 6.21, these squares are aligned in a square grid, and edges are routed along an edge grid. Section 6.5.4 will sketch this approach; it extends the network flow technique by Tamassia (1987) for 4-planar graphs that is described in Sections 6.5.2 and 6.5.3.

Papakostas and Tollis (1997c) present an algorithm where the size of any box for a vertex is less than twice the degree of the vertex. Their approach is a generalization of the pairing algorithm presented in the previous section, with boxes instead of vertices being placed on the grid. Outgoing edges leave a box on the top side, incoming edges enter a box on the left or right side. No edges are leaving or entering at the bottom side. Each edge has exactly one bend and the area bound for the algorithm is $m \times \frac{m}{2}$.

In the context of this section, we concentrate on a general framework for generating orthogonal drawings for graphs of high degree that was presented in Biedl et al. (1997a) and Biedl (1997). This *three-phase method* distinguishes the phases vertex placement, edge routing, and port assignment; in addition, there are preprocessing and postprocessing steps. In the preprocessing step, the graph is transformed into a *normalized graph*, i.e., a connected graph without reflex edges and without vertices of degree one. If the input graph is not connected, the connected components are drawn separately. In the first phase, vertices are represented as points and are placed on grid positions. In the second phase, edges are routed between vertices. The drawing obtained after these two phases is called a *sketch*; at this stage, it is not a valid drawing because the edges are routed with overlaps. From this sketch a drawing is produced by adding rows and columns to the drawing, such that vertices are enlarged to boxes. Furthermore, each edge is assigned a port at its vertices such that there is no intersection between any two edges on the same side. Since reflex edges and vertices of degree one were removed in the preprocessing step, they are reinserted in a postprocessing step. As a last step, drawings of the connected components are combined. It should be noted that this framework allows it to handle various kinds of constraints, like constraints on the position of vertices.

An example for this approach is the following algorithm for directed graphs: Place the n vertices of the input graph in *general position* on an $n \times n$ grid, i.e., no two vertices are placed on the same grid line. Edges are routed such that they always leave a vertex on the left or right side, and that they always enter a vertex at its top or bottom. Since the vertices are in general position, exactly one bend is needed to draw any edge. So if $e = (v_i, v_j)$ is an edge directed from v_i to v_j, then we place a temporary bend in the row of v_i and the column of v_j. Next, generate the boxes for the vertices and consider a row r. This row contains one vertex v, and some number of bends. Let $l(v)$ denote the number of edges emerging from the left side of v, and $r(v)$ the number of edges emerging from the right side of v. Analogously, let $b(v)$ and $t(v)$ denote the number of edges leaving the bottom or top side of v. We add $\max\{r(v), l(v), 1\} - 1$ bends above the row of v. We distribute these bends among these rows such that no two edges on one side cross as shown in Figure 6.16. It is clear that the drawings generated by this algorithm have height and width at most m, leading to an area bound of $m \times m$.

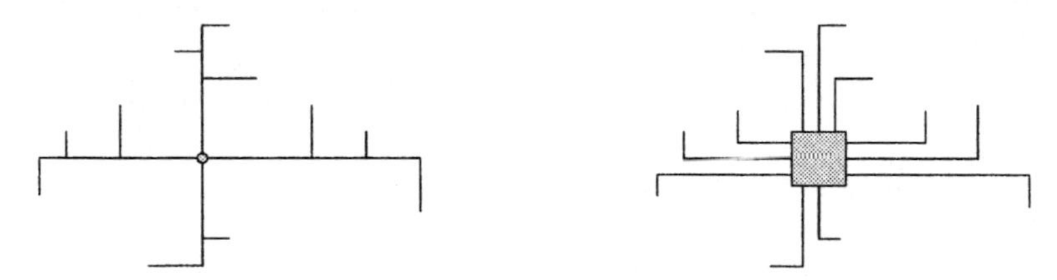

Fig. 6.16. An example for port assignment.

The worst case for the above area bound arises when all edges of the given graph leave the vertex on the same side. The algorithm by Biedl and Kaufmann (1997) achieves a better area bound by balancing the edges across the sides of a vertex box. This can be done by an appropriate placement in the initial phase of placing vertices. For a description, we use the following notation:

Definition 6.14. *Let G be a directed graph and let $\{v_1, \ldots, v_n\}$ be an arbitrary vertex order. An edge directed from v_i to v_j is called* good *if $i < j$ and* bad *otherwise. A predecessor (successor) v_i of v_j is* good *if the edge connecting v_i and v_j is good. We denote by $indeg^{good}(v_j)$ and $indeg^{bad}(v_j)$ the number of good and bad incoming edges of v_j. Similarly, we define $outdeg^{good}(v_j)$ and $outdeg^{bad}(v_j)$.*

Now the rows for the vertices are computed in ascending vertex order. If a vertex v has no good predecessor, then we create a new row at an arbitrary place. Otherwise, we add a row close to the median of the good predecessor rows. Vertex v is placed in this new row. Similarly, a column for each vertex is computed, proceeding in reverse order, and considering good successors instead of good predecessors.

The following lemma shows how vertex order and edge orientation affect the number of edges connected to each side.

Lemma 6.15. *For each vertex, we have the bounds $\lfloor outdeg(v)/2 \rfloor \le r(v)$, $l(v) \le \lceil outdeg^{good}(v)/2 \rceil + outdeg^{bad}(v)$ and $\lfloor indeg(v)/2 \rfloor \le t(v)$, $b(v) \le \lceil indeg^{good}(v)/2 \rceil + indeg^{bad}(v)$.*

Proof. Consider $b(v)$. By the placement of bends, any bend at the top of v belongs to an incoming edge of v. By the placement of vertices, at most half (rounded up) and at least half (rounded down) of the good predecessors are below v. Nothing can be said about the place of the bad predecessors. The result follows for $b(v)$. The proofs for the other three sides are similar.

It follows from the above lemma that the number of bad predecessors and successors of a vertex v should be minimized in order to reduce the size of a vertex box. The next lemma shows that there always exists a vertex order and an edge orientation such that there is a low number of bad predecessors and successors.

Definition 6.16. *A vertex order together with an edge orientation is called* polar-free almost acyclic, *if*
(a) $indeg(v) \geq 1$ *and* $outdeg(v) \geq 1$ *for all* $v \in V$, *and*
(b) $indeg^{bad}(v) \leq 1$, *if* $indeg^{good}(v) \geq 0$ *then* $indeg^{bad}(v) = 0$, *and*
(c) $outdeg^{bad}(v) \leq 1$, *if* $outdeg^{good}(v) \geq 0$ *then* $outdeg^{bad}(v) = 0$.

Lemma 6.17. *Let G be a simple graph without vertices of degree zero or one. Then G has a polar-free almost acyclic order and orientation. It can be found in O(m) time.*

For a proof of this lemma see Biedl and Kaufmann (1997).

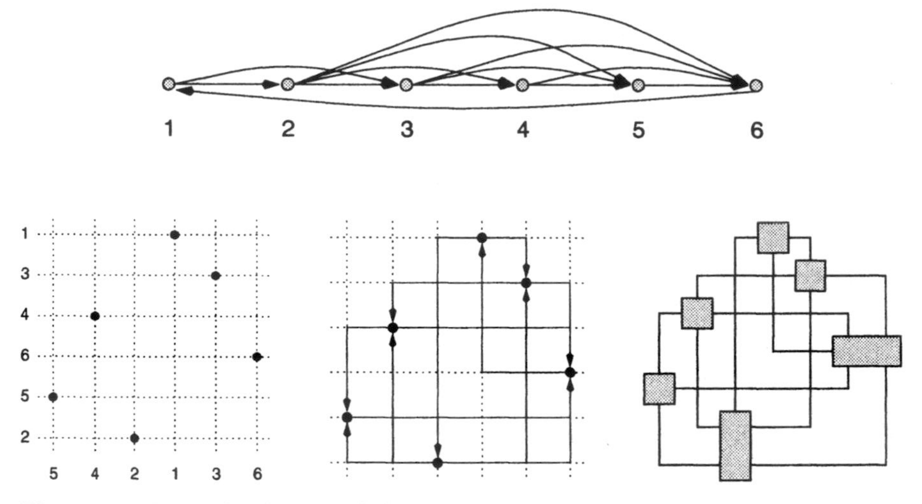

Fig. 6.17. Example of a run of the algorithm with polar-free almost acyclic order and orientation of the input graph.

Theorem 6.18 (Biedl and Kaufmann 1997). *Let G be a simple and connected graph. Then G has an orthogonal drawing in an* $\frac{n+m}{2} \times \frac{n+m}{2}$*-grid with one bend per edge. The drawing can be found in O(m) time.*

Proof. We only show the claim for the height, the claim for the width is similar. Suppose that G has no vertices of degree one. After the vertex placement, we have n rows. For each vertex v, we add $\max\{r(v), l(v), 1\} - 1$ rows. Thus, the height is $\sum_{v \in V} \max\{1, r(v), l(v)\}$. By Lemma 6.15 and the conditions on the polar-free almost acyclic order, we have $r(v), l(v) \leq \lceil \frac{outdeg(v)}{2} \rceil$. Thus, $\sum_{v \in V} \max\{1, r(v), l(v)\} \leq \sum_{v \in V} \frac{outdeg(v)}{2} + 1 = \frac{m+n}{2}$.

It remains to be shown that vertices of degree one can be inserted in the drawing without violating the area bound. Consider a vertex v of degree 1 in the postprocessing phase. First, one row is added above the top side of the neighbor vertex of v. Then a new column is generated, such that the width of the neighbor vertex increases by one and no adjacent edges of the neighbor vertex cross this new column. Then v is placed in the new row and column. No bend is needed to connect this vertex to its neighbor.

The vertex order and edge orientation can be found in $O(m)$ time as shown in Lemma 6.17. By using the data structure of Dietz and Sleator (1987), the algorithm can be implemented in $O(m)$ time. Thus, the overall complexity of the algorithm is $O(m)$.

An interactive version of the algorithm can be found in Biedl and Kaufmann (1997). A version of this algorithm that considers constraints can be found in Wiese and Kaufmann (1998).

6.4.6 A Divide-and-Conquer Approach

Now we describe a divide-and-conquer approach that originates from VLSI design. We already noted in Section 6.3 that there is a close relationship between graph drawing and VLSI design. While the first considers vertices and edges, the latter deals with transistors and wires. Clearly, there is a correspondence between the respective objects, but it is not without any problems, as wires and transistors need a certain amount of area. Thus, an important parameter for VLSI design is the *minimum feature size* λ, which is the width of the narrowest wire that can be manufactured. The smallest transistor that can be manufactured is a square with edge length λ and area λ^2. Further difficulties may arise from the fact that a general graph may have arbitrary degree, whereas a transistor can only have a limited number of connections. We resolve these difficulties by restricting us to graphs with vertices of a degree bounded by a constant, and by further assuming that vertices occupy only a constant area of silicon.

The VLSI model used here is similar to that of Thompson (1980), where wires have unit width and only two wires may cross at a point. Vertices are represented by little boxes that are placed on a rectangular grid, so that each box lies within a grid square. Edges run horizontally and vertically, one per grid square, except that an edge running horizontally may cross one running vertically. See Figure 6.18 for an example.

Layouts in this model are *sliceable*. That is, a horizontal or vertical line can be used to bisect the layout, the pieces can be moved apart, and the severed wires can be reconnected to realize the original graph connections. Slicing can be used to introduce a new edge in an existing layout as follows: perform two vertical and two horizontal cuts through the layout to separate the vertices. Separate the pieces by a grid unit, and reconnect the severed edges across the gaps in order to connect the vertices. If the length of the original layout was L and the width W, the new layout has length at most $L + 2$ and width at most $W + 2$.

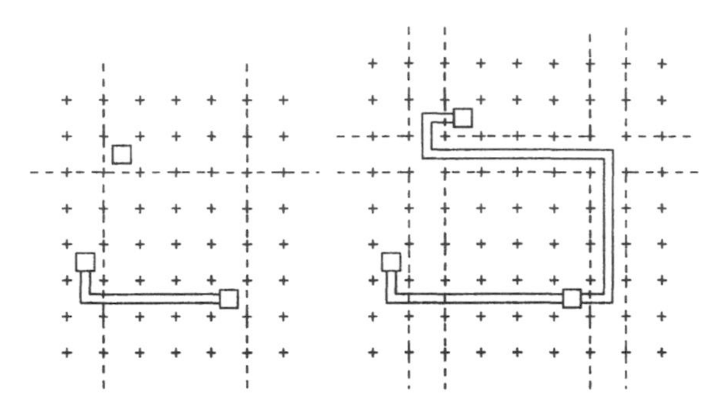

Fig. 6.18. An example for slicing and edge routing.

This property can be exploited to generate layouts for graphs in a divide-and-conquer approach. First, the graph is divided into unconnected components by removing edges. Then these components are laid out separately and the removed edges are inserted by slicing. Note that there may be crossings inserted in the slicing process, which implies that the produced drawing may be non-planar.

The quality of the layout produced by this approach depends on how many edges have to be removed to obtain unconnected components that differ in size by only a constant. In the worst case we have to remove $O(n)$ edges to obtain two unconnected components with this property, but there exist classes of graphs that can be separated by removing fewer edges. These classes of graphs are characterized by *separator theorems*. It is crucial that the classes are closed under the subgraph relation, i.e., each subgraph of a graph in such a class is again in the class. An example is the class of planar graphs, as it is closed under taking subgraphs. On the other hand, the class of trees is not closed under the subgraph relation.

Definition 6.19 ($f(n)$-separator). *Let S be a class of graphs that is closed under the subgraph relation. An $f(n)$-separator Theorem for S is a theorem of the following form: There exist constants α_s and c_s where $0 < \alpha_s \leq \frac{1}{2}$ and $c_s > 0$, such that if G is a graph with n vertices in S, then by removing at most $c_s f(n)$ edges, G can be partitioned into disjoint subgraphs G_1 and G_2*

having αn and $(1-\alpha)n$ vertices respectively, where $\alpha_s \leq \alpha \leq 1-\alpha_s$. The set of removed edges is called the cut set of the bisection.

Lipton and Tarjan (1970) showed that any planar graph with n vertices can be divided into two subgraphs of approximately the same size by removing $O(\sqrt{n})$ vertices in $O(n)$ time. When removing k nodes to split a graph with maximum degree d, at most $d \cdot k$ edges are removed from it. Since we are only considering graphs with bounded degree, and planar graphs are closed under the subgraph relation, this means that planar graphs have a $\sqrt{n}$-separator theorem. It is easy to see that forests have a 1-separator, see Valiant (1981) for a proof.

The analysis of the algorithm is quite complicated and results in solving recurrence equations. Leiserson (1980) and Valiant (1981) showed independently that graphs belonging to a graph class with a $\sqrt{n}$-separator can be drawn using $O(n \log^2 n)$ area and that graphs belonging to a graph class with a 1-separator can be drawn using $O(n)$ area.

Corollary 6.20. *Let G be a planar graph with n vertices and bounded degree. There is a layout of G, such that the area occupied by G is $O(n \log^2 n)$.*

For trees with maximum degree 4 one can use a method that is different from slicing for edge routing, which avoids crossings, but guarantees the same area bound (Valiant, 1981).

Corollary 6.21. *Let T be a tree with n vertices and bounded degree. There is a layout of T, such that the area occupied by T is $O(n)$. If T has maximum degree less than or equal to 4, there exists a planar layout with the same area bound.*

6.5 Flow-Based Methods

6.5.1 Drawing Graphs with Few Bends

We have seen in the preceding sections that over the years, orthogonal drawings of graphs have received a large amount of attention: there have been many different approaches to finding such drawings with desirable properties, like a small number of bends in the rectilinear paths representing edges. Unfortunately, one of the results by Formann et al. (1990) shows that it is $\mathcal{NP}$-complete to decide whether a planar graph with maximum degree 4 (a "4-planar graph") has an orthogonal drawing without any bends. The crux of the reduction is the fact that finding the right order of edges around each vertex in a drawing is difficult.

This difficulty arises when we are only given the information provided by vertices and edges, but have to find an optimal embedding – as in the hardness proof by Formann et al. It should be noted that for the special case of 3-planar graphs, it was shown by Di Battista et al. (1998a) that it is possible to optimize the number of bends in a drawing in polynomial time. This implies that it is the existence of degree 4 vertices in a graph that makes the problem difficult. As it was shown by Didimo and Liotta (1998), there are algorithms with polynomial complexity if the number of these vertices is bounded; the running time is exponential in the number of degree 4 vertices.

For many graphs that need to be drawn, the embedding is fixed, so the $\mathcal{NP}$-hardness result does not apply. In fact, it was shown by Tamassia (1987) that for a fixed embedding of a 4-planar graph, there is a nice combinatorial algorithm that computes a drawing with the smallest possible number of bends. We spend the rest of this section describing the idea of Tamassia's algorithm, and sketch some implications and extensions. For simplicity of notation, we do not distinguish between combinatorial objects (e.g., edges in a graph) and the geometric objects representing them (e.g., edge segments in a drawing).

6.5.2 A Network for Angles

Suppose we are given a planar graph $G = (V, E)$, and a fixed embedding of G, described by a clockwise order $\langle e_1, \ldots, e_{d(v)} \rangle$ of edges around any vertex $v \in V$. Let F be the set of faces in this fixed embedding, with f_0 being the exterior face. In any orthogonal drawing of G, the angles $\phi(e_1, e_2)$ between adjacent edge segments e_1 and e_2 are multiples of $\frac{\pi}{2}$. We use the notation $\psi(e_1, e_2) = 2\phi(e_1, e_2)/\pi$, write $E(v)$ for the (ordered) set of edge segments adjacent to a vertex $v \in V$, and $\Psi(f)$ for the set of angles in a face f. Then we can write the following conditions on these angles:

$$\sum_{e_i \in E(v)} \psi(e_i, e_{i+1}) = 4 \qquad \text{for all } v \in V. \tag{6.4}$$

$$\sum_{p(e_i, e_{i+1}) \in \Psi(f)} \psi(e_i, e_{i+1}) = 2|\Psi(f)| - 4 \quad \text{for all } f \in F \setminus \{f_0\}. \tag{6.5}$$

$$\sum_{p(e_i, e_{i+1}) \in \Psi(f_0)} \psi(e_i, e_{i+1}) = 2|\Psi(f_0)| + 4. \tag{6.6}$$

$$\psi(e_i, e_{i+1}) \geq 1. \tag{6.7}$$

Note the correspondence of these conditions to the linear relaxation described in Section 6.2.

Each angle p_i occurs exactly twice in this system of conditions, once at a vertex or bend in an edge, and once as an angle of a face. If we think of the size of these angles as amount of flow of some entity, we can introduce a network as follows:

1. There is a "source" node n_v for each vertex $v \in V$.
2. There is a "sink" node n_f for each face $f \in F$.
3. There is an arc $a_{v,f}$ from node n_v to node n_f if vertex v is incident to face f.
4. There is a source s, connected to all nodes n_v, and a sink t, connected from all nodes n_f.
5. For any two adjacent faces $f_1, f_2 \in F$, there are two arcs a_{f_1,f_2} and a_{f_2,f_1}.

Flow among these arcs has the following significance:

The sink node allocates angles to the vertices; as described, any vertex has a total sum of angles summing up to 4, so we fix the flow $x_{s,v}$ to this amount. A flow of $x_i \geq 1$ units on arc $a_{v,f}$ indicates that the angle p_i incident to vertex v and face f has size x_i. By requiring flow conservation at each node n_v, we make sure that condition (6.4) is valid. The lower bound (6.7) guarantees that each angle is positive. An example is shown in Figure 6.19.

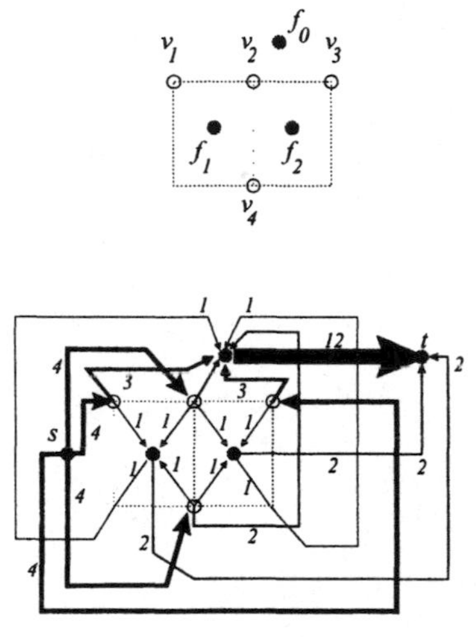

Fig. 6.19. An orthogonal drawing of a graph (top); flows in the corresponding network, with the amount of flow for each arc indicated by numbers and arc thickness (bottom). At the bottom, original edges are grey, and only arcs with positive flow are shown.

Furthermore, a flow of x_i units on arc $a_{v,f}$ indicates that the angle p_i incident to vertex v and face f has size x_i. A bend in an edge between two faces f_1 and f_2 creates a "reflex" angle of 3 on one side, and a "convex" angle of 1 on the other. In conditions (6.5) and (6.6), any angle in $P(f)$ contributes an angle of 2 to the face balance. The difference is accounted for by a flow of one unit from the face with the reflex angle to the face with the convex angle. This leaves a total amount of $2|V(f)| - 4$ for the angles at the set of vertices $V(f)$ incident to faces $f \in F \setminus \{f_0\}$, and $2|V(f_0)| - 4$ for the set of angles $V(f_0)$ incident to the exterior face f_0. By requiring this flow on the arcs $a_{f_i,t}$, and requiring flow conservation, conditions (6.5) and (6.6) are kept valid.

In addition to the above nodes and arcs, the following capacities and costs are defined:

1. Any arc $a_{v,f}$ from node n_v to node n_f gets capacity 4, and cost 0.
2. Any arcs a_{f_1,f_2} gets unbounded capacity, and cost 1 per unit of flow.

These capacities arise from the fact that no angle is larger than 4, but any edge can have an unbounded number of bends. Since our objective function is the total number of bends, and any unit of flow in an arc a_{f_1,f_2} corresponds to one bend, the cost function is chosen this way.

6.5.3 Optimal Flow in the Network and Implications

It is clear from our above discussion that any feasible drawing of G corresponds to a feasible flow in the network. Furthermore, if there is a feasible flow, the integrality of the arc capacities implies that there is a feasible solution where the flow on each arc is integral. Using the methods described in Section 6.3, it is straightforward to derive a feasible orthogonal representation for this flow. We have shown in Section 6.3 that any feasible orthogonal representation can be used to construct a feasible drawing in linear time. This leaves it to find a minimum cost feasible flow in the network, a classical problem of combinatorial optimization. See the book by Ahuja et al. (1993) for an overview over the basic algorithmic approaches. We summarize:

Theorem 6.22 (Tamassia 1987). *For a fixed embedding of a planar graph G with n vertices, an orthogonal drawing with a minimum number of bends can be found in time that is required for finding a minimum cost flow on a network with $O(n)$ arcs and $O(n)$ vertices.*

In 1987, the resulting complexity was $O(n^2 \log n)$. Currently, the best running time is $O(n^{\frac{7}{4}}\sqrt{\log n})$, using an improved network flow algorithm by Garg and Tamassia (1997).

There are a number of consequences and extensions. Any modification of a feasible flow that leads to a flow of reduced cost can be interpreted as a local improvement of a drawing. See Figure 6.20 for a number of examples, where the improvement of the flow is performed by identifying a cycle of negative cost in a reduced cost network. (Considering these types of local improvements in flow networks is a standard approach.) Any unit of flow along an arc in a negative cycle implies that we should increase an angle by a single multiple of $\frac{\pi}{2}$ at the expense of another. In the figure, such flow is indicated by the places where the cycle crosses an edge, a vertex, or a bend in the drawing. Performing these changes along the full cycle reduces the total number of bends along the encountered places, while all parts inside and outside of the cycle keep the same angles. As shown in the figure, this corresponds to a "rotation" of the inside against the outside.

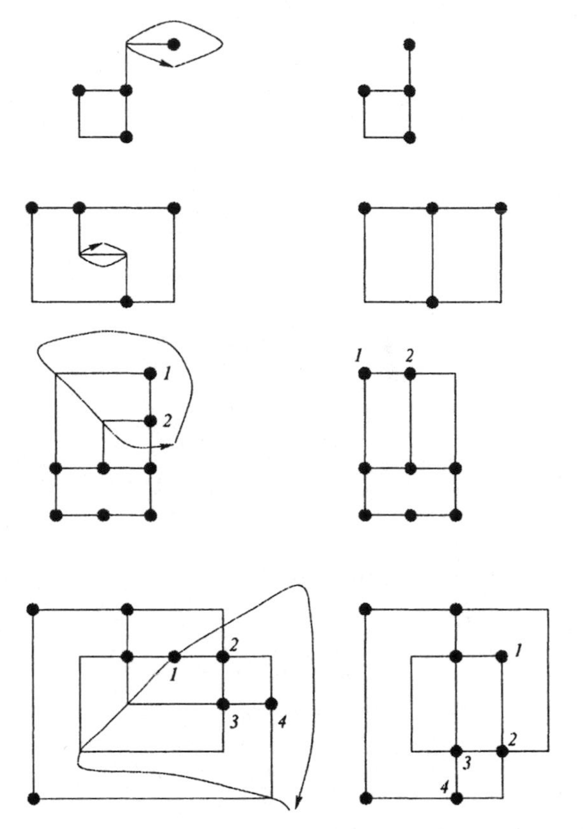

Fig. 6.20. Four examples: orthogonal drawings with an improving cycle in the corresponding flow (left); improved drawings (right).

Furthermore, it is straightforward to extend the ideas of Tamassia to flow networks for other types of grids. One example is the treatment of drawings in a hexagonal grid, where angles are multiples of $\frac{\pi}{3}$. However, these type of drawings allow angles of size $\frac{\pi}{3}$ or $\frac{2\pi}{3}$ at a bend; in order to use the network flow approach, angles of the first type have to be considered to carry twice the cost of angles of the latter type. Another issue is the question of finding a feasible drawing for a given flow (treated in Section 6.3 for the orthogonal case), which is unresolved, as the number of degrees of freedom in a hexagonal grid is different from the geometric dimension. See the paper by Tamassia (1987).

6.5.4 Kandinsky

If we need to draw planar graphs with maximum degree beyond 4, drawing vertices as points must create overlap, as the degree of a vertex exceeds the number of different orthogonal directions. As we described in Section 6.4.5, one possible remedy is to draw nodes not as points, but as boxes. A variant that allows it to make use of flow optimization techniques is given by the KANDINSKY model, which was first introduced by Fößmeier and Kaufmann (1995). Here, all vertices are given as identical $k \times k$ squares, with the size k determined appropriately. These squares are aligned on a square grid of size $(2k-1) \times (2k-1)$, and edges are routed as axis-parallel paths along the grid lines running through boxes. See Figure 6.21 for an illustration.

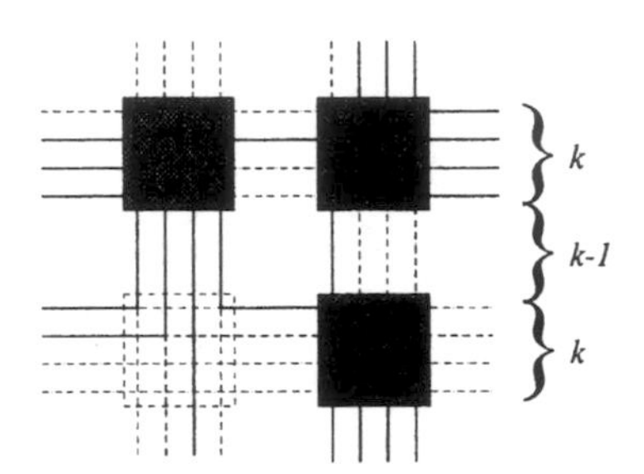

Fig. 6.21. The basic layout of vertices and edges in the KANDINSKY model.

Obviously, there are a number of technical issues that have to be taken care of, among them the choice of grid size, and more generally compaction methods that are modified from the steps described in Section 6.3. Details can be found in the original paper by (Fößmeier and Kaufmann, 1995), and in the thesis (Fößmeier, 1997b). Here we concentrate on the modifications of the flow network that have to be performed.

The key modification of the flow network for bend minimization arises from the fact that for a vertex of degree larger than four, there have to be edges leaving in the same direction. Clearly, neighboring edges of this type enclose an interior angle of 0; for the network described above, only positive angles are feasible. This can be fixed by allowing a flow of -1 to represent a zero angle, which can be interpreted as a flow in the opposite direction.

While this fix takes care of the angles, it creates another problem: since these flows do not incur any cost, it is possible to shift flow until the overall cost is zero. Thus, a minimum cost flow does not correspond to a feasible drawing.

However, if we exclude parallel edges (which can be done in a preprocessing step), any pair of edges enclosing an angle of 0 at one vertex must connect this vertex to two different vertices. This means that at some point, the two edges cannot continue to run in parallel. Because of the underlying grid in the Kandinsky model, this can only occur when one of the two edges bends. Thus, an angle of 0 forces a bend in one of the enclosing edges. We can charge this forced bend for the zero angle to the vertex – see Figure 6.22. If f is the face with the zero angle at vertex v, with neighboring faces g and h, then the forced bend can be interpreted as a unit of flow from g or h to v, at a cost of 1.

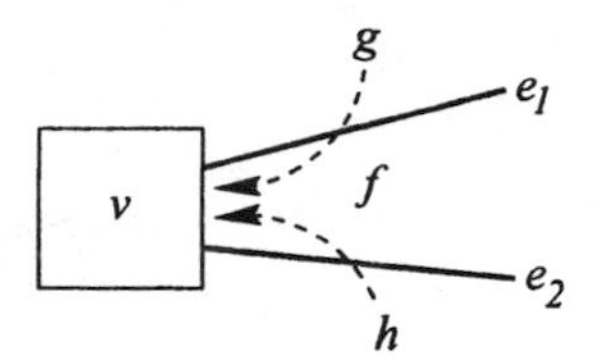

Fig. 6.22. Auxiliary arcs at a vertex.

After this modification, there are new issues that need to be taken care of: we only want to charge the cost for a zero angle once, and if we charge the arc from g to v, we must not charge the arc from h to v. This can be resolved by using the modified network for the flow between v, f, g, and h as shown in Figure 6.23, using additional edges with cost $2c + 1$ and $-c$. (Note that for clarity in Figure 6.23, the reference to vertex v in the labeling of the auxiliary nodes H is omitted.) In particular, the arc from a face f to an incident vertex v with edges e_i and e_j is represented by a path formed by the following arcs:

- Arcs with capacity 1 and cost $2c + 1$ from f to an auxiliary node $H_i^{v,fg}$ and $H_j^{v,fh}$. Here, g and h are the faces separated from f by e_i and e_j.
- Arcs with capacity 1 and cost $-c$ from $H_i^{v,fg}$ to $H_i^{v,gf}$, and vice versa.
- Arcs with capacity 1 and cost 0 from $H_i^{v,fg}$ to an auxiliary node H_g^v.
- Arcs with capacity 1 and cost 0 from H_g^v if v lies on the boundary of g.

This introduces cycles with negative cost into the network, while a flow that is feasible for a drawing must be decomposable into partial flows from s to t. Thus, we are no longer dealing with a classical minimum cost flow problem. However, using minimum cost flow algorithms based on augmentations along shortest paths from s to t still yields the desired result that an optimal flow corresponds to a feasible drawing with a minimum number of bends.

6.5.5 Constraints and Extensions

There are many algorithms that arise from the basic ideas described in the previous section. Some of them are able to consider a variety of constraints by using integer programming methods. See (Eiglsperger et al., 2000) for a description.

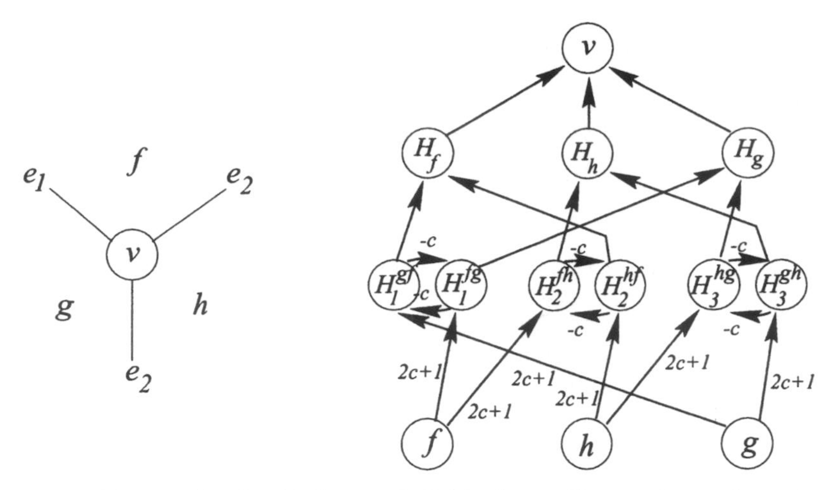

Fig. 6.23. The neighborhood of a graph (left); the auxiliary nodes and arcs in the flow network (right).

There are other graph drawing problems where the objective is closely related to what we described in the previous sections, but may yet have some differences. We give a few pointers to and descriptions of these issues.

While flow models minimize the total number of bends, it may very well be that the objective is to minimize the maximum number of bends in any edge. As Fößmeier et al. (1996) demonstrated, it is possible to construct a drawing of a planar graph in the Kandinsky-model, such that no edge has more than one bend. This allows it to minimize the total number of bends under this restriction: the constraint can be enforced by imposing an upper limit of 1 on the capacities of the face-to-face arcs in the flow network. Similarly, other upper bounds on the number of bends for individual edges can be handled.

There have been efforts to use flow models for bend minimization if the embedding is not fixed. Clearly, serious difficulties are to be expected, as we pointed out at the beginning of this section: since the problem for a non-fixed embedding is $\mathcal{NP}$-hard, such approaches can only be expected to lead to heuristics, or we may have to accept a worst-case running time that is not polynomial. As we already mentioned above, there is an algorithm by Didimo and Liotta (1998) for finding a drawing with a minimum number of bends for non-fixed embedding that is only exponential in the number of degree four nodes. Extending earlier work by Bertolazzi et al. (1997), the algorithm proceeds by a branch-and-bound search. Branching steps correspond to local modifications of the current graph embedding; the full set of possible embeddings is maintained by using a special data structure called an SPQ*R-tree. The four different types of tree nodes (S, P, Q*, and R) correspond to different types of triconnected components that allow a limited number of local modifications. The flow method by Tamassia is used as a subroutine. Experiments seem to indicate that the algorithm may be practical for test graphs of up to 200 nodes, since the distribution of degree four nodes in the test graphs tends to keep running times significantly below the worst case estimates.

It has been attempted to use flow techniques even for nonplanar graphs, by making use of sufficient a priori knowledge of the location of edge crossing. Details are quite tricky and technical; see the thesis (Fößmeier, 1997b).

6.6 Compaction

Compaction is the process of changing a given orthogonal layout, so that either the area, the total edge length, or the maximum edge length decreases. In this section, we will focus on compaction techniques that maintain the topology and shape properties of the input.

Most of the algorithms described in this section have their roots in VLSI design, but have been adapted to solve the compaction problems in graph drawing. In addition, we present two recent developments originating in the area of graph drawing. An overview of the techniques in VLSI layout is given in Lengauer (1990, Chap. 10) and in LaPaugh (1998, Sect. 23.3). Before describing the compaction techniques we state the underlying compaction problems in a formal way, and we discuss their complexity in Section 6.6.1. One-dimensional algorithms attack the problems by dividing them into two separate subproblems for the horizontal and vertical direction. They are covered in Section 6.6.2: we focus on the compression-ridge technique and the graph-based compaction strategies. Finally, Section 6.6.3 is dedicated to optimal methods.

6.6.1 Problems and Their Complexity

Depending on the various aesthetic criteria we want to optimize, there are several versions of compaction problems originating in the topology-shape-metrics approach. The input of the third phase within this paradigm is an orthogonal representation H. Such a representation may have been produced by the flow-based methods of Section 6.5. By introducing auxiliary vertices, we may assume that H is simple. The task is to find an orthogonal grid embedding respecting the shape of H with either minimum total edge length, minimum area, or minimum length of the longest edge. We will refer to these problems as COMP_{sum}, COMP_A, and COMP_{max}, respectively. It is also possible to state the problems for existing orthogonal grid embeddings, especially as a postprocessing step for drawings as produced in Section 6.3. In this case the task is to change the coordinates of vertices and bends, but not the angles formed by the edge segments. These formulations may seem somewhat restrictive in the case of existing drawings (in a more general scenario, it may be possible to introduce or remove bends), but even then, they may serve as a local improvement step.

Almost all variants of the two-dimensional compaction problem in VLSI design are $\mathcal{NP}$-hard. In most cases, the corresponding proofs are reductions of the problem 3PARTITION (see Garey and Johnson (1991)). They exploit the fact that in VLSI problem formulations, wires may be allowed to swap their connection points at components. Such a swap corresponds to a change of the embedding and is not allowed in graph drawing problem formulations. These changes are crucial for the reductions and cannot be used in a setting with fixed topology.

For a long time it was conjectured that the compaction problems (i.e., COMP_{sum}, COMP_A, and COMP_{max}) for a fixed embedding are also $\mathcal{NP}$-hard. Recently this was proven by Patrignani (1999a). His proof for COMP_A is based on a reduction of the satisfiability problem SAT (see Garey and Johnson (1991)). Given a formula ϕ, there is an orthogonal representation $H_A(\phi)$ with n_A vertices and m_A edges, and area at most $(9n_A+2)(9m_A+7)$, if and only if ϕ is satisfiable. The reductions for COMP_{sum} and COMP_{max} are similar.

6.6.2 One-Dimensional Compaction Methods

Due to the large sizes of instances, research in VLSI design has focused on one-dimensional methods. Only one dimension may be changed at a time; the other dimension is fixed. We will refer to the restricted, one-dimensional compaction problems as $\text{COMP}^1_{\text{sum}}$, COMP^1_A, and $\text{COMP}^1_{\text{max}}$, respectively.

After a compaction step, the layout is changed: alternating the direction and performing another step results in an iterative process. However, at each step the decisions are purely local, and compaction in one direction may prevent greater progress in the other direction. Furthermore, the layout may be blocked in both dimensions, but still be far away from an optimal solution (see Figure 6.24 for an example).

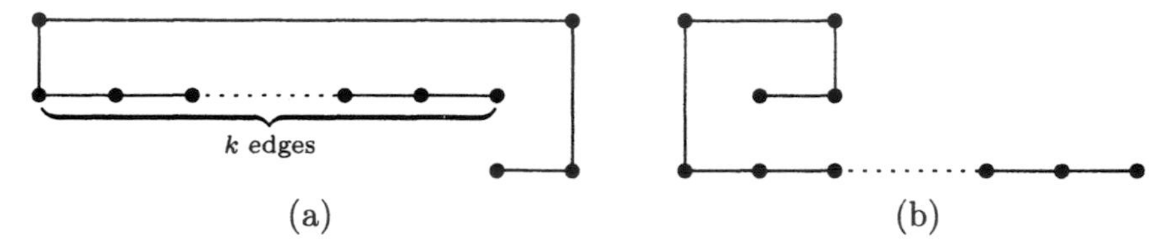

Fig. 6.24. (a) Both directions are blocked, the total edge length is $2k+5$. (b) An optimal layout with edge length $k+6$.

Originating in VLSI design, the *compression-ridge method* searches the layout for cuts that divide it into two parts and pass through regions of empty space. For fixed embedding the "empty space" corresponds to edges that are longer than the minimum length of one unit. If such a cut has been found, its edges can be shortened by at least one unit and the resulting grid embedding is still feasible.

We sketch the method from (Dai and Kuh, 1987), adapted to different scenarios in the area of graph drawing. All cuts are found as an interpretation of a maximum flow in a network N that depends on the initial drawing. A compaction step in x-direction for the example introduced in Section 6.3 (page 130) is shown in Figure 6.25. Compaction in y-direction can be explained similarly. First the layout is dissected into horizontal stripes. This corresponds to the dissection process described in Section 6.3 with the restriction that only artificial edges of horizontal direction are allowed. A modification of the dissection method still runs in linear time; the result is a drawing with internal faces of rectangular shape. Now the network N can be constructed as follows: each rectangular face f corresponds to a node $n(f)$ in N. In addition, there are two nodes s and t for the outer face; s at the top of the drawing, t at the bottom. Arcs are directed downwards: for each horizontal edge e separating an upper face f from a lower face g, there is an arc $a_e^+ = (n(f), n(g))$ and an arc $a_e^- = (n(g), n(f))$. The capacity of a_e^+ is the length of e minus one. This corresponds to the maximal possible shortening of e. A capacity of ∞ is assigned to the opposite arc a_e^-, accounting for possible elongations of e. The maximum flow from s to t in this network corresponds to the shortening that can be applied to obtain a minimum width drawing; thus, we get a layout of optimal horizontal width. Each compaction step has running time $O(n \log n)$, the bottleneck being the computation of a maximum flow problem in N. (By construction, the network N is planar and linear in size of the original graph.)

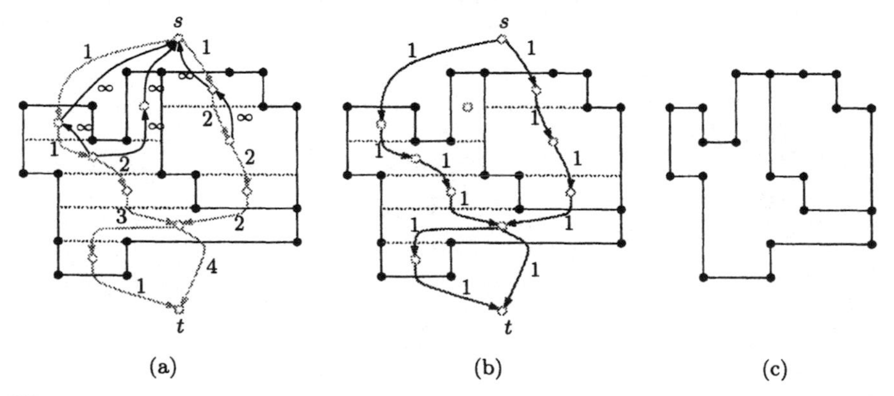

Fig. 6.25. The compression-ridge method in graph drawing: (a) the network N for x-compaction of Figure 6.9 (d) (only some of the unbounded upward arcs are shown); (b) the maximum flow in N; (c) the drawing after the horizontal compaction step.

So-called *graph-based compaction methods* represent a different and more efficient approach: Two *layout graphs* – one for each direction of the compaction – encode the visibility properties between maximally connected vertical and horizontal paths in the given grid embedding. These paths are also referred to as *bars* in Di Battista et al. (1999), *maximal chains* in Bridgeman et al. (1999), or *segments* in Klau and Mutzel (1999b).

Definition 6.23. *A horizontal* segment *is a maximally connected component in* (V, E_h), *the subgraph of* G *containing only the horizontal edges. Similarly, we define vertical segments in* (V, E_v). *The sets* S_h *and* S_v *refer to the horizontal and vertical segments, and the set* $S = S_h \cup S_v$ *refers to all segments. A vertex* v *lies on the two unique segments* $\mathrm{hor}(v) \in S_h$ *and* $\mathrm{ver}(v) \in S_v$.

The directed layout graphs $D_x = (V_x, A_x)$ and $D_y = (V_y, A_y)$ are built as follows: the node set V_x of the horizontal graph D_x corresponds to the set of vertical segments S_v. A similar construction applies to D_y, here $V_y = S_h$.

For an arc set A let $\mathrm{trans}(A)$ be the transitive hull of A. Geometric relations between the segments define the arc sets in the digraphs: whenever a horizontal segment s_i is to the left of another horizontal segment s_j, we want to find a directed path between s_i and s_j. We characterize the vertical relationships analogously. More formally, we want to have

$$\mathrm{trans}(A_x) = \{(s_i, s_j) \mid s_i \text{ is to the left of } s_j\} \quad \text{and} \tag{6.8}$$

$$\mathrm{trans}(A_y) = \{(s_i, s_j) \mid s_i \text{ is below } s_j\} \ . \tag{6.9}$$

Any sets with properties (6.8) and (6.9) can be used as the arc sets of the layout graphs. Figure 6.26 shows two layout graphs for the running example in this section with arc sets produced by a sweep-line method.

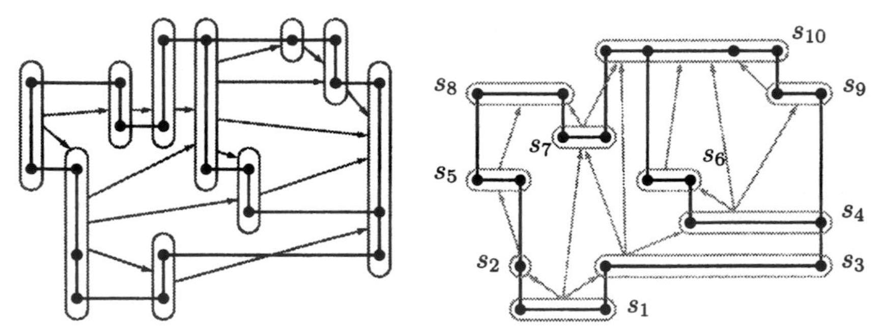

Fig. 6.26. The directed layout graphs D_x (left) and D_y (right).

Each arc corresponds to a distance constraint for a pair of segments. Since the visibility properties must be maintained in one-dimensional compaction (recall that the coordinates of the other direction are fixed) an arc (s_i, s_j)

describes the fact that all the vertices of s_j must be assigned a greater coordinate than the one for vertices of s_i. Hence, the task of one-dimensional compaction in x-direction reduces to computing a topological numbering for the nodes in D_x. Similarly, a vertical compaction step corresponds to a topological numbering in D_y. Since the layout graphs are acyclic by construction, such an order Φ can be computed in time $O(|V_x| + |A_x|)$ or $O(|V_y| + |A_y|)$, e.g., with the longest path method. It is easy to show that D_x and D_y are planar; thus, $|V_x|$, $|V_y|$, $|A_x|$, and $|A_y|$ are in $O(n)$. Therefore, the running time for computing the topological numbering Φ is linear in the size of the original graph.

We illustrate the method by performing vertical compaction on the example graph. Consider D_y in Figure 6.26. The longest path method results in the following topological numbering $\Phi : S_h \to \mathbb{Z}$.

$$\begin{array}{lllll} \Phi(s_1) = 0 & \Phi(s_2) = 1 & \Phi(s_3) = 1 & \Phi(s_4) = 2 & \Phi(s_5) = 2 \\ \Phi(s_6) = 3 & \Phi(s_7) = 2 & \Phi(s_8) = 3 & \Phi(s_9) = 3 & \Phi(s_{10)} = 4 \ . \end{array}$$

The new vertical coordinate of a vertex v is just the topological number of hor(v). Setting all y-coordinates in this manner results in the compacted drawing with minimum possible height in a one-dimensional setting, as shown in Figure 6.27 (a).

Note, however, that this method tends to push vertices as far to the bottom as possible (or to the left when performing a horizontal compaction step). Each segment gets its minimum topological number. For segments lying on the longest path tree, this is the optimal assignment; other segments should rather be placed closer to their neighbors than to the bottom or left margin. In Figure 6.27 (a) this has no negative influence on the aesthetics, but Figure 6.27 (b) shows an example where this is the case. Though the drawing has minimum width and height, the bottom edge is drawn longer than its one-dimensional minimum length of one grid unit. In addition to the unpleasant drawing, these edges might prevent the following compaction step in the other direction from a better performance.

Minimizing the total one-dimensional edge length corresponds to minimizing the difference between the topological numbers. In the area of VLSI this problem is known as *wire balancing*. For vertical compaction, the corresponding optimization problem is

$$\begin{array}{lll} \min & \sum_{(v,w)\in E_v} \Phi(\text{hor}(w)) - \Phi(\text{hor}(v)) & (6.10) \\ \text{s.t.} & \Phi(s_j) - \Phi(s_i) \geq 1 \qquad \text{for all } (s_i, s_j) \in A_y \ . & \end{array}$$

Note that (6.10) is the same problem as the layer assignment for layered drawings of graphs (see Section 5.3). The optimization problem can be seen as the dual of a flow problem, as shown by the following steps. Let $\Delta_{\text{ver}}(s)$ denote the *vertical degree* of a horizontal segment $s \in S_h$, defined as

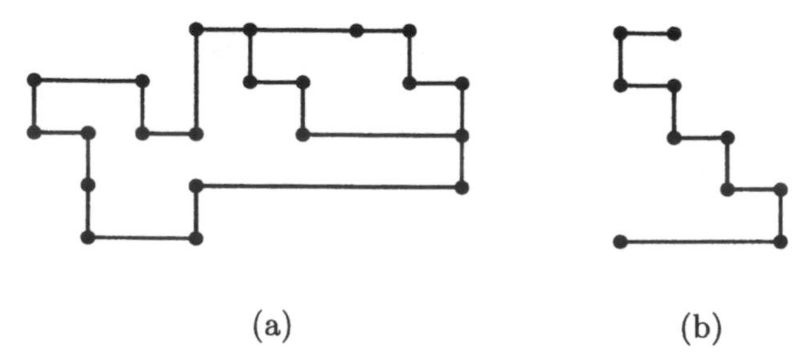

Fig. 6.27. The graph-based compaction method with longest-path computations: (a) the example from Figure 6.9 after a vertical compaction step; (b) the method does not lead to optimal layouts with respect to the one-dimensional compaction problem $\mathrm{COMP}^1_{\mathrm{sum}}$.

$$\Delta_{\mathrm{ver}}(s) = |\{(v,w) \in E_v \mid \mathrm{hor}(v) = s\}| - |\{(u,v) \in E_v \mid \mathrm{hor}(v) = s\}|. \quad (6.11)$$

Then $\sum_{s \in S_h} \Delta_{\mathrm{ver}}(s) = 0$, and the optimization problem (6.10) becomes

$$\begin{aligned} \min \quad & \sum_{s \in S_h} \Delta_{\mathrm{ver}}(s)\Phi(s) && (6.12) \\ \text{s.t.} \quad & \Phi(s_j) - \Phi(s_i) \geq 1 \qquad \text{for all } (s_i, s_j) \in A_y. \end{aligned}$$

The dual of (6.12) is

$$\begin{aligned} \max \quad & \sum_{a \in A_y} \Psi(a) && (6.13) \\ \text{s.t.} \quad & \sum_{a=(s,t)} \Psi(a) - \sum_{a=(r,s)} \Psi(a) = \Delta_{\mathrm{ver}}(s) && \text{for all } s \in S_h, \\ & \Psi(a) \geq 0 && \text{for all } a \in A_y. \end{aligned}$$

The objective function of this problem can be written as

$$\min \sum_{a \in A_y} -\Psi(a),$$

leading to the standard form of a minimum cost flow problem. This implies that there is a polynomial-time method for finding an optimal solution for the one-dimensional compaction problems COMP^1_A and $\mathrm{COMP}^1_{\mathrm{sum}}$.

6.6.3 Optimal Compaction Methods

There are several methods for finding an optimal solution for two-dimensional compaction problems of the types $\mathrm{COMP}_{\mathrm{sum}}$, COMP_A, and $\mathrm{COMP}_{\mathrm{max}}$ that we have introduced in Section 6.1. Both the algorithms by Kedem and Watanabe (1984); Watanabe (1984), and by Schlag et al. (1983) are based on a branch-and-bound approach and originate in VLSI design. Recently, there have been developments in the area of graph drawing: Bridgeman et al. (1999) specify a class of orthogonal representations for which optimal drawings with respect to area and total edge length can be found employing the methods from Section 6.3. Klau and Mutzel (1999b) present an optimal branch-and-cut approach for minimizing the total edge length for a given orthogonal representation; they also characterize classes of representations for which their algorithm runs in polynomial time.

The algorithm by Kedem and Watanabe (1984); Watanabe (1984) is based on a translation of the two-dimensional compaction problem COMP_A into a nonlinear mixed integer programming formulation that is solved by a branch-and-bound algorithm. They express the problem as minimizing the nonlinear area function under a set of linear and nonlinear constraints. Their formulation, however, sacrifices the general statement of COMP_A as defined in Section 6.1 and considers only a subset of the feasible solutions in order to achieve a better running time. In the following we sketch the branch-and-bound algorithm.

A vector of decision variables d determines the interaction of components (remember that components correspond to vertices of a graph drawing instance). In the given formulation, only two positions are possible for a pair of components, coded as an entry in the 0/1-vector d. Each combination corresponds to a different relative placement of the component pair – either a horizontal or a vertical constraint is active. The entries in d correspond to arcs in the layout graphs, i.e., a fixed d specifies two (possibly infeasible) one-dimensional compaction problems. This formulation has the drawback of being able to handle only two-way choices instead of four possible relative placements. Though the area of the computed layout is optimal for a given partial order, it may not be optimal for an instance of COMP_A as formulated in Section 6.1. The authors propose a postprocessing step to determine where the partial order of elements has to be swapped, but they do not present a method that guarantees an overall optimal solution.

A fixed set of relative positioning decisions – corresponding to a node in the branch-and-bound tree – results in two one-dimensional problems. They are solved using the graph–based longest path method described in Section 6.6.2. If the subproblem is infeasible, the tree of problems can be cut at this node. If the node is a leaf and a feasible solution is found that is better than the previous global upper bound, the bound is updated. Otherwise, a solution may cause an update of the local lower bound. If the latter becomes greater than the global upper bound, an optimal solution cannot be found below the current node, and again the tree is cut at this point.

In general, an optimal solution for the restricted problem can be found in short time by using this method. However, the proof of optimality may be very time-consuming.

A different branch-and-bound approach was proposed by Schlag et al. (1983). They give a characterization of feasible layouts in terms of satisfiability of a special Boolean expression.

If the distance constraints between two components i and j in the layout process are not fulfilled, the pair is called a *violation*. Four constraints $c_{ij}^1, c_{ij}^2, c_{ij}^3$, and c_{ij}^4 define the relative placement of elements i and j. A violation can be seen as the nonemptiness of the intersection between the two rectangles R_{ij} and j, shown in Figure 6.28. The rectangle R_{ij} contains element i; at each of the four sides it is enlarged by the appropriate minimum distance to element j.

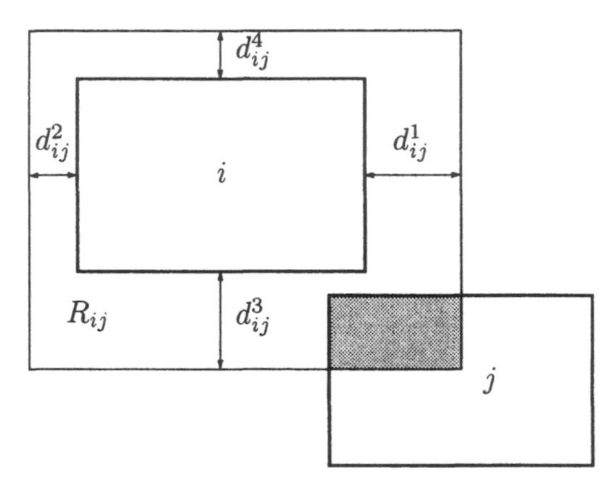

Fig. 6.28. A violation formed by elements i and j.

With each constraint $c \in \bigcup_{1 \le i < j \le n} \{c_{ij}^1, c_{ij}^2, c_{ij}^3, c_{ij}^4\}$ and each layout P, we also associate a logical variable. The variable is "true" if the constraint is fulfilled in P and "false" otherwise. Then a legal layout, i.e., a feasible orthogonal grid embedding, is characterized by the following properties:

- It satisfies the base constraints determining the sizes of the elements.
- For every c the formula F is "true", where

$$F = \bigwedge_{i,j} \left(c_{ij}^1 \vee c_{ij}^2 \vee c_{ij}^3 \vee c_{ij}^4\right) \quad . \tag{6.14}$$

For a practical application, the size of set F is too big. The basic idea of the two-dimensional compaction algorithm is to start with $F = \{\}$, and to obtain a so-called *smashing* by solving the system of inequalities with the longest path method from Section 6.6.2. A smashing is a possibly illegal layout that respects only the set of constraints in the current set F. Then the algorithm determines a violation (i, j) in the smashing by means of a rectangle intersection algorithm, searching for situations like in Figure 6.28. Once such a pair (i, j) is found, a branching step is performed: in each of the four subproblems, a different constraint c_{ij}^k ($k \in \{1, \ldots, 4\}$) is added to the set of active constraints F. Then the algorithm calls the procedure recursively with each of the four different sets $F + \{c_{ij}^1\}, \ldots, F + \{c_{ij}^4\}$.

An optimal solution is obtained like in the algorithm by Kedem and Watanabe. If the subproblem is a leaf, the generated constraints are either inconsistent, or the layout is legal and may become the new upper bound. Computations at inner nodes in the branch-and-bound tree have the following effects: illegal subproblems and problems exceeding the global upper bound cause the algorithm to cut the tree at the current node; otherwise, the objective value of the smashing becomes the new local lower bound.

This concludes the overview of optimal VLSI methods for two-dimensional compaction problems. Though not applicable to the typically huge instances of VLSI problems, the algorithms can be useful for the compaction of orthogonal grid embeddings. The rest of this section is dedicated to recent developments in the area of orthogonal graph drawing.

As seen in Section 6.3, there is a class of orthogonal representations where the two-dimensional compaction problems can be solved to optimality–these are representations with faces not containing forbidden sub-shapes, resulting in inner faces of rectangular shape. The work by Bridgeman et al. (1999) studies the class of orthogonal representations and introduces so-called *turn-regular orthogonal representations* to devise polynomial-time heuristics for the compaction problems in graph drawing.

An orthogonal representation H is *turn-regular*, if it does not contain opposite angles inside of a face. A pair of angles is *in opposition*, if it forms one of the 18 configurations shown in Figure 6.29.

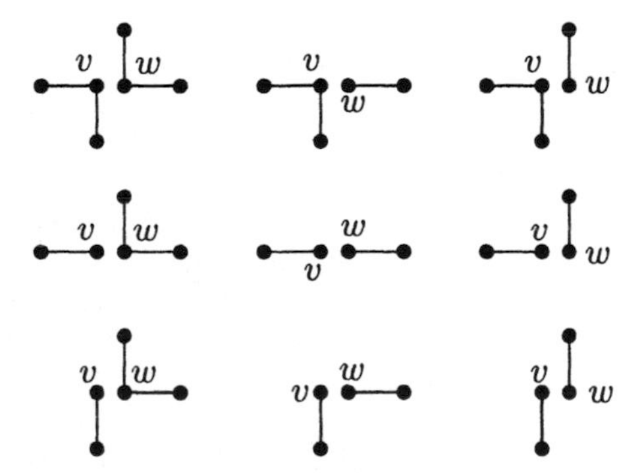

Fig. 6.29. Nine configurations for a pair of opposite angles inside of a face. The other nine can be obtained by 90° rotation.

Turn-regularity of an orthogonal representation can be tested in linear time. The authors show that the relative position of every pair of vertices is defined, if and only if the underlying orthogonal representation is turn-regular. This implies that a drawing that respects all the relative positions is a feasible orthogonal grid embedding.

For a turn-regular representation, the networks introduced in Section 6.3 can be used in order to obtain a minimum area drawing in linear time. Furthermore, the $O(n^{7/4}\sqrt{\log n})$ time algorithm introduced in Garg and Tamassia (1995a) can compute the drawing with minimum total edge length within this optimal area.

Based on this theoretical background, the following compaction heuristic can be applied to any orthogonal representation H: First, H is tested for turn-regularity with a linear-time algorithm. If the test is positive, an optimal drawing can be computed in polynomial time. Otherwise, the heuristic turns the non-regular faces into regular ones. Techniques similar to the dissection method presented in Section 6.3 can be used for this purpose, e.g., it is possible to insert straight artificial edges between pairs of opposite vertices. In general, the drawings resulting from this dissection method are better than the ones from Section 6.3, but are still far away from an optimal solution. Figure 6.30 shows a drawing for the example from Section 6.3, constructed with this heuristic method.

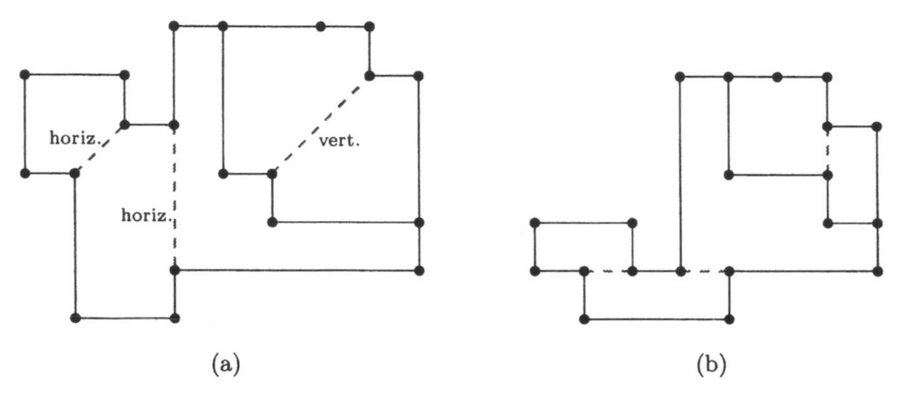

Fig. 6.30. The heuristic based on turn-regularity applied to the example from Figure 6.9: (a) pairs of opposite angles have been linked by either horizontal or vertical artificial edges. The resulting representation is turn-regular and compactable in polynomial time.

The method by Klau and Mutzel (1999b) solves the problem of minimizing the total or maximal edge length for a given orthogonal representation. It makes use of a necessary and sufficient condition for all feasible solutions of a given instance of the compaction problem. This condition is based on existing paths in so-called *constraint graphs.* This pair of graphs is similar to the layout graphs defined in Section 6.6.2. As in one-dimensional graph-based compaction, nodes in these graphs represent the segments (see Definition 6.23), and arcs characterize relative positioning relations.

Figure 6.31 shows an example of a pair of constraint graphs. The arcs specify exactly the relative relationships known from the given simple orthogonal representation H. Each edge in H determines the relative position of two segments in every feasible orthogonal grid embedding for H. Pairs of constraint graphs whose arc sets consist of all such arcs are also called *shape descriptions*.

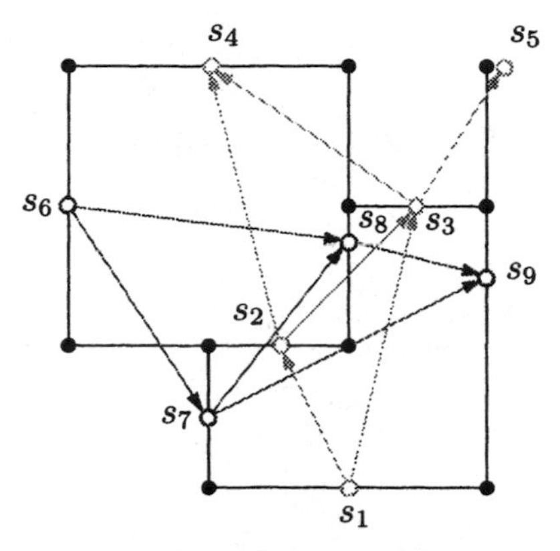

Fig. 6.31. A pair of constraint graphs that is a shape description. Each segment is limited by two horizontal and two vertical segments. The left limit of segment s_3 is $l(s_3) = s_8$, its right limit is $r(s_3) = s_9$. The bottom and top limits of s_3 are the segment itself, i.e., $b(s_3) = t(s_3) = s_3$.

The optimal compaction method is based on the following observations:

- The arcs of a shape description are contained in the layout graphs of every drawing that reflects the given shape.
- Most frequently, the information in a shape description σ is not enough to produce a feasible orthogonal grid embedding. Respecting only the relative positioning constraints encoded in σ may lead to crossings and overlapping edges. If this is not the case, however, we call such a pair of constraint graphs *complete*.
- In general, there are many possibilities for extending a shape description to a complete pair of constraint graphs.

Let $u \xrightarrow{*} v$ denote the existence of a directed path from u to v. The following is a precise characterization of complete pairs of constraint graphs in terms of paths that must be contained in the arc sets: a pair of graphs is complete if and only if both arc sets are acyclic and for every pair of segments $(s_i, s_j) \in S \times S$, one of the following four conditions holds:

$$
\begin{aligned}
&1.\quad r(s_i) \xrightarrow{*} l(s_j), \qquad &&3.\quad t(s_j) \xrightarrow{*} b(s_i), \\
&2.\quad r(s_j) \xrightarrow{*} l(s_i), \qquad &&4.\quad t(s_i) \xrightarrow{*} b(s_j).
\end{aligned}
\tag{6.15}
$$

In this definition, $l(s), r(s), b(s)$, and $t(s)$ denote the limits of a segment s as introduced in Figure 6.31. If one of the conditions applies we also call the pair of segments *separated.*

We can now express a one-to-one correspondence between these complete extensions and feasible orthogonal grid embeddings. For each simple orthogonal drawing with shape description σ, there exists a complete extension τ of σ and vice versa: every complete extension τ of a shape description σ corresponds to a simple orthogonal drawing with shape description σ.

Hence, the compaction task can be seen as the search for a complete extension of the given shape description leading to minimum total edge length (or minimum maximal edge length). One way to characterize the set of complete extensions is by means of an integer linear program (ILP). We introduce a binary variable x_{ij} for each arc (s_i, s_j) that may be part of some extension of the given shape description $\sigma = \langle (S_v, A_h), (S_h, A_v) \rangle$. If (s_i, s_j) is contained in the extension, the corresponding variable x_{ij} is one; otherwise, it is zero. We refer to the set of arcs in σ by $A = A_h \cup A_v$ and to the set of potential additional arcs by A^+. In addition, there is a variable $c_s \in \mathrm{Z}$ for each segment $s \in S$ denoting the coordinate of s. This yields the following ILP:

$$\min \sum_{e \in E_h} c_{r(e)} - c_{l(e)} + \sum_{e \in E_v} c_{t(e)} - c_{b(e)} \tag{6.16}$$

subject to

$$x_{r_i,l_j} + x_{r_j,l_i} + x_{t_j,b_i} + x_{t_i,b_j} \geq 1 \qquad \forall (s_i, s_j) \in S \times S \qquad (6.16.1)$$

$$c_j - c_i \geq 1 \qquad \forall (s_i, s_j) \in A \qquad (6.16.2)$$

$$c_j - c_i - (M+1)x_{ij} \geq -M \qquad \forall (i,j) \in A^+ \qquad (6.16.3)$$

$$x_{ij} \in \{0,1\} \qquad \forall (i,j) \in A^+ \qquad (6.16.4)$$

Inequalities (6.16.1) model the characterization of separation, i.e., the existence of necessary paths in an extension as required by conditions (6.15). In this formulation, r_i is short for the segment $r(s_i)$; the same abbreviation applies to all other limits. Inequalities (6.16.2) force the coordinates to obey the distance rules coded by the arcs in the underlying shape description. The same must hold true for the potential additional arcs: whenever a variable x_{ij} has value 1, we want an inequality of type (6.16.2); otherwise, there should be no restriction on the coordinate variables. This situation is modeled by inequalities (6.16.3) with the help of a big constant M. The authors show that in a feasible solution, the corresponding arc sets are acyclic and the entries of the coordinate vector c integral.

Like the one-to-one correspondence between complete extensions and feasible orthogonal grid embeddings, there is a one-to-one correspondence between feasible solutions of the ILP and complete extensions of the given shape description.

To formulate COMP_{max}, i.e., the minimization of the longest edge, the ILP has to be slightly modified. Only a linear number of inequalities have to be added and the objective function must be changed for that purpose.

For the class of turn-regular orthogonal representations defined in Bridgeman et al. (1999), there is only one complete extension of the corresponding shape description. In a preprocessing phase, the algorithm extends the given shape description as far as possible by adding arcs when there is only one possibility of meeting the four conditions (6.15). In case of complete constraint graphs, the integer linear program decomposes into two separate one-dimensional compaction problems that can be solved with the algorithm from Section 6.6.2, which is optimal in the one-dimensional case. Figure 6.32 shows a drawing for the example graph that is optimal with respect to total edge length, constructed with the branch-and-cut algorithm. It should be noted that the algorithm performs quite well on medium-sized instances, despite of its exponential worst-case time complexity.

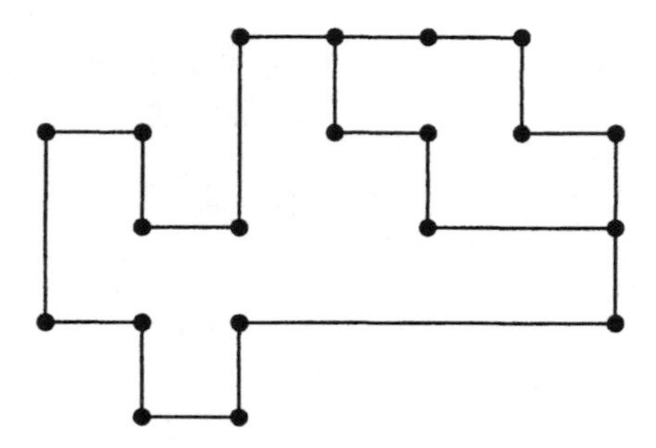

Fig. 6.32. An optimal solution of the two-dimensional compaction problem produced by the branch-and-cut algorithm (example from Figure 6.9).

To conclude this section, observe that the aesthetic criteria "area" and "edge length" may contradict each other, even when the corresponding orthogonal representation is turn-regular and the underlying shape description is complete. Unlike the rectangular case, optimality of the two criteria does not always coincide (see Figure 6.33).

6.7 Improving Other Aesthetic Criteria

In this section we present efficient postprocessing routines for improving other aesthetic criteria. Here the focus is on efficiency rather than optimality. In addition to decreasing area and edge length, these techniques aim at reducing the number of bends, the number of crossings, and the sizes of vertices.

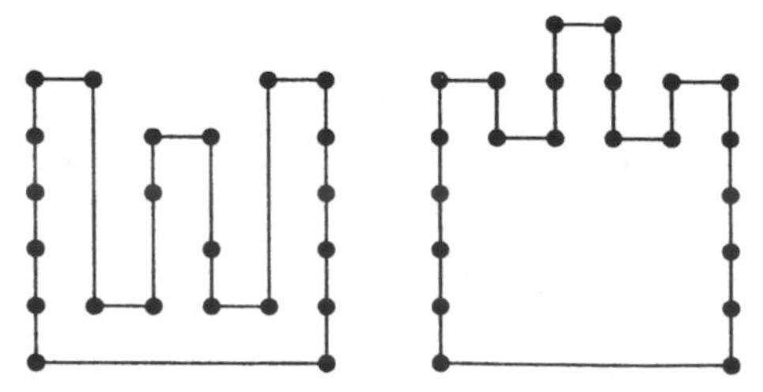

Fig. 6.33. Underlying shape description and orthogonal representation are complete and turn-regular, respectively. Nevertheless, minimum area (left) and minimum total edge length (right) exclude each other. (The example is taken from Patrignani (1999a).)

Few efforts have been made in this direction. The bend-stretching transformations by Tamassia and Tollis (1989) as presented in Section 6.4 fall in this category. They may reduce the number of bends in an orthogonal drawing. The *refinement algorithm* by Six et al. (1998) provides an additional set of elementary transformations to reduce the number of crossings. A more complicated and less efficient approach is given by Fößmeier et al. (1998). They also consider changing the sizes of vertices in order to get smaller drawings and fewer bends.

The goal of the algorithm by Six et al. (1998) is to find an efficient way for obtaining a better drawing in terms of area, number of bends, number of crossings, and total edge length. To improve area and edge length, they use the linear-time one-dimensional compaction method described in Section 6.6.2. The number of bends is reduced by using the bend-stretching transformations introduced by Tamassia and Tollis (1989) that are described in Section 6.4.2.

In addition, Six et al. (1998) consider the following configurations that can be removed in order to increase readability of the orthogonal drawing.

1. U-Turns are three consecutive edge segments forming two 90° angles (as shown in Figure 6.34 (a)).
2. Poorly placed degree two vertices are those which are neither on a bend nor distributed evenly in the drawing (Figure 6.34 (b)).
3. Self-crossings occur between two edges that are incident to the same vertex. The authors distinguish between near and far self-crossings (Figure 6.34 (c)).
4. A stranded vertex has only one neighbor that is placed far away (as shown in Figure 6.34 (d)).

A preprocessing phase constructs a so-called *abstracted graph* G' by deleting vertices with degree at most two. In some cases, the following simple procedures can repair the configurations of Figure 6.34: if a U-turn is found, the algorithm checks whether the middle segment can be moved towards the ends of the "U". If the necessary space is available, this operation can save cross-

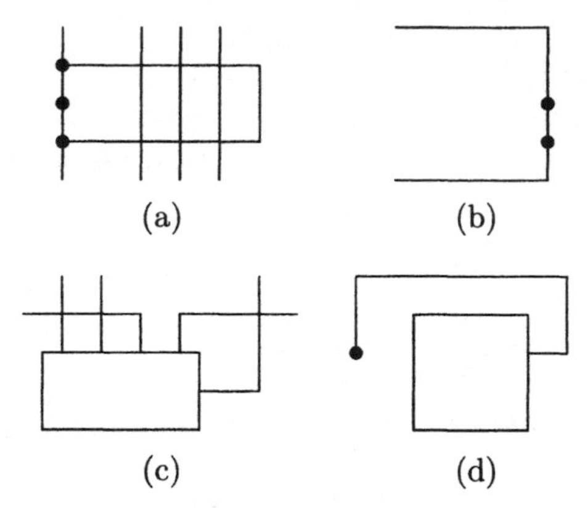

Fig. 6.34. The four additional configurations considered in Six et al. (1998): (a) a U-turn; (b) poorly placed degree two vertices; (c) self-crossings (near and far); (d) a stranded vertex.

ings and reduce total edge length. The bends on edges e that represent chains of degree two vertices are redistributed, so that either they lie on bends of e, or they are in the middle of an edge segment. Another operation removes near self-crossings by swapping the affected edges. For far self-crossings, the procedure tries to reroute the edges. Finally, stranded nodes are placed as close as possible to their neighbors. The authors give an $O(n+m)$ time bound for their refinement algorithm.

Another approach to postprocessing is the *4M-algorithm* by Fößmeier et al. (1998). It consists of the four operations *moving*, *matching*, *morphing*, and *merging*.

The *moving* operation is a one-dimensional compaction method similar to the compression ridge method presented in Section 6.6.2. The authors introduce a *moving line* (corresponding to an s-t flow) to cut the drawing into two parts. They propose a depth-first search to find this line more efficiently. Moving is shown in Figure 6.35 (a).

Matching resembles the third bend-stretching transformation in Tamassia and Tollis (1989). The aim is to save bends by moving vertices to the geometric places of bends. The technique, however, is different. Analogous to the moving line in the preceding operation, a *matching line* is used for finding these configurations. A theoretical characterization of matching lines can be found in Tamassia (1987) and in Di Battista et al. (1999). Figure 6.35 (b) illustrates the matching operation.

The *morphing* procedure saves bends by changing the size of a vertex v, drawn as a box. This operation is the inverse to shrinking vertices by introducing bends, which is used to get an orthogonal drawing from a visibility representation, described in Section 6.4.2. The basic idea of morphing is to expand v in direction of a close bend b so that the geometric representation of v covers b. Then the operation changes the box of v to its smallest possible size. Figure 6.35 (c) demonstrates an application of a morphing step. There are many cases, however, where this operation is not applicable: vertices may grow too big or overlap other parts of the drawing. Again, a *morphing line* is used for finding configurations where the operation can be performed successfully.

The last operation in the 4M-algorithm is *merging*. This operation aims at reducing the sizes of vertices drawn as a box and is illustrated in Figure 6.35 (d). Merging is a combination of inverse morphing and matching. First it introduces a bend b by resizing a vertex v in order to place a neighbor w of v on the position of b. As a result, either the width or the height of vertex v decreases by one grid unit.

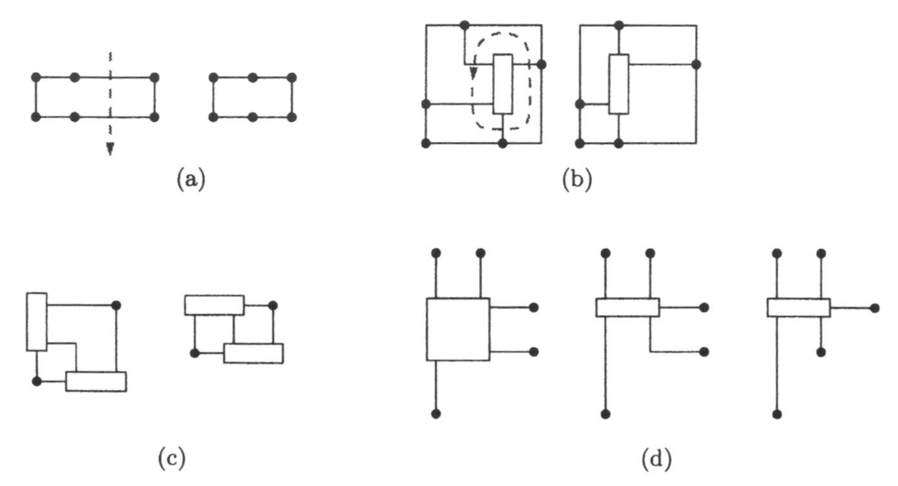

Fig. 6.35. The 4M-algorithm: (a) moving; (b) matching; (c) morphing; (d) merging.

Different variants of the 4M-algorithm run in $O(n^2)$ or $O(n \log n)$ time.

6.8 Conclusions and Open Problems

We have described a number of models and methods for orthogonal drawings of graphs. As we saw in the beginning, these problems are somewhat related to the issue of angles in drawings.

Drawing edges as axis-parallel paths makes it relatively easy to give combinatorial descriptions of these drawings. This allows it to use combinatorial arguments for getting a first drawing, as well as methods from mathematical programming for improving it. This basic approach of discretization can also be applied by using other grids; however, the resulting combinatorial issues may be harder to resolve.

Many of the performance guarantees of the presented heuristics still leave a gap between the number of bends in a drawing that can be achieved and the number of bends that may be necessary. It is conceivable that some of these gaps will be narrowed or closed.

Another elegant application of methods from mathematical programming is given by the polynomial flow-based algorithms for bend minimization. Like in some other cases, an immediate application is restricted to the relatively small class of planar graphs with maximum degree 4, but there are some extensions to cope with other models of orthogonality. A possible approach is to draw vertices as boxes; doing this in a specific manner leads to the KANDINSKY model, but others are possible.

The same applies to local improvement of the metric quality of a drawing, i.e., compaction. If we are dealing with a graph that has vertices of degrees exceeding 4, it easy to perform graph-based compaction on orthogonal drawings where vertices of high degree are represented as boxes. Crossings can be modeled as virtual vertices of degree four. The algorithms can be changed so that they can also process input drawings in Kandinsky-style, i.e., drawings with different grids for vertices and edges. There are many more variations of the problems that can be formulated elegantly so that a solution is found using the methods presented in this chapter.

We conclude by listing some of the open problems concerning the quality of orthogonal drawings:

- Can we give good approximation algorithms for drawing a 4-planar graph with few bends if the embedding is *not fixed*?
- Can we extend these approximation algorithms to general planar graphs?
- Are there approximation algorithms for classes of non-planar graphs?
- Are there more classes of orthogonal representations for which appropriate compaction algorithms find the optimal drawings in polynomial time?
- Several algorithms in this chapter operate in a fixed embedding or fixed shape setting. How do the optimal drawings with respect to the aesthetic criteria change if the embedding and/or the shape may be changed?
- To date, no approximation algorithms exist for the compaction problems. It would be very interesting to have efficient heuristics with a good performance guarantee.
- Can some of the ideas for planar drawings be extended to three dimensions?

Some aspects of the last question are discussed in Chapter 7.

7. 3D Graph Drawing

Britta Landgraf

7.1 Introduction

There is a large number of effective methodologies and algorithms for the creation of aesthetically pleasing graph drawings in two dimensions. However, representing graphs in three dimensions offers various benefits. The extra dimension gives greater flexibility for placing the vertices and edges of a graph and crossings can be always avoided. On the other hand new challenges arise: current output media have a two-dimensional nature and can only provide a limited resolution and display area. Thus, the resulting drawings become complex and difficult to survey. These disadvantages can be weakened by the use of navigational operations such as rotation, shifting and zooming. These operations enable an effective use of screen space and allow users to resolve ambiguities in large graphs while maintaining their overall mental map. The possibility of changing the viewpoint in 3D will also diminish the relevance of edge crossing in the (two-dimensional) screen representation of the graph.

The most commonly implemented 2D algorithms can be grouped into the following three categories: physical simulations, layering, and orthogonal graph drawing. Examples for physical-based methods for drawing undirected graphs are force-directed algorithms. The main idea of layering-based methods is to partition the nodes into layers and order the nodes within the layers, such that edge crossings are reduced. These algorithms are particularly well suited to draw directed acyclic graphs. An orthogonal graph drawing places all vertices at grid points, i.e., at points whose coordinates are all integer. The edges are represented by sequences of contiguous segments of grid lines, i.e., axis-parallel line segments determined by the grid points. Edge routes are allowed to contain bends, but are not allowed to cross or to overlap. Whereas the first two methods can be extended naturally to 3D (Ostry, 1996), orthogonal graph drawing in three dimensions requires mostly new algorithms. Therefore, the focus of this paper is on 3D orthogonal graph drawing.

This chapter is organized as follows. Section 7.2 describes the special aspects that are to be considered for physical-based methods in 3D. In Section 7.3 the extension of the layering approach to 3D is briefly described. Section 7.4 presents some orthogonal graph drawing algorithms for graphs of maximum degree six. Section 7.5 summarizes the results for 3D orthogonal graph drawing of high-degree graphs. Finally, Section 7.6 addresses the problem of finding good viewpoints for 3D straight-line graph drawings.

M. Kaufmann and D. Wagner (Eds.): Drawing Graphs, LNCS 2025, pp. 172-192, 2001.

7.2 Physical Simulation

Spring embedder (Kumar and Fowler, 1994; Sim, 1996) is a heuristic algorithm based on a physical model. As described in Section 4.2, a spring embedder works by replacing nodes by mutually repulsive charges and the edges by springs that attract connected nodes. The idea is that a minimum energy state of the system should correspond to a good layout.

The extension of the spring embedder from two to three dimensions is straightforward, because the algorithm make no particular assumption on the number of dimensions: one adds simply the third coordinate in the calculation of the forces. Figure 7.1 shows a three-dimensional spring embedding of the complete graph K_6.

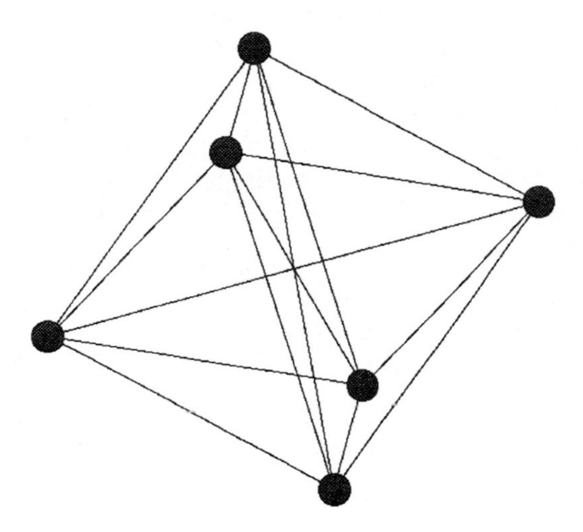

Fig. 7.1. 3D spring embedding of K_6.

Many refinements and extensions of the basic algorithm have been suggested. However, all these and the basic algorithm have the disadvantage that they frequently get stuck in local minima, i.e., states where the energy is locally minimal but not globally minimal. Another problem with this method is the difficulty of adding more sophisticated forces like ones that can deal with edge-densities or edge-crossings.

The *simulated annealing* (Davidson and Harel, 1996; Sim, 1996; Monien et al., 1995) algorithm uses randomness to overcome the problem of ending up in local minima. A detailed description of simulated annealing can be found in Section 4.3. Most of the extensions of the simulated annealing algorithm from 2D to 3D are straightforward, e.g., perturbing a point within a sphere instead of a circle. However, the choice of components for the energy function that reflect the desired aesthetics of the final graph differs significantly, because the aesthetic criteria themselves differ. For instance, the possibility of changing the viewpoint in 3D diminishes the relevance of edge crossings.

Simulated annealing tends to be very flexible but has the disadvantage that the cooling must be very slow to enforce uniform and symmetric layouts. It needs about 10 times as many iterations as normal spring embedders. Experiments have shown that the combination of both spring embedding and simulated annealing can be useful: one moves the nodes in direction of the forces, but adds a small random force. With a certain probability, moves are accepted that would increase the global energy.

An algorithm that is based on such a combination of the spring-embedder and the simulated annealing approach is GEM (graph embedder) (Frick et al., 1995). It contains several heuristics to speed up convergence, including local temperatures, the attraction of vertices towards the barycenter of their neighbors, and the detection of oscillations and rotations.

GEM-3D (Bruß and Frick, 1996) is the three-dimensional version of GEM. Since GEM contains nothing that is inherently two-dimensional, the extension requires essentially only an adaptation to 3D geometry, e.g., the notion of opening angles is extended to opening cones. The most difficult part is the detection of a rotation, as there are infinitely many planes of rotations. GEM-3D considers three alternatives to detect rotations:

1. Consider only rotations in the projections of the last and current movement vector onto the coordinate planes.
2. Count the number of 90° angles.
3. Use a global cooling schedule instead of rotation detection.

Unfortunately, the paper by Bruß and Frick (1996) contains no comparison of the quality of the produced drawings for these three alternatives.

7.3 Layering

Sugiyama et al. (1981) introduced an effective layer-based method for 2D drawings of directed graphs. The algorithm proceeds in four stages:

1. Make the graph acyclic.
2. Assign vertices to layers, i.e., partition the vertices of the directed graph into an ordered sequence of subsets in such a way that the edges have directions consistent with the subset ordering. Introduce dummy vertices to avoid "long edges", i.e., edges which traverse one or more layers.
3. Permute the vertices within the layers to reduce the number of crossings.
4. Reduce the number of bends by readjusting the position of vertices on each layer.

For a detailed description of the algorithm the reader is referred to Chapter 5.

Most of these stages for the 2D algorithm can be applied identically to 3D drawing: the graph must be made acyclic and the vertices must be assigned to layers. The barycenter heuristic is applicable to 3D by computing the average positions for both coordinates of the planes defining the layers. As in the 2D case, collisions can occur if two or more vertices in a layer have the same set of adjacent vertices. Resolving these collisions by separating the vertices to a predescribed distance or by altering the vertex positions in the primary layer is more difficult in 3D than in 2D, because the direction of separation must be defined as well as the separation distance to avoid introducing further collisions. A simpler method for this stage of the layout algorithm is to start with a barycentric heuristic layout, and then to refine by applying a constrained spring algorithm (Ostry, 1996).

Ostry (1996) also describes a 3D extension of layering which reduces the perceptual problems caused by apparent edge and vertex overlap. This approach satisfies the additional 3D aesthetic constraint that vertices should be placed on simple surfaces. The most appropriate surfaces for layered drawings are the cone and the cylinder. Then the vertices in a layer lie on a circle. Basically, the drawing has the overall form of a 2D layered drawing wrapped around a cone or a cylinder.

Another common approach for visualizing hierarchical structures is a cone tree (Robertson et al., 1993). In a cone tree, each subtree is associated with a cone, such that the vertex at the root of the subtree is placed at the apex of the cone, and its children are circularly arranged around the base of the cone. The cone tree can be oriented top-bottom or left-right. In the latter case the tree is also called a *cam tree*. Usually the cone structure is transparently shaded to allow visualization of all nodes, even those that would normally be blocked by other cones or nodes. In addition, permanent rotation can be helpful to enable a view of all nodes. A user can select a node by clicking on it. When a new node has been selected, the node and its parent path to the root are brought into focus by rotating the tree.

A third technique for 3D hierarchical graph drawing combines 2D drawings with a lifting transformation, i.e., first a 2D non-upward representation of a directed acyclic graph is created, and then the vertices are lifted along a third dimension. The lifting height of the vertices reflects the hierarchy. This method is used in GIOTTO3D (Garg and Tamassia, 1996a), which constructs the 3D drawing of a directed acyclic graph G in three phases. In the first phase, GIOTTO3D constructs a 2D non-upward drawing of G in the XY-plane. For this a variation of GIOTTO is used. GIOTTO transforms a graph into a planar graph by replacing each crossing with a fictitious vertex and then constructs an orthogonal drawing, using flow networks to minimize the number of edge bends and the total area of the graph. In the second phase Z-coordinates are assigned to the vertices and to the bends of the edges, such that their placement reflects the hierarchy. The purpose of the third phase is to increase the visual appeal of the drawing by drawing vertices as spheres and edges as Bezier tubes. In addition, a footprint of the 3D drawing, i.e., a projection of the graph to the XY-plane, assists the user in understanding the hierarchy-independent connectivity information of the graph.

7.4 3D Orthogonal Drawings of Graphs of Maximum Degree Six

An 3D orthogonal graph drawing places all vertices at grid points, i.e., at points whose coordinates are all integers. The edges are represented by sequences of contiguous segments of grid lines, i.e., axis-parallel line segments determined by the grid points. Edge routes are allowed to contain bends, but are not allowed to cross or to overlap. Because each grid point lies at the intersection of three grid lines, any graph that admits an 3D orthogonal graph drawing has vertex degree at most 6. By representing vertices by boxes it is possible to construct 3D orthogonal drawings for graphs of arbitrary degree (see Section 7.5).

The most common proposed measures for determining the quality of an orthogonal drawing are the bounding box volume, i.e., the volume of the smallest grid-box containing the drawing, and the maximum number of bends per edge. Using straightforward extensions of the corresponding 2-dimensional NP-hardness results, optimizing any of these criteria is NP-hard (Eades et al., 1996).

Optimizing both the volume and the number of bends per edge are conflicting goals. Minimizing the number of bends in a graph drawing often increases the bounding box volume. Table 7.1 shows this trade-off for the 3D orthogonal drawing algorithms for graphs of maximum degree six, presented in this section.

The COMPACT drawing algorithm of Eades et al. (1996b) requires the least volume at the expense of more bends per edge. This volume bound is tight. Kolmogorov and Bardzin (1967) showed that no algorithm can produce asymptotically more compact drawings.

The 3-BENDS algorithm of Eades et al. (1996b) and the INCREMENTAL algorithm of Papakostas and Tollis (1997a,b) establish an upper bound of 3 for the number of bends per edge route.

It is unknown if the upper bound of 3 for the maximum number of bends per edge route is tight. In fact, Wood (1998a,b) showed that for every maximum degree five graph a 3D orthogonal drawing having at most 2 bends per edge route exists.

Figure 7.2 shows some 3D orthogonal graph drawings of K_6, the complete graph on 6 vertices.

Often it is much easier for a human to recognize the structure of a graph in a spring embedding than in a 3D orthogonal drawing, although the former is not crossing-free. The main reason for this seems that bend minimization is the most important aesthetic criterion for diagram readability. For an

Table 7.1. Upper bounds for 3D orthogonal drawing algorithms for n-vertex graphs of maximum degree six.

Algorithm	Volume	Max Bends
COMPACT drawing algorithm of Eades, Symvonis and Whitesides (Eades et al., 1996b) and its refinements (Eades et al., 2000)	$O(n^{3/2})$ $O(n^2)$ $O(n^{5/2})$	7 6 5
algorithm of D. R. Wood for maximum degree six graphs (Wood, 1998a)	$2.37n^3$	4
3-BENDS algorithm of Eades, Symvonis and Whitesides and its refinements(Eades et al., 1996b; Wood, 2000)	$27n^3$ $8n^3$ $n^3 + o(n^3)$	3 3 3
INCREMENTAL algorithm of Papakostas and Tollis Papakostas and Tollis (1997a,b)	$4.66n^3$	3
2-bend algorithm of D. R. Wood for maximum degree five graphs (Wood, 1998b)	n^3	2

example compare the drawings of K_6 in Figure 7.2 with the spring embedding of K_6 in Figure 7.1. Only the REDUCE FORKS algorithm (Patrignani and Vargiu, 1997) produces a drawing which can be recognized just as easily. Furthermore, the REDUCE FORKS algorithm seems to produce the most compact drawing. Unfortunately, there are no proven bounds for the REDUCE FORKS algorithm. In experimental tests (Patrignani and Vargiu, 1997) the REDUCE FORKS algorithm produced drawings with an average of less than 2.5 bends per edge route and with bounding box volume $0.6n^3$. These tests involved only graphs with average degree 4. Wood (1998a) showed, that, even for the 12-vertex graph $K_{6,6}$, the REDUCE FORKS algorthm introduces a 9-bend edge and has more volume that the Papakostas and Tollis algorithm. In the experimental study of Di Battista et al. (1998b) the REDUCE FORKS algorithm only produces drawings with the least volume for graphs with approximately < 35 vertices. For larger graphs the COMPACT drawing algorithm of Eades et al. (1996b) performs better.

In summary the existing 3D orthogonal graph drawing algorithms, when applied to large graphs, do not produce drawings appropriate for visualisation purposes. So the value of orthogonal graph drawing – apart from special applications like VLSI design – lies at present mainly in the theoretical analysis of 3D graph drawing algorithms, whereas spring embedding, simulated annealing, and hierarchical approach are more applicable in practice.

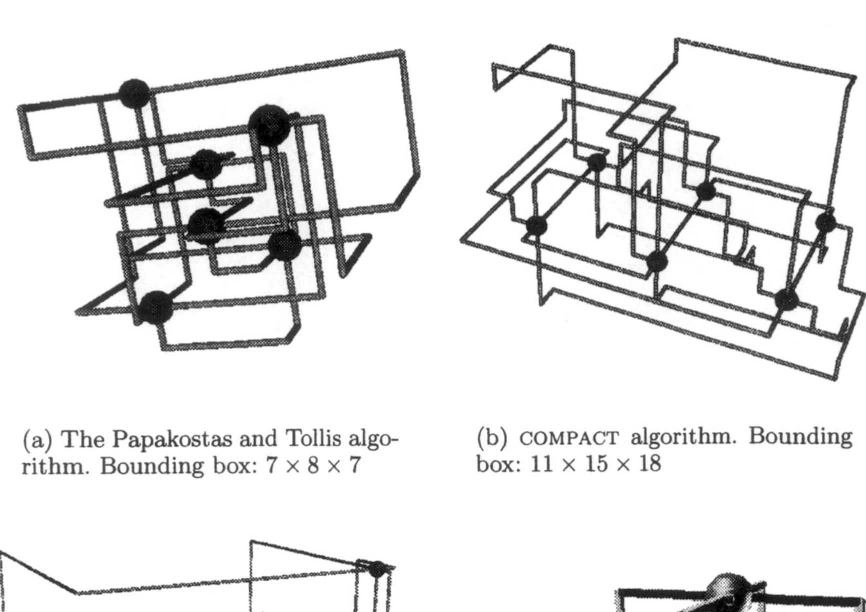

(a) The Papakostas and Tollis algorithm. Bounding box: $7 \times 8 \times 7$

(b) COMPACT algorithm. Bounding box: $11 \times 15 \times 18$

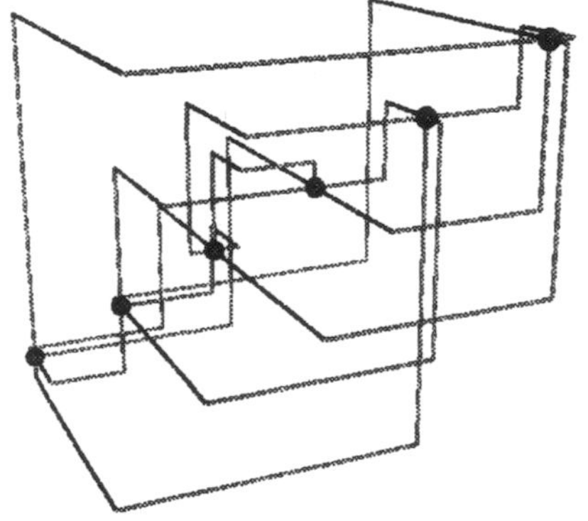

(c) 3-BENDS algorithm. Bounding box: $16 \times 15 \times 16$

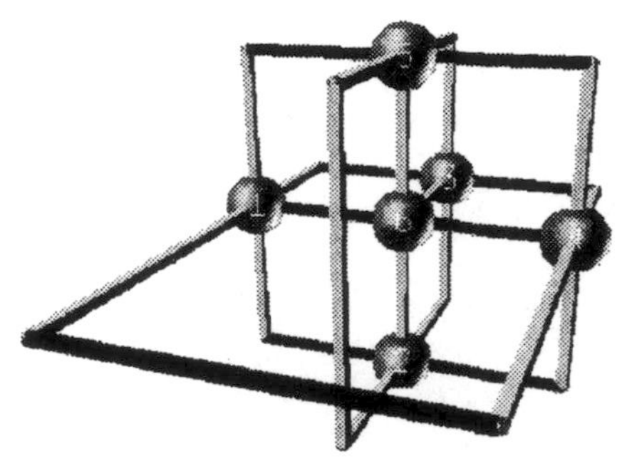

(d) REDUCE FORKS algorithm. Bounding box: $2 \times 3 \times 2$

Fig. 7.2. 3DCube's (Patrignani and Vargiu, 1997) Snapshots of K_6.

7.4.1 Approaches to 3D Orthogonal {Point}-Drawing

In the following, all algorithms take an input graph $G = (V, E)$ of maximum degree at most 6 and $|V| = n$.

Compact Drawing Algorithm. This section briefly describes the COMPACT drawing algorithm of Eades et al. (1996b), which produces a grid drawing having at most 7 bends per edge, maximum edge length $16\sqrt{n} - 7$, and bounding box dimensions $(3\lceil\sqrt{n}\,\rceil + 2) \times 5\lceil\sqrt{n}\,\rceil \times (8\lceil\sqrt{n}\,\rceil - 6)$.

The COMPACT drawing algorithm is based on a preprocessing step using basic graph theory to construct a directed graph G' whose underlying undirected graph contains G. Then the preprocessing algorithm computes a partition of the arcs of G' into three arc-disjoint cycle covers, denoted C_{red}, C_{blue} and C_{green}. A *cycle cover* of a directed graph is a spanning subgraph that consists of directed cycles.

Each of these cycle covers is arranged within different areas of the drawing, so that no differently colored arcs can cross. To obtain a drawing for G, the algorithm routes the undirected edges of G according to the routes for the corresponding directed arcs of G'. Arcs which were inserted for the construction of G', but do not arise from edges of G, are simply not drawn.

All vertices of G are placed in an array of 5×5 squares in the plane $Z = 0$. The vertices of each directed cycle of C_{red} are placed successively in a snake-like fashion by following the cycle. Then the arcs of these cycles are routed completely in the plane $Z = 0$ using at most 6 bends per route.

The blue colored arcs are routed above the plane $Z = 0$. The route for an arbitrary arc (v, w) of C_{blue} consists of 8 segments, as shown in Figure 7.3.

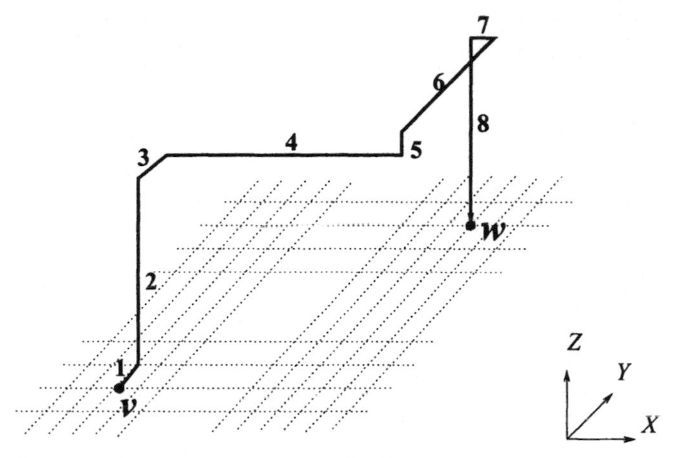

Fig. 7.3. The 7-bend route of arc (v, w) (Eades et al., 1996b).

The segments of type 1,3,5 and 7 have length 1. These unit length segments, the arrangement of the vertices in the square array, and a suitable selection of the Z-coordinate of segments of type 4 and 6 ensure that no segments of different routes can overlap or cross.

The arcs of C_{green} are routed like the arcs of C_{blue}, but on the other side of the plane $Z = 0$.

Eades et al. (2000) refine the COMPACT drawing algorithm to explore the trade-offs between the number of bends per edge and the dimension of the bounding box of the drawing. For this purpose they eliminate successively the unit-lengths segments from the routes of the arcs in C_{green} and C_{blue}.

For each eliminated segment the maximum number of bends per edge reduces by one while the lengths of the bounding box side parallel to this segment increases by a factor of $O(\sqrt{n})$, see Table 7.1.

A detailed discussion of the trade-offs and the refinements of the COMPACT drawing algorithm can be found in Eades et al. (2000). Furthermore, the authors present an algorithm which draws a maximum degree 4 graph in a $O(n) \times O(n) \times O(1)$ bounding box with at most 3 bends per edge.

3-Bends Algorithm. The 3-BENDS algorithm of Eades et al. (1996b) constructs an 3D orthogonal drawing with at most 3 bends per edge, maximum edge length $9(n-1)+2$, and bounding box dimensions $(3n-2) \times (3n-3) \times (3n-2)$. It is based on the same preprocessing algorithm as the COMPACT drawing algorithm described in Section 7.4.1 to obtain a directed graph G' together with three arc-disjoint color classes. It places the vertices of G on the diagonal of a $3n \times 3n \times 3n$ cube according to an arbitrary ordering. Each pair a, b of vertices in G' determines a cube $C(a,b)$ with $p_a = (3a, 3a, 3a)$ and $p_b = (3b, 3b, 3b)$ at opposite corners. The colored arc (a,b) is routed along the edges of this cube or with an offset of one unit near the path of cube edges; the particular choice depends on the color of the arc and on whether the predecessor and successor of b corresponding to that color are positioned both on the same side of b on the diagonal. See Figure 7.4.

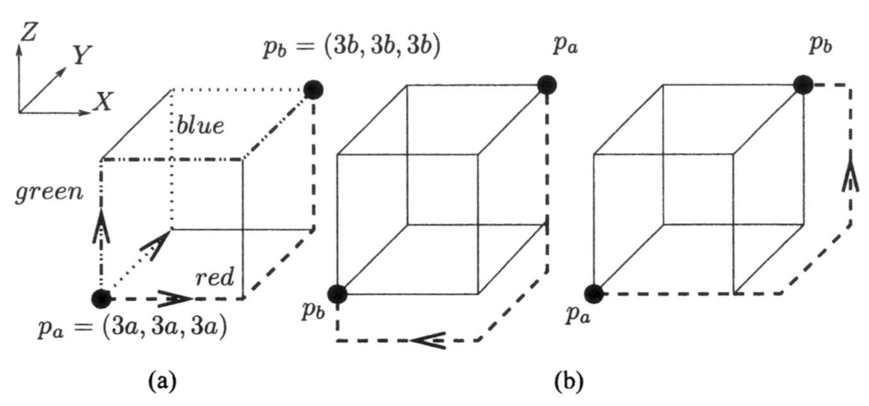

Fig. 7.4. 3-bends routing along (a) or near (b) the path of cube edges.

Colored paths of cube edges on the same cube get close to one another only in the vicinity of the ends of the paths. Routing one unit near the path of cube edges enables the use of all six directions in the endpoints without conflicts. Therefore there are obviously no illegal intersections of routes.

By deleting empty grid-planes in drawings produced by the 3-BENDS algorithm the volume can be improved from $27n^3$ to $8n^3$.

Incremental Orthogonal Graph Drawing Algorithm of Papakostas and Tollis. The INCREMENTAL algorithm of Papakostas and Tollis (1997a,b) produces a 3D orthogonal drawing with volume at most $4.66n^3$, and at most 3 bends per edge in linear time. This slighthly outperforms the 3-BENDS algorithm of Eades et al. (1996b) with regard to the volume of the drawing and with same run time (Eades et al., 2000). As opposed to this algorithm, which places all vertices before the routing, the algorithm of Papakostas and Tollis operates interactively, i.e., vertices arrive and enter the drawing on-line. Thus, the bounds of the representation must be increased only if necessary. The decision about where a new vertex will be routed depends entirely on the free directions around the adjacent vertices. Placing a new vertex and routing its incident edges has several cases. The advantage of the algorithm of Papakostas and Tollis is the incremental mode of operation, whereas the 3-BENDS algorithm is superior in the elegance and the ease of implementation.

2-Bends Drawing of Maximum Degree Five Graphs. For an n-vertex m-edge graph with maximum degree 6 the algorithm of Wood (1998a) produces drawings with bounding box volume at most $2.37n^3$ and with a total of $7m/3$ bends, using no more than 4 bends per edge route. The resulting drawing is in *general position model*, i.e., no grid plane intersects any two vertices.

For maximum degree five graphs the bounding box has volume n^3 and each edge route has two bends. Furthermore Wood has given 2-bend 3D orhtogonal drawings of the 6-regular multi-partite graphs (Wood, 1998b). This raises the problem whether there is a 2-bend orthogonal drawing for every maximum degree six graph, which is still open at present.

Due to space limitations only an outline of the algorithm for maximum degree five graphs without a proof of correctness is given here.

The 3-BENDS algorithm of Eades et al. (1996b) positions the vertices along the diagonal of a cube according to an arbitrary ordering. Wood uses an approximately balanced ordering to place the vertices along the diagonal of a cube. The use of an approximately balanced ordering essentially guarantees that the number of predecessors and successors for each vertex in the graph are distributed more evenly. Wood uses also a 3-coloring of the arcs of the directed graph G' to move the vertices and to route the edges according to their color class. For routing an edge $\{v, w\}$ with two bends the color $I \in \{X, Y, Z\}$ of the arc (v, w) is interpreted as direction for the start segment and the color of its reversal arc (w, v) as direction of the end segment. From the diagonal each vertex is moved in up to two dimensions dependent on its number of predecessors and successors and its color. A balanced vertex v remains unmoved and the positive (respectively negative) directions are assigned to the successor (predecessor) arcs of v. At an unbalanced vertex v, say with more successors than predecessors, the positive directions can be assigned to at most three successor arcs of v. The remaining successor arcs (v, w) must be assigned a negative direction. To do so v is moved past w in the relative I-ordering if (v, w) has the color $I \in \{X, Y, Z\}$.

Edge crossings are resolved by swapping the directions of the starting segments of the crossing edges. A swapping operation may create new edge crossings. However it reduces the sum of the lengths of the middle segments of the two edge routes involved. This sum is bounded below, so a finite number of swaps suffice to create a crossing-free 3D orthogonal drawing. An example for re-routing intersecting edges is shown in Figure 7.5.

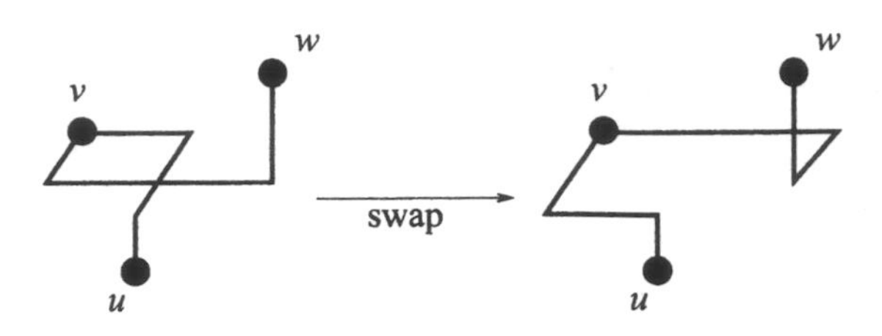

Fig. 7.5. Re-routing intersecting edges Wood (1998a).

7.5 3D Orthogonal Drawings of Graphs of Arbitrary Degree

Representing vertices as points enables only crossing-free 3D orthogonal drawings for graphs of maximum degree six. As in the two-dimensional case (see Section 6.4.5) it is possible to construct valid 3D orthogonal drawings for graphs of higher degree by representing vertices as three-dimensional boxes.

Currently there are only few results for 3D orthogonal {box}-drawings. Papakostas and Tollis (1997a,b) present an algorithm to embed any graph in a 3D grid of volume $O(m^3)$ with at most 2 bends per edge. Biedl (1998) presents three approaches for creating 3D drawings. Biedl et al. (1997b, 1999) study 3D orthogonal drawings of the complete graph K_n, and hence for any simple graph, established lower bounds. Wood (1999a) presents an algorithm for producing orthogonal drawings in any number of dimensions. Furthermore Wood (1999b) introduces an algorithm for 3D orthogonal drawings of arbitrary degree n-vertex-m-edge multigraphs with $O(m^2/\sqrt{n})$ bounding box volume and 6 bends per edge route. In this paper Wood also discuss many open problems in 3D orthogonal graph drawing.

The results of these papers suggest a trade-off between cube-like appearance of the vertex boxes and bounding box volume. For instance the improved version of the first approach of Biedl (1998), *edge-lifting*, yields drawings with a bounding box volume that asymptotically matches the lower bound, but the vertex boxes are $1 \times 1 \times \frac{4}{3}n^{1.5}$-boxes and therefore may be disproportionally large. In the second approach, called *half-edge-lifting*, the surface area of each vertex box is proportional to the degree of the vertex, but the boxes are also highly degenerate. With a slight modification of this approach, the boxes become cubes at the cost of an increase in volume. Edge-lifting as well as half-edge-lifting are essentially two-dimensional, because they are created by starting with a 2D orthogonal drawing and lifting it into 3D. Hence one sees the 2D input drawing when looking at the final 3D drawing from the top. The third approach, called the *three phase method*, and the algorithm of Papakostas and Tollis have a more natural 3D appearance, but result in drawings with higher bounding box volume.

From a truly three-dimensional drawing one expects that their vertices should be displayed more or less as cubes. For this purpose Biedl has introduced three models which describe the restriction of the size of vertex boxes. The *unlimited growth model* imposes no restrictions on the dimension of the vertices. In the *degree-restricted model* the surface area of each vertex v is proportional to its degree v, but there is no restriction on the shape of a vertex. In the *cube model* the box of v must be a cube whose surface area is proportional to the degree of v.

The main disadvantage of the unlimited growth model is that vertices are not recognized as points. The main advantages of this model are that bends can often be saved by stretching a vertex to cover a bend, and that it yields very small volumes. Hence the use of this method is of a theoretical nature to explore worst-case upper bounds.

Table 7.2 summarizes the above results and shows the trade-off between the shape of a vertex and the bounding box volume. In this table Δ denotes the maximum degree of a graph.

7.5.1 Bounds for 3D Orthogonal {Box}-Drawings

Biedl et al. (1997b) have established lower bounds for the volume and the number of bends of 3D orthogonal graph drawings. Their focus has been on the complete graph K_n since any simple graph G with n vertices is a subgraph of K_n. Therefore, upper bounds for K_n yield upper bounds for all other simple graphs on n vertices, and no simple graph on n vertices can yield larger lower bounds than K_n. For drawings of K_n, they prove a lower bound of $\Omega(n^{2.5})$ on the volume and a lower bound of $\Omega(n^2)$ on the total number of bends.

A Lower Bound on the Volume: To show that the minimum possible volume for a 3D orthogonal drawing of K_n is $\Omega(n^{2.5})$ one considers a drawing of K_n in a $X \times Y \times Z$-grid and distinguishes three cases that describe the distribution of the vertices in the space: in case 1 one assumes that there exists a line l that intersects many vertices. Case 2 assumes that no such line exists, but a plane that intersects many vertices. Case 3 treats the remaining situation that no plane intersects many vertices. As an example we sketch the treatment of case 2.

Case 2: Assume that no grid-line intersects as many as $\frac{1}{16}n$ vertices, but there exists a plane p_z that is parallel to the XY-plane and intersects at

Table 7.2. Upper bounds for 3D orthogonal drawing algorithms for graphs of arbitrary degree.

Algorithm	Volume	Max Bends	Model
INCREMENTAL algorithm (Papakostas and Tollis, 1997a,b)	$O(m^3)$	2	degree-restricted
edge-lifting (Biedl, 1998)	$O(n^3)$	1	unlimited growth
modified edge-lifting (Biedl, 1998)	$O(n^{5/2})$	3	unlimited growth
half-edge-lifting (Biedl, 1998)	$O(n^3)$	2	degree-restricted
improved half-edge-lifting (Biedl, 1998)	$O(nm\sqrt{\Delta})$	2	cube-model
three-phase-method (Biedl, 1998)	$O(n^2m)$	2	degree-restricted
modified three-phase-method (Biedl, 1998)	$O((nm)^{3/2})$	2	cube-model

least $\frac{1}{4}n$ vertices. Let p_x be a plane that is parallel to the YZ-plane and intersects the X-axis at the point $(x, 0, 0)$. By assumption, each p_x intersects p_z in a line that intersects fewer than $\frac{1}{16}n$ vertices. $x = x_0$ is now selected in such a way that it is the largest integer value, such that fewer than $\frac{1}{16}n$ vertices intersect p_z to the left of p_{x_0}, i.e., they have X-coordinates less then x_0. Thus the number of vertices that intersect p_z and that lie to the right of p_{x_0+1} is at least $\frac{1}{16}n$. Since in K_n all nodes are connected, there are at least $\left(\frac{1}{16}n\right)^2$ edges between the vertices on the left and the vertices on the right of p_{x_0}, so $YZ \geq \left(\frac{1}{16}n\right)^2$. If one applies the same argument to the y-direction and considers that by assumption $XY \geq \frac{1}{4}n$, it follows that $XYZ = \sqrt{YZ \cdot XZ \cdot XY} \geq \frac{n^{5/2}}{512}$.

For all sufficiently large n, the bound given by case 2 is the smallest of the three. Hence any drawing of K_n has volume $\Omega(n^{2.5})$.

A Lower Bound on the Bends: It was shown by Fekete and Meijer (1999) that K_n has no bend-free drawing for $n > 183$. This can be used to prove that any drawing of K_n has a total number of bends $\Omega(n^2)$:

Let c be an integer such that any 3D orthogonal drawing of K_c has a bend. For $n > c$ the graph contains $\binom{n}{c}$ copies of a K_c. Each of these copies must have a bend. Any edge belongs to exactly $\binom{n-2}{c-2}$ of these copies of K_c.

Hence the number of edges with a bend must be at least $\binom{n}{c}/\binom{n-2}{c-2} \geq \frac{n^2}{c^2}$ for $n \geq c$.

7.5.2 Approaches to 3D Orthogonal {Box}-Drawings

Incremental Orthogonal Graph Drawing Algorithm of Papakostas and Tollis. Papakostas and Tollis (1997a,b) present an incremental orthogonal drawing algorithm for graphs of arbitrary degree. They map vertices to grid boxes with surfaces proportional to their degrees, i.e., they consider the degree-restricted model. New vertices are placed outside of the current drawing, such that the *general position property* is kept, i.e., there is no plane parallel to one of the three base planes containing grid points of two different boxes in the current drawing. This rule enables the crossing-free edge routing with two bends. Edges that are adjacent to a vertex are attached to the surface of its box at grid points. As a result of edge routing, edges may require attachment to specific sides of incident boxes. If there are no available grid points on that side, the box must be increased. Since the box of every vertex may grow in various different ways in the course of the drawing process, the resulting drawing follows the degree-restricted model and cannot follow the cube model. In addition, with each enlargement of a box new planes in the current 3D drawing are inserted to accommodate the size of the box. This affects the coordinates of some ports and bends which shift by one unit along the X-, Y- or Z-axes, but the general shape of the drawing remains the same. The produced drawing has two bends per edge and a bounding box volume of $O(m^3)$.

Figure 7.6 shows a sample 3D orthogonal drawing of K_5 produced by this algorithm. The box numbers denote the vertex insertion order.

Lifting-Based Approaches. Biedl (1998) presents two lifting-based approaches to create an 3D orthogonal drawing. *Edge lifting* and *half-edge lifting* start with a semi-valid 2D orthogonal drawing, i.e., no vertices overlap and no edge crosses a vertex, but edges may overlap each other. Then this drawing is split into valid orthogonal drawings. These crossing-free drawings are placed into different Z-planes. The difference between both methods lies in the splitting technique. In edge lifting, an arbitrary partition can be used, whereas half-edge lifting first splits the 2D drawing into two drawings, where one drawing contains all horizontal edge segments and the other drawing contains all vertical edge-segments. In edge lifting every vertex must be extended to all Z-planes to get a drawing of the underlying graph. In half-edge lifting every vertex is extended only to those Z-planes that contain an incident edge. In addition, at every bend of an edge in the original 2D drawing a Z-segment is added to connect the two endpoints of the horizontal and the vertical segment incident to this bend. Consequently, the vertices in drawings produced by edge lifting follow the unlimited growth model, whereas half-edge lifting yields drawings in the degree-restricted model.

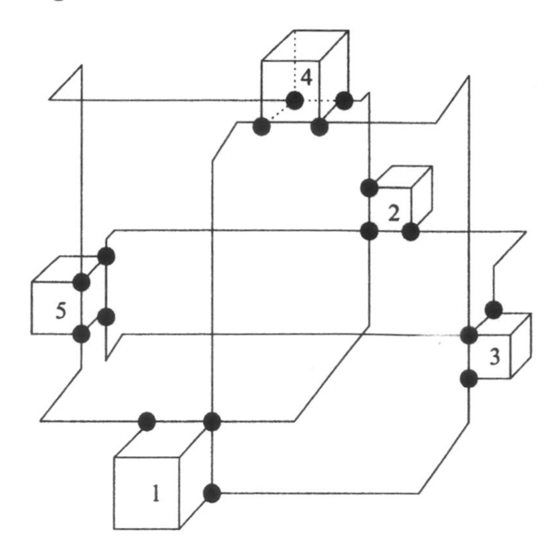

Fig. 7.6. 3D orthogonal drawing using boxes to represent vertices (Papakostas and Tollis, 1997a).

The edge lifting method is used in Biedl et al. (1997b, 1999) to get constructions that achieve lower bounds for the volume of the bounding box and the total number of bends under the restriction that the drawing has at most k bends per edge route. For $k = 1$ or $k = 2$ the constructions produce drawings for K_n in $O(n^3)$ volume. For $k \geq 3$ an unconstrained construction is given that asymptotically yields the optimum volume. For $k = 1$ and $k = 2$ it is an open problem whether the lower bound for the volume is attainable. All constructions match the lower bound on the number of bends.

In order to achieve a volume as small as possible, the vertices and edges must be placed skillfully and one needs an efficient splitting into crossing-free drawings. The more bends per edge are admissible the more space-saving the edges can be nested into one another. For $k = 1$ Biedl et al. (1997b, 1999) use nested triangular edges, for $k = 2$ nested rectangular edges, and for $k = 3$ L-shaped or Γ-shaped edges that are diagonally arranged. Figure 7.7 shows examples for these edge sets after splitting into crossing-free drawings and Figure 7.8 shows the 3D orthogonal graph drawing of K_6 produced by these constructions. The pictures in the latter figure was created with OrthoPak (Closson et al., 1998).

For $k = 1$ and $k = 2$ the constructions can still be improved with respect to the volume by a skillful combination of two drawings of $K_{\frac{n}{2}}$. However, in both cases the achieved volume $O(n^3)$ does not match the lower bound. For $k = 3$ the construction asymptotically yields the optimum volume $O(n^{2.5})$. Since this construction generates at most 3 bends on any edge, it is valid for each $k \geq 3$.

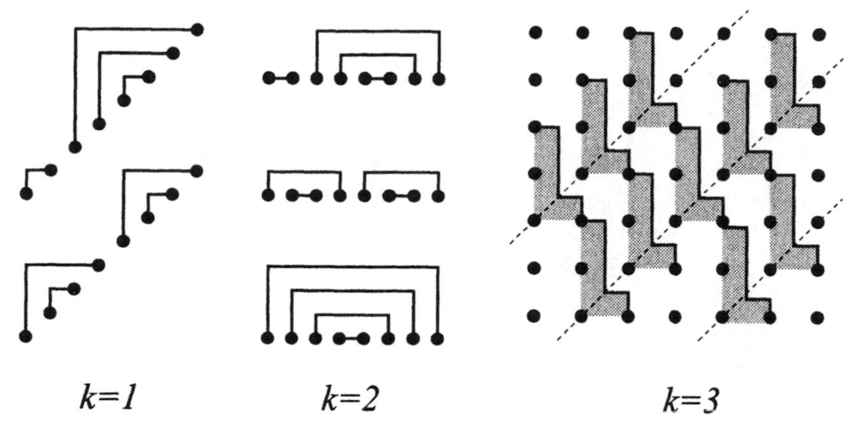

Fig. 7.7. Edge set examples (Biedl et al., 1997b, 1999).

As already mentioned in the introduction to this section, the half-edge lifting approach is in the degree-restricted model like the algorithm of Papakostas and Tollis, but it yields better results regarding the bounding box volume. On the other hand, the drawings created with this approach are essentially two-dimensional.

Three-Phase Method. Creating truly 3D drawings is straightforward using the *three-phase method* (Biedl, 1998) which is an extension of the corresponding method in 2D (see Chapter 6).

In the first phase, *vertex placement*, vertices are drawn as points, not as boxes. They are placed in 3D arbitrarily in XY-general position, i.e., every X-plane and every Y-plane intersects at most one vertex. This condition is weaker than the general position property used in the algorithm of Papakostas and Tollis described in Section 7.5.2, since several vertices may be placed in the same Z-plane.

In the second phase the *edge routing* with 2 bends is done using directed Z-routes. Directed Z-routes are those cube routes, i.e., routes along the edges of a cube, for which the middle segment is parallel to the Z-axis. Biedl indicates two further subclasses of cube routes that can be used in order to produce a crossing-free drawing: the colored cube routes from the 3-BENDS algorithm of Eades et al. (1996b), compare Section 7.4.1, and the *shortest-middle routes*. An edge is said to be routed using shortest-middle routes, where the middle segment is the shortest segment. However, the last two types of routes require that the vertices are placed in general position, i.e., every grid-plane intersects at most one vertex, so that compared to the INCREMENTAL algorithm of Papakostas and Tollis, no substantial improvement of the volume is to be expected.

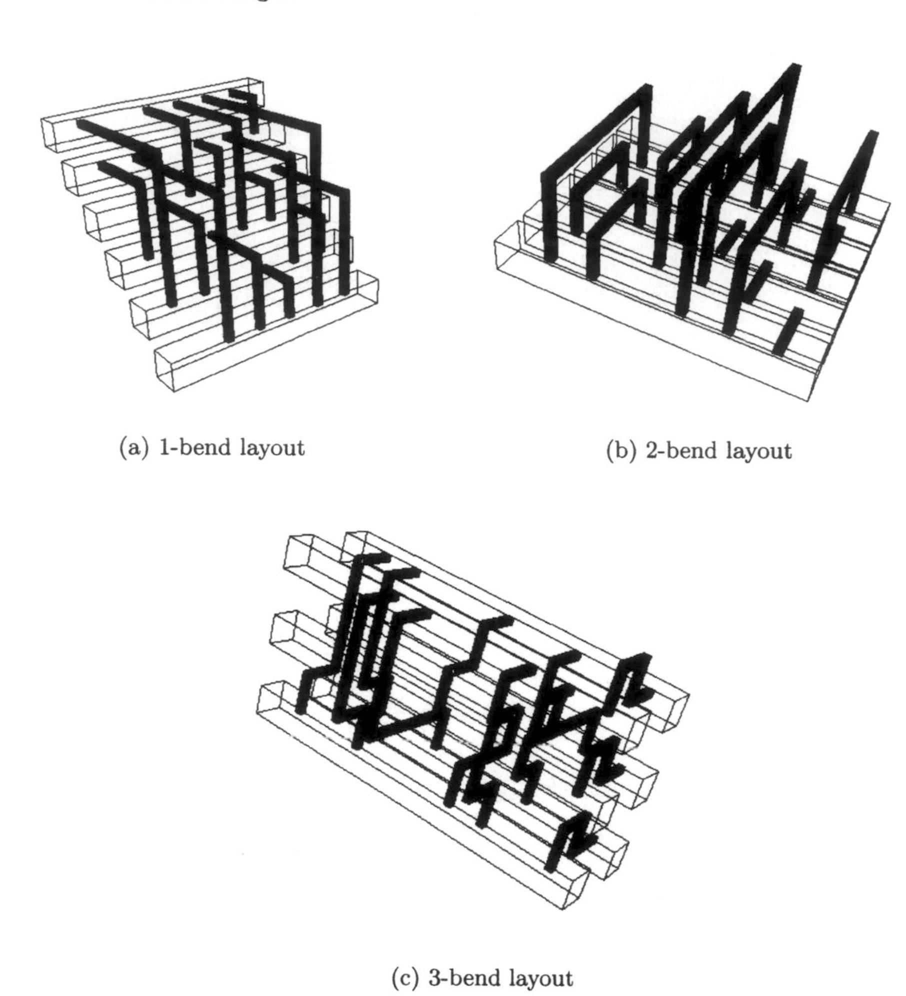

(a) 1-bend layout

(b) 2-bend layout

(c) 3-bend layout

Fig. 7.8. OrthoPak's Snapshots of K_6.

In the third phase, *port assignment*, each grid plane is replaced by sufficiently many grid planes. Each vertex is replaced by a grid box that is the intersection of the planes inserted for the respective vertex. The edges are re-assigned to ports of the vertex boxes, such that all overlaps and crossings are removed. For this purpose the edges attached to one side are split into four groups, depending on their direction of continuation. Then one assigns sufficiently many ports to each group, such that no edges of two different groups could possibly cross. Next the edges in each group are sorted by the coordinates of the next bend and assigned to a port of their group. By this sorting the attached segments are arranged in such a way that they pass one above the other, as is shown in Figure 7.9 for edges attached to a Y-side of a vertex-box.

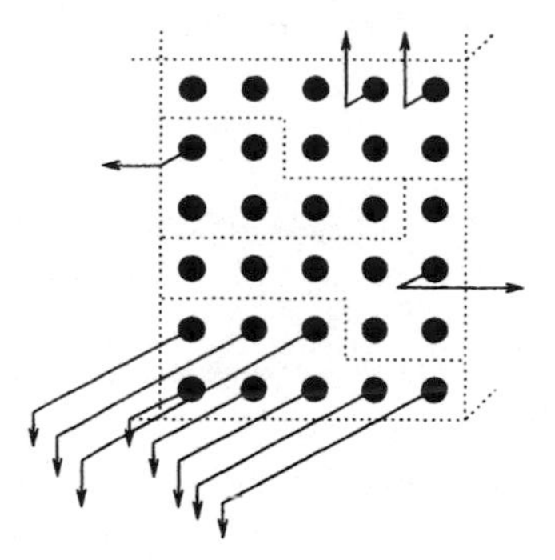

Fig. 7.9. Three-phase method: Edge assignment to ports.

With a suitable vertex placement one can draw every normalized graph, i.e., graphs that are simple, connected and have no nodes of degree 1, with the three-phase method so that the resulting drawing has 2 bends per edge, and either

- lies in an $n \times n \times m$-grid box with vertex contained in a $1 \times 1 \times (d_G(v)/2+1)$-grid box, i.e., the drawing is in the degree-restricted model,
  or
- lies in an grid box of side length $n + 2\sqrt{nm}$ and each vertex is contained in a cube of side-length $2\lceil \sqrt{d_G(v)/2}\, \rceil$, i.e., the drawing is in the cube model.

Multi-dimensional Orthogonal Graph Drawing. Wood (1999a) investigate the general position model for D-dimensional ($D \geq 2$) orthogonal drawing of arbitrary degree graphs. Some of the ideas of his algorithm for 3D orthogonal {point}-drawing, described in section 7.4.1, can be transferred to D dimensions: representing vertices by D-dimensional hyperboxes instead of using points, using a multi-dimensional balanced vertex layout to determine a layout in general position, and using a D-coloring for routing edges. Edge crossings can be eliminated again by swapping the direction of the starting segments of the crossing edges. New for hyperbox-drawings in contrast to {point}-drawings is the problem how to assign ports for each edge route so that no two edges routed on the same face can intersect. This is done by an algorithm similar to the port assignment in the three-phase method in Biedl (1998). The edges of a face are arranged in groups according to their direction of continuation and within a group the edges are assigned to ports in increasing order of the length of the first segment of the route. For a detailed description of this layout-based algorithm and a routing-based algorithm the reader is referred to Wood (1999a).

7.6 Viewpoints

Because of the two-dimensional nature of current output media, a 3D graph drawing must be transformed by projection into a 2D image that can be rendered on a computer screen or paper. In order to keep the loss of information as small as possible, a projection should be used that shows as much as possible of the 3D image. The most important parameter of a projection is the viewpoint, i.e., an observer position together with a direction of view. A starting point for the definition of good viewpoints is the preservation of the abstract graph of a drawing under projection. For instance, Kamada and Kawai (1988) consider viewpoints as good that preserve the shape information of a wire-frame drawing, and exclude viewpoints for which edges appear collinear. Bose et al. (1996) presented several models to describe the quality of a viewpoint. In one model they define good viewpoints as those from which the image of a 3D drawing appears monotonic. These viewpoints are particularly important for viewing hierarchical graphs. They also propose a model that relates the quality of a viewpoint to the number of edge crossings in the 2D projection of a 3D wire-frame drawing. The third model of Bose et al. (1996) preserves the depth-order of a wire-frame drawing, permitting only viewpoints that yield regular projections, i.e., projections under which no three 3D points map to the same 2D point. The end-points of edges count as two points.

Webber (1997, 1998) extend some of these models and define a good viewpoint as one that yields a projection in which no item hides another one and no false incidences are suggested. One distinguishes between four types of such *occlusions*, see Figure 7.10:

vertex-vertex: A pair of vertices from the 3D graph drawing map to a single vertex in the 2D image.

vertex-edge: A vertex of the 3D graph drawing maps to an internal point of an edge in the 2D image.

edge-vertex: Similar to vertex-edge occlusions: the edge appears in front of the vertex.

edge-edge: There are two cases of edge-edge occlusions. A *crossing occlusion* occurs when a pair of 3D edges map to a pair of 2D edges that cross at a single internal point. It is insignificant, since the relational information is not effected by projection. On the other hand, a significant *edge-edge occlusion* occurs when two 3D edges map to a pair of 2D edges that share a continuous sequence of points. The line segment shared by the occluding edges is not to be detected in the projection. Therefore some relational information is lost.

An occlusion point is a viewpoint that generates an occlusion.

Webber (1997, 1998) presents two measures for the quality of a viewpoint under orthographic parallel projection and develops algorithms to find best viewpoints under these models. In the *rotational separation* measure, the

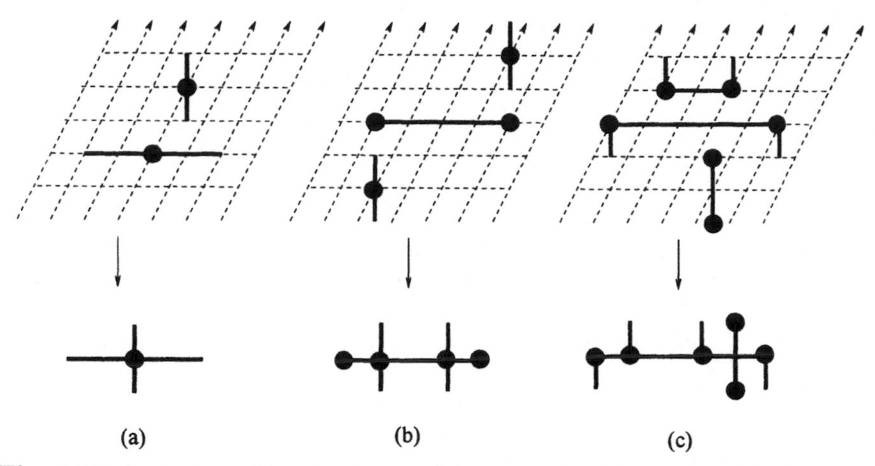

Fig. 7.10. Occlusions: (a) vertex-vertex (b) vertex-edge/edge-vertex (c) edge-edge.

quality of a viewpoint is defined to be the angle measured between viewpoint directions to the nearest occlusion point. The best viewpoints under this model are those for which this angle is maximized. When interactively viewing a 3D graph, this means that the angle by which the user can rotate the drawing without causing an occlusion is maximized. For an example of a best viewpoint for two occlusions see Figure 7.11 (a).

In the *observed separation* measure, the quality of a viewpoint is defined to be the shortest Euclidean distance between the projections of two elements that cause an occlusion for an arbitrary viewpoint. The best viewpoints under this measure are those for which the Euclidean distance is maximized between the closest pair of points in the resulting image. This implies that the level of detail required to allow discrimination between elements in the image is minimized, see Figure 7.11 (b).

The rotation separation measure is more suitable for the interactive display of 3D graph drawings, while the observed separation measure is more suitable for static displays. However, the complexity of the algorithms based on the above measures is too high to be useful in interactive applications. For instance, to determine the rotational separation diagram, which is needed to determine the quality of a given viewpoint, can require $O(|G|^4 \log |G|)$ time (in terms of the size of the graph G, i.e., $|G| = |V| + |E|$) in the worst-case. Webber (1997, 1998) developed two classes of faster algorithms that find approximative good viewpoints, using two distinct heuristic approaches. The first class of algorithms tests trial viewpoints until a given termination criterion is reached. Various methods can be used to choose a trial viewpoint, for example, random selection within a circle, centered at the initial viewpoint. Similarly one has different possibilities for the selection of the termination criterion. Ideally, a new viewpoint should be found in no more time than it takes to render the 3D graph drawing. Thus it is reasonable to terminate

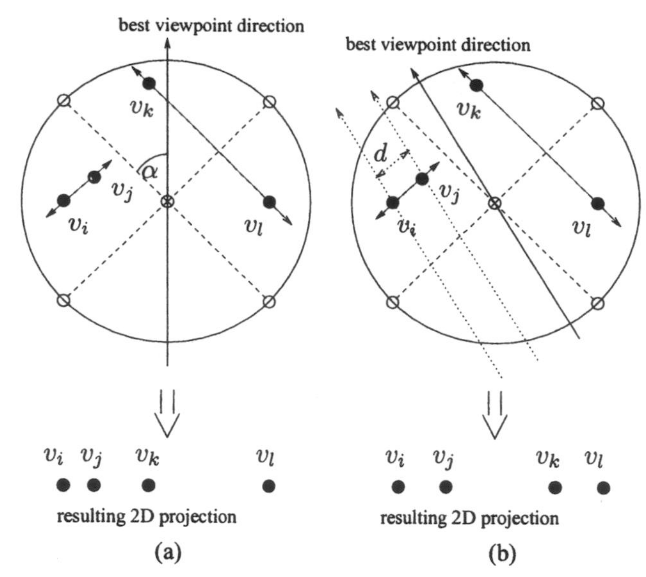

Fig. 7.11. Example of a best viewpoint for two occlusions under: (a) rotation separation measure (b) observed separation measure.

the calculation after this time. Alternatively, one can choose the maximal time that a user is willing to wait for a new viewpoint. By animating this algorithm, the time can be extended. The second class of algorithms is based on the force-directed approach (see Section 7.2). Roughly spoken, the force applied on the current viewpoint depends on the occlusion points.

Experimental results show that both the iterative improvement and force-directed approaches result in useful algorithms for finding reasonably good viewpoints.

8. Drawing Clusters and Hierarchies

Ralf Brockenauer and Sabine Cornelsen

Large graphs such as WWW connection graphs or VLSI schematics cannot be drawn in a readable way by traditional graph drawing techniques. An approach to solve this problem is, for example, fish-eye representation, which allows one to display a small part of the graph enlarged while the graph is shown completely (see e.g. Formella and Keller (1995)). Another way is drawing only a part of the graph. The method presented here is clustering, that is grouping the vertex set.

Apart from the use of clustering to draw large graphs, already clustered graphs occur in applications such as statistics (see e.g. Godehardt (1988)), linguistics (see e.g. Batagelj et al. (1992)) or divide and conquer approaches. To visualize these structures it is also important to find a method of drawing clustered graphs in an understandable way.

In this chapter, Section 8.1 gives an overview of several terms occurring in connection with clustering in the literature. Section 8.2 presents a few main methods of finding good clusters. The following three sections introduce some graph drawing algorithms for clustered graphs. Beginning with the special case of planar drawing methods in Section 8.3, Section 8.4 works on general graphs with a hierarchical structure, called compound graphs, and Section 8.5 deals with arbitrarily clustered graphs using force-directed methods. Finally, Section 8.6 shows a case study for drawing partially known huge graphs.

8.1 Definitions

The usage of the term clustering is not determined uniquely in the literature. In this chapter several terms concerning clustering are defined to give an overview. We only consider vertex clustering, but it is worth mentioning that clustering with respect to edges can be of interest as well. A method to do this can be found in Paulish (1993, Chapter 5).

Clustering of graphs means grouping of vertices into components called clusters. Thus, clustering is related to partitioning the vertex set.

Definition 8.1 (Partition). *A (k-way)* partition *of a set C is a family of subsets $(C_1, \ldots, C_k)$ with*

- $\bigcup_{i=1}^{k} C_i = C$ *and*
- $C_i \cap C_j = \emptyset$ *for* $i \neq j$.

The C_i are called parts. *We refer to a 2-way partition as a* bipartition.

Now, we can define one of the most basic definitions of clustered graphs.

M. Kaufmann and D. Wagner (Eds.): Drawing Graphs, LNCS 2025, pp. 193-227, 2001.

Definition 8.2 (Clustered Graph). *A* clustered graph *is a graph with a partition* $(C_1, \ldots, C_k)$ *on the vertex set. The* C_i *are called* cluster.

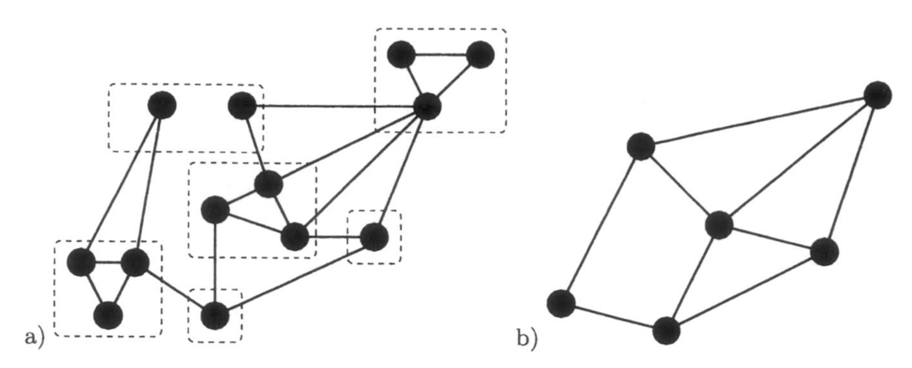

Fig. 8.1. a) A clustered graph. Clusters are framed with rectangles. b) Quotient graph of the clustered graph in a).

Sometimes (e.g. Alpert and Kahng (1995)) the term clustering is only used for large $k \in \theta(n)$ where n is the number of vertices. The expression clustered graph is also used to denote the quotient graph defined by the partition.

Definition 8.3 (Quotient Graph). *For a partition* $(C_1, \ldots, C_k)$ *on the vertex set of a graph* $G = (V, E)$*, the quotient graph* $\mathcal{G} = (\mathcal{V}, \mathcal{E})$ *is defined by* shrinking *each part into a single* node, *i.e.*

- $\mathcal{V} = \{C_1, \ldots, C_k\}$ *and*
- $(C_i, C_j) \in \mathcal{E} \iff i \neq j$ *and* $\exists v \in C_i, w \in C_j \quad (v, w) \in E$.

The elements of $\mathcal{V}$ *are called* nodes.

The quotient graph of the clustered graph in Figure 8.1 a) is shown in Figure 8.1 b). A drawing of the quotient graph is also called a *black-box drawing* (Paulish, 1993) of the corresponding clustered graph. Another way of shrinking subgraphs into a single node is proposed by Lengauer (1990). He calls the construction hierarchical graph. Lengauer used this structure to find faster algorithms, e.g., for planarity testing (Lengauer, 1989) or connectivity testing, for large graphs.

Definition 8.4 (Hierarchical Graph). *A* hierarchical graph *is a finite sequence* $\Gamma = (G_1, \ldots, G_k)$ *of graphs* G_i *called* cells. *The vertex set of the cells is divided into* pins *and* inner vertices. *The set of inner vertices, again, is divided into* terminals *and* non-terminals. *Each non-terminal has a* type. *The type of a non-terminal in* G_i *is a cell* G_j *with* $j < i$. *The degree of a non-terminal* v *is the number of pins of its type* G *and the neighbors of* v *are bijectively associated with the pins of* G.

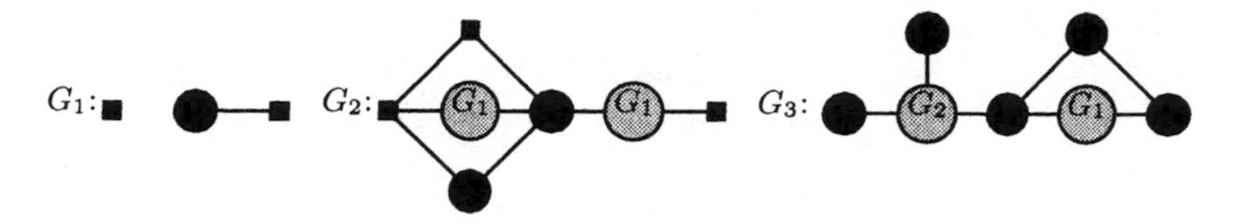

Fig. 8.2. Example for a hierarchical graph. Pins are drawn as rectangles, inner vertices as circles and non-terminals are shaded grey. The bijection between the neighbors of a non-terminal and the associated pins is given via the position in the figure.

An example for a hierarchical graph is shown in Figure 8.2. Note, that the cells need not to be connected. A hierarchical graph represents a graph which is obtained by *expansion*. It is a substitution mechanism that glues pins of a cell to neighbors of non-terminals the type of which is this cell. Note that if $\Gamma = (G_1, \ldots, G_k)$ is a hierarchical graph, so is any *prefix* $\Gamma_i = (G_1, \ldots, G_i)$, $1 \leq i \leq k$.

Definition 8.5 (Expansion). *The* expansion $G(\Gamma)$ *of a hierarchical graph* $\Gamma = (G_1, \ldots, G_k)$ *is obtained recursively as follows:*

$k = 1$: $G(\Gamma) = G_1$

$k > 1$: *For each non-terminal* v *of* G_k, *let* v *be of type* G_j. *Delete* v *and its incident edges and insert a copy of* $G(\Gamma_j)$ *by identifying pins of* $G(\Gamma_j)$ *with their associated vertex in* G_k.

Thus, a hierarchical graph is a clustering of the expansion graph and clusters can include other clusters. This can be illustrated as a tree – the *hierarchy tree*. The expansion and the hierarchy tree of the hierarchical graph in Figure 8.2 are shown in Figure 8.3.

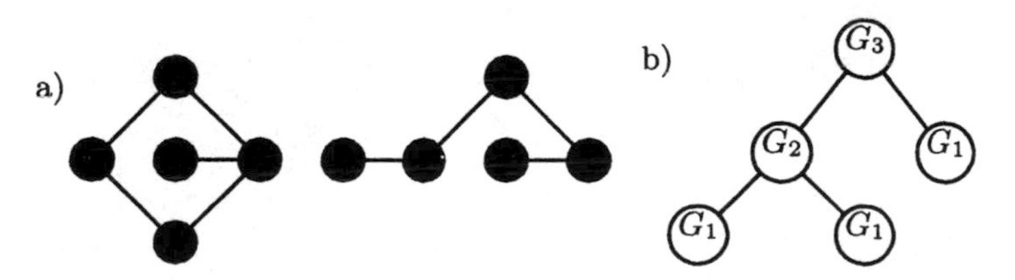

Fig. 8.3. Example for the a) expansion and b) inclusion tree of the hierarchical graph in Figure 8.2.

A quite similar concept introduced by Feng et al. (1995) are hierarchical clustered graphs.

Definition 8.6 (Hierarchical Clustered Graphs). Hierarchical clustered graphs $C = (G, T)$ *consist of a graph* $G = (V, E)$ *and a rooted tree*

T such that the leaves of T are exactly V. Vertices of T are called nodes. *Each node ν of T represents the* cluster $V(\nu)$ *of leaves in the subtree of T rooted at ν. T is called the* inclusion tree *of C. An edge e is said to be* incident *to a cluster $V(\nu)$, if $|e \cap V(\nu)| = 1$.*

An example of a hierarchical clustered graph is shown in Figure 8.4 a). Algorithms for drawing planar hierarchical clustered graphs are presented in Section 8.3. A way of generalization is to allow clusters to intersect. Combined with a hierarchical structure this leads to compound graphs presented by Sugiyama and Misue (1991).

Definition 8.7 (Compound Graphs). *A* compound (directed) graph *is a triple $D = (V, E, I)$ such that $D_a = (V, E)$ is a (directed) graph and $D_c = (V, I)$ is a directed graph. The elements of E are called* adjacency edges, *those of I* inclusion edges.

Thus, $(v, w) \in I$ means that v includes w. Of course, this interpretation only makes sense if the directed graph D_c is acyclic. An example of a compound graph is shown in Figure 8.4 b). In Section 8.4 a drawing algorithm for compound graphs is given in the special case where D_c is a rooted tree and D_a is a directed graph, such that no vertex is adjacent to an ancestor. A hierarchical clustered graph can be seen as a compound graph where D_c is a rooted tree and adjacency edges are only incident to leaves. Note the different meaning of V.

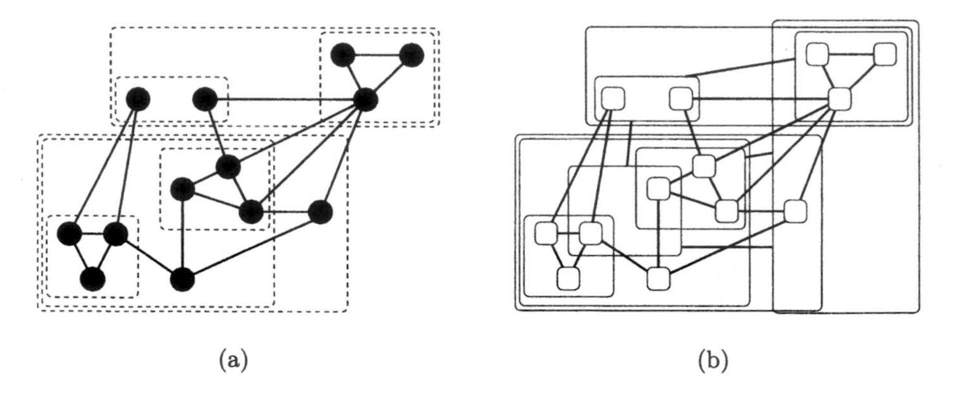

(a) (b)

Fig. 8.4. a) A hierarchical clustered graph. The inclusion tree is drawn by the inclusion representation. b) A compound graph. Inclusion edges are drawn as including rectangles.

8.2 Clustering Methods

There is a large number of clustering approaches. This section will only mention some of them to give an idea of the main methods. A good overview is given in Alpert and Kahng (1995) or Jain and Dubes (1988). The approaches introduced in this section that are not cited separately can be found there as well. Some partitioning methods concerning VLSI are also summarized in Lengauer (1990).

Clustering is a type of classification. This classification can be *extrinsic* or *intrinsic*. Extrinsic classification uses category labels on the objects and clusters are defined by these categories. For example, take trade relation between world wide spread companies and countries as clusters. On the other hand intrinsic classification is only based on the structure of the graph. In the following discussion, we consider intrinsic classification.

There are two main goals in clustering graphs. The first, which has applications, for example, in VLSI, parallel computing and divide-and-conquer algorithms, is to partition a graph into clusters of about the same size and with as few edges connecting the clusters as possible. The second one, which is a method used in statistical applications, is to explore the structure of the data. Thus, the number of clusters is not fixed.

8.2.1 k-Way Partition

The first goal can be formalized as the *Min-Cut k-Way Partition* (cf. Lengauer (1990) p. 253). In the course of the following passage, let $G = (V, E)$ be a graph with vertex-weight $c : V \to \mathrm{N}$ and edge-weight $w : E \to \mathrm{N}$.

Definition 8.8 (Min-Cut k-Way Partition). *Given a fixed $k \in \mathrm{N}$ and $b(i), B(i) \in \mathrm{N}$ for $i = 1, \ldots, k$, find among all k-way-partitions $(C_1, \ldots, C_k)$ of V which satisfy $b(i) \le c(C_i) \le B(i)$ for all $i = 1, \ldots, k$ one that minimizes the weight of the partition*

$$w(C_1, \ldots, C_k) = \frac{1}{2} \sum_{i=1}^{k} \sum_{\substack{e \in E \\ |e \cap C_i| = 1}} w(e) .$$

Exact cluster size balance is achieved by setting $b(i) = c(V)/k - \epsilon$ and $B(i) = c(V)/k + \epsilon$, where $\epsilon > 0$ may be necessary to obtain a solution at all.

Move-Based Approaches. Unfortunately, the multiway partition problem is $\mathcal{NP}$-complete even in the special case of bipartition. A classical good graph bipartitioning heuristic was introduced by Kernighan and Lin (1970). A natural local search method for solving this problem is to start with an initial bipartition and to exchange pairs of vertices across the cut, if doing so improves the cut-size. To reduce the danger of being trapped in local minima, Kerninghan and Lin modified the search, proceeding in a series of passes.

During each pass of the algorithm, every vertex moves exactly once. At the beginning of a pass each vertex is unlocked. Iteratively the pair of unlocked vertices with the highest gain is swapped, where the *gain* of vertices $v_1 \in C_1, v_2 \in C_2$ is defined by

$$w(C_1, C_2) - w((C_1 \cup \{v_2\}) \setminus \{v_1\}, (C_2 \cup \{v_1\}) \setminus \{v_2\})$$

that is the decrease in cut-weight that results from the pair swap. Then, both swapped vertices are locked. The swapping process is iterated until all vertices become locked. The bipartition with the lowest cut-weight observed during the pass is the initial bipartition for the next pass. The algorithm terminates when a pass fails to find a solution with lower weight than its initial bipartition.

Maintaining a sorted list of gains, the complexity of this algorithm is in $O(n^2 \log n)$. Fiduccia and Mattheyses modified the algorithm of Kerninghan and Lin to permit an $O(|E|)$ implementation. The main difference is that a new bipartition is derived by moving a single vertex either from C_1 to C_2 or from C_2 to C_1 instead of exchanging two of them. Therefore, the algorithm must violate the exact cluster size balance constraint. The solution is permitted to deviate from an exact bipartition by the size of the largest vertex. The algorithm of Fiduccia and Mattheyses can also be extended to $k > 2$.

There are several functions combining cluster size balance and minimization of cut weights within a single objective. One of them is the *Ratio Cut Partition* proposed by Wei and Cheng (1991) for $k = 2$. Their approach has several generalizations for arbitrary k. One of them is presented by Roxborough and Sen (1997).

Definition 8.9 (Ratio Cut Partition). *Find among all k-way-partitions $(C_1, \ldots, C_k)$ of V one that minimizes*

$$R(C_1, \ldots, C_k) = \frac{w(C_1, \ldots, C_k)}{c(C_1) \cdot \ldots \cdot c(C_k)} .$$

Finding a ratio cut partition is also $\mathcal{NP}$-complete. In Wei and Cheng (1991) a heuristic based on the algorithm of Fiduccia and Mattheyses is proposed to find good bipartitions. In Figure 8.5 an example is shown where the ratio cut partition is a much more intuitive one than the exact cluster size balanced min-cut bipartition.

Spectral Methods. For a graph $G = (\{v_1, \ldots, v_n\}, E)$ with enumerated vertex set, a partitioning solution can be represented in terms of vectors and matrices.

Definition 8.10 (Characteristic Vector). *For a graph G with a given a k-way partition $(C_1, \ldots, C_k)$ the* characteristic vector *for cluster C_h is the n-dimensional vector $\mathbf{x}_h = (x_{1h}, \ldots, x_{nh}) \in \{0,1\}^n$ with $x_{ih} = 1$ if and only if $v_i \in C_h$. The $n \times k$ matrix X with column h equal to $\mathbf{x}_h$ is the* assignment matrix *of the partition.*

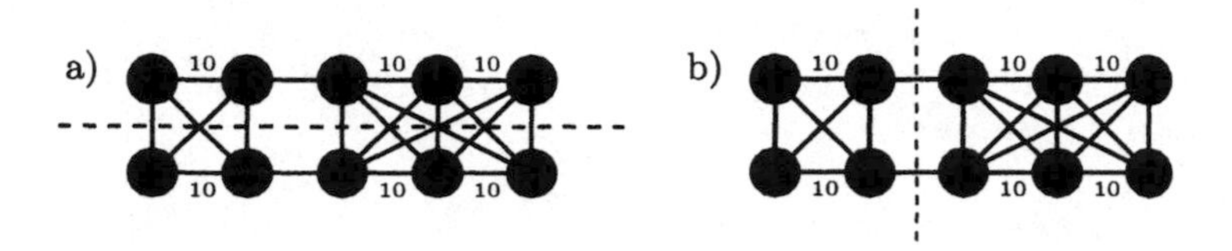

Fig. 8.5. a) Exact cluster size balanced min-cut bipartition and b) ratio cut bipartition.

Let $A = (a_{ij})$ be the adjacency matrix of an undirected graph G and $D = (d_{ij})$ the degree matrix, that is $d_{ij} = \deg(v_i)$ for $i = j$ and zero otherwise. The *Laplacian matrix* of G is defined as $L = D - A$. Since L is a symmetric matrix

- all eigenvalues of L are real and
- there is a basis of the n-dimensional space of mutually orthogonal eigenvectors of L.

Since $\mathbf{x}^T L\mathbf{x} = \frac{1}{2}\sum\sum a_{ij}(x_i - x_j)^2 \geq 0$ for all $\mathbf{x} \in \mathrm{R}^n$, matrix L is positive semi-definite and thus, all eigenvalues are non-negative. Furthermore, the columns of L add to zero and we get

- the smallest eigenvalue of L is 0 with corresponding eigenvector $(1, \ldots, 1)$. (The multiplicity of 0 as an eigenvalue of L is equal to the number of connected components of G.)

Now, let $\mathbf{x}$ denote the characteristic vector for one part of a given bipartition (C_1, C_2) in a connected undirected graph, then we can express the value of the cut defined by the bipartition by

$$w(C_1, C_2) = \frac{1}{2}\sum_{i=1}^{n}\sum_{j=1}^{n} a_{ij}(x_i - x_j)^2 = \mathbf{x}^T L\mathbf{x}\,.$$

Allowing non-discrete solutions, a normalized eigenvector μ to the smallest positive eigenvalue minimizes $\mathbf{x}^T L\mathbf{x}$ among all normalized $\mathbf{x}$. Although a non-discrete solution for $\mathbf{x}$ is meaningless, this result suggests heuristically finding the discrete solution closest to μ. Given cluster size constraints $|C_1| = m_1$ and $|C_2| = m_2$, the closest discrete solution is obtained by placing the m_1 vertices with the highest coordinates of μ in C_1 and the rest in C_2 or vice versa.

This approach is known as *spectral bipartitioning*. Unfortunately, it can be arbitrarily worse than optimal, as illustrated by the following example: Let G be the graph shown in Figure 8.6 in which two $n/4$-cliques are connected, each by a single edge, to an $n/2$-clique. Spectral bipartition will cut $K_{n/2}$ into equal halves, cutting $(n/4)^2$ edges. But the optimal cluster size balanced bipartition has weight 2.

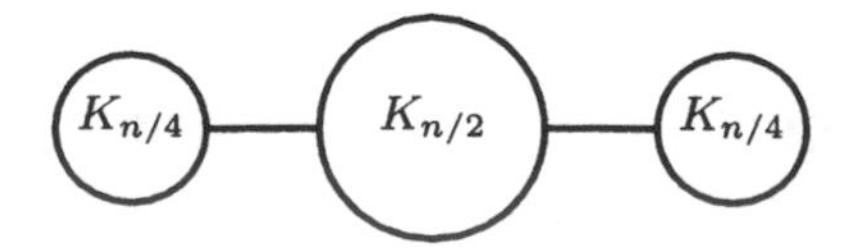

Fig. 8.6. Example of bad spectral bipartition.

To improve the result, Frankle and Karp proposed to find an characteristic vector that is close to a linear combination of the eigenvectors of the d smallest eigenvalues. A summary of spectral clustering including an extension to $k > 2$ is given in Alpert and Kahng (1995) Section 4.

8.2.2 Structural Clustering

We now turn to the second goal of graph clustering which tries to identify certain intuitive properties ρ such as cliques or connectivity. In contrary to the last section, the number of clusters is not given. One important algorithm used to solve this problem is *agglomerative clustering*, which starts with the n-way partition and constructs iteratively a k-way partition from the $k+1$-way partition.

For a complete graph $G = (V, E)$ with edge weight $w : E \to \mathrm{R}$ and no two edges having the same weight, the iterative step is defined in more detail, for example, by Hubert (see Jain and Dubes (1988) p. 63). For a given *threshold* d let $G_d = (V, \{e \in E; w(e) \leq d\})$. For a pair of clusters (C_r, C_s) in the $k+1$-way partition, define

$$Q_\rho(C_r, C_s) = \min\{d; \text{the subgraph induced by } C_r \cup C_s \text{ in } G_d \text{ is either complete or has property } \rho\}\,.$$

Merge cluster C_p and C_q if

$$Q_\rho(C_p, C_q) = \min_{r,s} Q_\rho(C_r, C_s)\,.$$

Each specification of property ρ defines a new clustering method. Every cluster must at least be connected. Some suitable graph properties $\rho(k)$ with integer parameter k are listed below:

k-edge-connectivity: All pairs of vertices are joined by at least k edge disjoined paths. 1-edge-connectivity is also called *single-link* property and n-edge-connectivity is called *complete-link* property.

k-vertex-connectivity: All pairs of vertices are joined by at least k vertex disjoined paths.

vertex degree k: A connected graph such that each vertex has at least degree k.

diameter k: All pairs of vertices are joined by a path of length at most k.

An agglomerative algorithm constructs a hierarchical clustering. The above mentioned algorithm is serial. To minimize the height of the inclusion tree, an alternative strategy is to find many good clusters to merge, then perform all merges simultaneously.

A special drawing of the inclusion tree of a hierarchical clustered graph, that reveals the order in which clusters are merged, is called *dendrogram*. Cutting a dendrogram horizontally creates a partition of the vertex set. Figure 8.7 gives two examples on a weighted K_5, one for single- and the other for complete-link property.

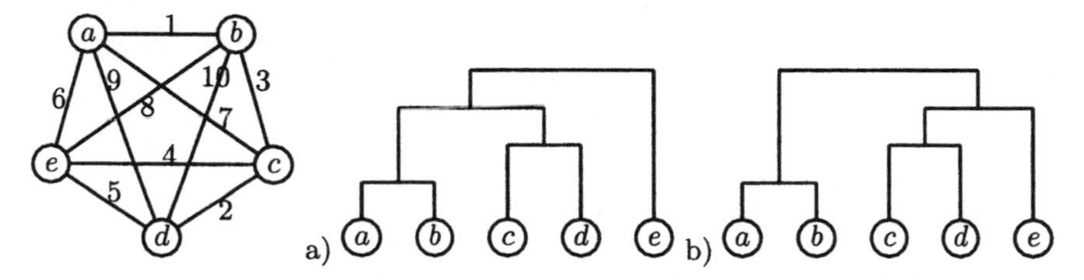

Fig. 8.7. Two dendrograms on a weighted K_5. a) single link, b) complete link.

8.2.3 Other Approaches

There are several other methods for clustering graphs. Duncan et al. (1998) presented a method to partition an already laid out graph along horizontal, vertical and diagonal lines. The approach of Sablowski and Frick (1996) is based on the successive identification of patterns in the graph. Other approaches consider constraints, e.g., some vertices should belong to different clusters. Nakano et al. (1997) presented a work on this topic.

Most of the clustering methods presented so far produce connected clusters. But it is important to know, that unconnected clusters might occur in practice. This can be seen immediately by considering extrinsic clustering, because it can be defined arbitrarily. But there are also some intrinsic classifications which produce unconnected components. They are, for example, used in social network analysis. The following two classifications presented by Wasserman and Faust (1994) are of this type.

Definition 8.11 (Structural Equivalence). *Two vertices v, w of a graph (V, E) are* structural equivalent *if and only if they have the same neighborhood, that is if and only if for all $u \in V$ holds*

$$(v,u) \in E \Longleftrightarrow (w,u) \in E$$
$$(u,v) \in E \Longleftrightarrow (u,w) \in E.$$

Thus, two vertices are structural equivalent if their rows and columns in the adjacency matrix are identical.

Definition 8.12 (Regular Equivalence). *Two vertices v and w of a graph (V, E) are* regular equivalent *($v \cong w$) only if they have equivalent neighborhoods, that is only if for all $u \in V$ holds*

$$(v, u) \in E \Longrightarrow \exists u' \in V \quad u' \cong u \ \wedge \ (w, u') \in E$$
$$(u, v) \in E \Longrightarrow \exists u' \in V \quad u' \cong u \ \wedge \ (u', w) \in E.$$

Structural equivalence is a special case of regular equivalence. Regular equivalence partitioning is not uniquely determined. The partition with the fewest equivalence classes that is consistent with the definition of regular equivalence is called the *maximal regular equivalence.* For example, in a tree where all leaves have the same height, taking the levels as equivalence classes yields the maximal regular equivalence on that tree.

8.3 Planar Drawings of Hierarchical Clustered Graphs

In her PhD thesis, Feng (1997) presented a characterization for planar connected hierarchical clustered graphs and introduced some algorithms for drawing them.

Definition 8.13 (Connected Hierarchical Clustered Graphs). *A hierarchical clustered graph $C = (G, T)$ is* connected, *if each cluster induces a connected subgraph of G.*

Definition 8.14 (Drawing of Hierarchical Clustered Graph). *A* drawing *$\mathcal{D}$ of a hierarchical clustered graph $C = (G, T)$ includes the drawing of the underlying graph G and of the inclusion tree T in the plane. Each vertex of G is represented as a point and each edge $\{v, w\}$ as a simple curve between $\mathcal{D}(v)$ and $\mathcal{D}(w)$. Each non-leaf node ν of T is drawn as a simple closed region $\mathcal{D}(\nu)$ bounded by a simple closed curve such that*

- *$\mathcal{D}(\mu) \subset \mathcal{D}(\nu)$ for all descendents μ of ν.*
- *$\mathcal{D}(\mu) \cap \mathcal{D}(\nu) = \phi$ if μ is neither a descendent nor an ancestor of ν.*
- *$\mathcal{D}(e) \subset \mathcal{D}(\nu)$ for all edges e of G with $e \subset V(\nu)$.*
- *$\mathcal{D}(e) \cap \mathcal{D}(\nu)$ is one point if $|e \cap V(\nu)| = 1$.*

The drawing of an edge e and a region $\mathcal{D}(\nu)$ have an *edge-region-crossing,* if $e \cap V(\nu) = \phi$ but $\mathcal{D}(e) \cap \mathcal{D}(\nu) \neq \phi$. Drawings where this occurs, are allowed, but they are not c-planar.

Definition 8.15 (c-Planar). *A drawing of a hierarchical clustered graph is* c-planar *(compound planar), if there are no crossing edges and no edge-region-crossings.*

For example, the drawing of the graph shown in Figure 8.4 on page 196 is c-planar.

Theorem 8.16 (Characterization of c-Planar Graphs). *A connected hierarchical clustered graph $C = (G, T)$ is c-planar if and only if there exists a planar drawing of G, such that for each node ν of T all vertices of $V - V(\nu)$ are in the outer face of the drawing of $G(\nu)$.*

Proof. Consider a clustered graph $C = (G, T)$ with a c-planar drawing $\mathcal{D}$. Suppose there is a node ν of T and a vertex $v \in V - V(\nu)$ which is not drawn in the outer face of $\mathcal{D}(G(\nu))$. Hence, any simple region that contains $\mathcal{D}(G(\nu))$ must also contain v, contradicting the c-planarity properties.

Consider now a planar drawing of G, such that for each node ν in T, $G - G(\nu)$ is drawn in the outer face of the drawing of $G(\nu)$. It remains to add cluster boundaries. Since $G(\nu)$ is connected for each ν in T, the outer face is bounded by a – not necessarily simple – cycle. Thus, cluster boundaries are constructed recursively, following T from bottom to top, along their external facial cycle.

Based on Theorem 8.16, Feng et al. (1995) developed an algorithm which constructs a *c-planar embedding* of a hierarchical clustered graph, that is a circular ordering of the incident edges ordered around each cluster. The algorithm applies the *PQ*-tree technique presented by Booth and Lueker (1976) and modified by Chiba et al. (1985) and takes time $O(n^2)$ under the additional condition, that each non-leaf node of T has at least two children.

It tries to embed the subgraph $G(\nu)$ induced by each cluster $V(\nu)$ recursively, following T from bottom to top. To guarantee the conditions in Theorem 8.16 for each *virtual edge* $e \in E$, that is an edge such that $e \cap V(\nu) = \{v_e\}$ has cardinality one, an additional vertex w_e and an edge $\{v_e, w_e\}$ is added to $G(\nu)$. Further one of the additional vertices is connected to all other additional vertices (see Figure 8.8).

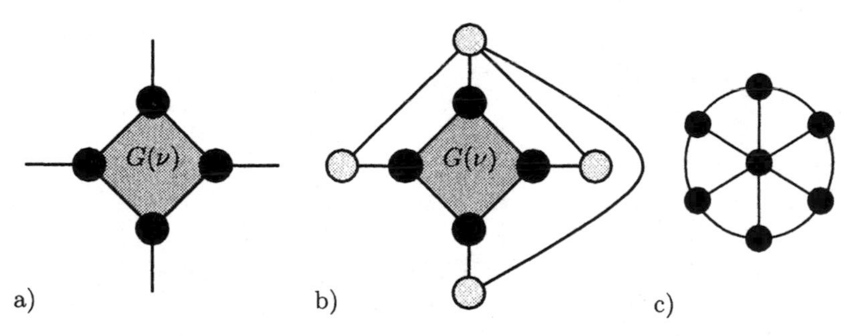

Fig. 8.8. a) Graph $G(\nu)$ with virtual edges is transformed into graph b). Additional vertices are shaded light grey. c) A wheel graph with 6 vertices on the rim.

To determine whether the embeddings of the children of a cluster ν can be combined to an embedding of $G(\nu)$, the graph $G(\mu)$ for each child μ of ν is replaced by a *representative graph* which is more or less constructed by replacing 2-connected components in $G(\mu)$ by *wheel graphs.* A wheel graph consists of a vertex called *hub* and a simple cycle called *rim*, such that the hub is connected to every vertex on the rim (see Figure 8.8 c)). They showed that a representative graph with given ordering of the virtual edges can always be embedded in such a way that the rims are in the outer face without changing the ordering of the virtual edges.

8.3.1 Straight-Line Drawings with Convex Clusters

For a given c-planar embedding of a connected hierarchical clustered graph $C = (G, T)$, Eades et al. (1996a) gave an algorithm to construct a drawing of C such that the edges of G are drawn as straight lines and the regions are convex. This drawing of C can be constructed in time $O(n^2 \log n)$ which is dominated by the time needed for constructing the convex hull of the clusters.[1]

The algorithm works as follows. First, graph G is triangulated. Then an *st*-numbering[2] of the vertices of G is computed such that vertices in the same cluster are numbered consecutively. Such a numbering is called *c-st-numbering*. These numbers are now used as a layer assignment – thus, there is one vertex per layer – and an algorithm for constructing planar straight-line drawings of layered graphs, which is also presented by Eades et al. (1996a), is applied to draw the graph. Since each cluster has consecutive layers, the convex hull of its vertices satisfies all conditions of a region in a c-planar drawing.

Apart from the construction of a planar straight-line drawing of layered graphs, the critical part of this method is the construction of the c-*st*-numbering. To ensure that the vertices of the same cluster are numbered consecutively, Eades, Feng and Lin used a top-down approach, ordering the children of the root of T first and thus having a lexicographical numbering on the clusters. To compute an order of the child cluster of ν, an *auxiliary graph* $F(\nu)$ is computed from $G(\nu)$ by shrinking each child cluster to a vertex. If ν is the root of T, an edge $\{s, t\}$ not belonging to any child cluster of T is chosen and an *st*-numbering is computed.

If ν is not the root, vertices σ and τ are added to the auxiliary graph $F(\nu)$. For a virtual edge $\{v, w\}$ with $v \in V(\nu)$, let μ be the lowest ancestor of ν with $w \in V(\mu)$ and k the number of the child cluster of μ which w belongs to. If $g(\nu) < k$ then edge $\{v, \tau\}$ is added to $F(\nu)$. Otherwise, edge $\{\sigma, v\}$ is added. In case $s \in V(\nu)$, the vertex representing the cluster containing s is set to be σ; similarly for t and τ. Now a $\sigma\tau$-numbering is computed.

The only thing missing now is that the auxiliary graphs are 2-connected. This is a consequence of the following lemma.

---

[1] In Eades et al. (1998), the time complexity is improved to $O(n^2)$.

[2] For the definition of *st*-numbering see Definition 2.9 on page 27.

Lemma 8.17. *For every non-root node ν of the inclusion tree of a connected c-planar hierarchical graph $C = (G, T)$ with triangulated G, the subgraph of G induced by $V \setminus V(\nu)$ is connected.*

Proof. Suppose that the subgraph of G induced by $V \setminus V(\nu)$ has $k \geq 2$ components denoted by $F_1, \ldots, F_k$. Since G is triangulated, it has a unique planar embedding. By Theorem 8.16, all vertices of $G - G(\nu)$ are in the same face of $G(\nu)$. Since G is connected, there is a face f of G such that its boundary contains an edge connecting $G(\nu)$ and F_i and also an edge connecting $G(\nu)$ and F_j for a $j \neq i$. Because G is triangulated, f also contains an edge connecting F_i and F_j, a contradiction.

Unfortunately, there are hierarchical clustered graphs such that any c-planar straight-line convex drawing strategy results in poor area requirement and angular resolution. Eades, Feng and Lin gave a family $C_n = (G_n, T_n)$ of clustered graphs which require area $\Omega(2^n)$ and have angles between two edges incident to a vertex in $O(1/n)$. A sketch of the recursive construction of the underlying graphs G_n can be seen in Figure 8.9. The root of T_n is adjacent to two nodes A and B, which have children $a_1, \ldots, a_n$ and $b_1, \ldots, b_n$, respectively.

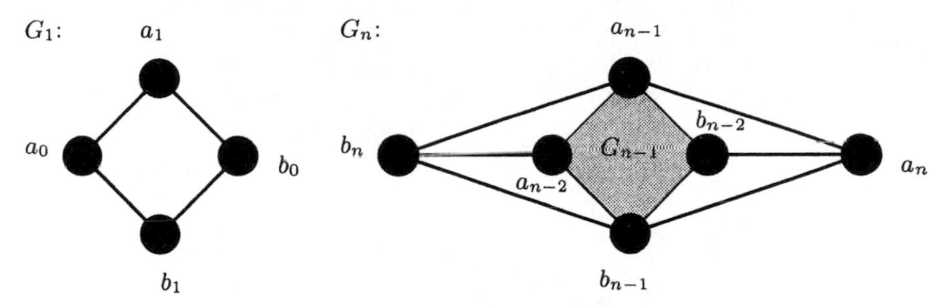

Fig. 8.9. Example for poor vertex and angular resolution.

8.3.2 Orthogonal Drawings with Rectangular Clusters

Eades and Feng (1997) gave an algorithm to construct a drawing of a hierarchical clustered graph $C = (G, T)$ with fixed embedding and degree at most 4 such that G is drawn orthogonal and the regions are rectangles. Using the constrained visibility representation, the algorithm takes time $O(n^2)$, the drawing space $O(n^2)$, and each edge has at most 3 bends.[3]

The algorithm works as follows: First triangulate G and compute a c-*st*-numbering as constructed in the previous section. Orient the edges from lower

[3] In Eades et al. (1999), the time complexity is improved to $O(n)$.

to higher numbers. Construct a directed graph G' from the oriented triangulation of G by adding four additional dummy vertices and replacing virtual edges for each cluster to ensure rectangular regions. Construct a constraint visibility representation of G' for a suitable set of non-intersecting paths. Construct an orthogonal drawing of G from the visibility representation of G' and finally reduce some bends.

How to construct a planar orthogonal drawing from a constraint visibility representation for non-clustered graphs is, for example, explained in Di Battista et al. (1999) Section 4.9 on page 130. A short introduction to this topic is also given in Section 6.4.2. So it remains to give the construction of G' and the additional constraints. Proceeding from the leaves to the root of T, for

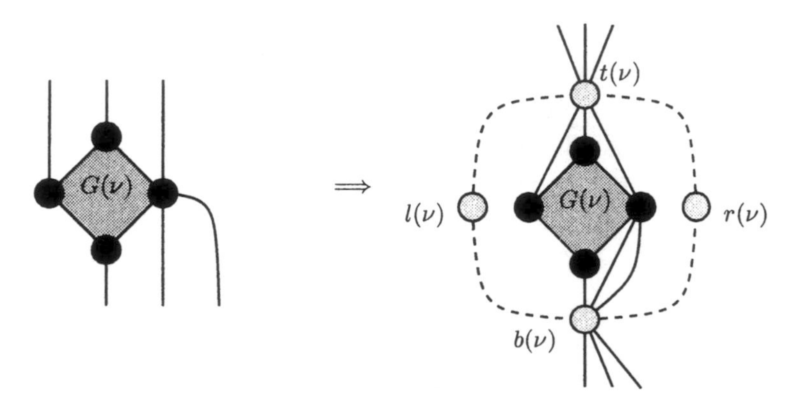

Fig. 8.10. Virtual edges are bunched together.

each non-leaf node ν of T add four dummy vertices denoted by $b(\nu)$ (bottom), $t(\nu)$ (top), $l(\nu)$ (left), and $r(\nu)$ (right) to $G(\nu)$ and split virtual edges (v, w) of $G(\nu)$ by a dummy vertex in the following way (illustrated in Figure 8.10):

- If $v \in V(\nu)$ replace (v, w) by $(v, t(\nu))$ and $(t(\nu), w)$.
- If $w \in V(\nu)$ replace (v, w) by $(v, b(\nu))$ and $(b(\nu), w)$.

Add edges $(b(\nu), r(\nu))$, $(r(\nu), t(\nu))$, $(b(\nu), l(\nu))$ and $(l(\nu), t(\nu))$ to $G(\nu)$.

For a node $\nu \neq s$ on the way from s to the root of T, let μ be the child of ν on this way. If $\mu \neq s$, add edge $(b(\nu), b(\mu))$, else add $(b(\nu), s)$. Similarly, for a node $\nu \neq t$ on the way from t to the root of T, let μ be the child of ν on this way. Add edge $(t(\mu), t(\nu))$ respectively $(t, t(\nu))$. By this construction, G' is a planar st-graph with $O(n)$ vertices and $O(n^2)$ edges.

Now, the alignment requirements in G' for the visibility representation, which is a set of paths ϕ, is specified. For each non-leaf node of T, the set ϕ contains the paths $(b(\nu), l(\nu), t(\nu))$ and $(b(\nu), r(\nu), t(\nu))$. Intercluster edges in G are replaced by paths in G'. These paths are also added to ϕ. Finally, some paths containing edges incident to vertices of G are added to ϕ to avoid unnecessary bends like in the non-clustered version. Thus, ϕ is a set of non-intersecting paths in the sense defined below and a constraint visibility representation can be computed.

Definition 8.18 (Set of Non-Intersecting Paths). *Two paths p_1 and p_2 of a planar graph G with given embedding are said to be* non-intersecting *if they are edge disjoint and there is no vertex v of G with edges e_1, e_2, e_3, and e_4 incident to v in this clockwise order around v, such that e_1 and e_3 are in p_1 and e_2 and e_4 are in p_2. A set of pairwise non-intersecting paths of G is called a* set of non-intersecting paths.

For each non-leaf node ν, the rectangle bounded by the drawing of the corresponding vertices $b(\nu)$ and $t(\nu)$ and the drawing of the paths $(b(\nu), r(\nu), t(\nu))$ and $(b(\nu), l(\nu), t(\nu))$ is defined to be $\mathcal{D}(\nu)$.

Having at most three bends per edge is as good as it gets: Eades and Feng gave a family $C_n = (G_n, T_n)$ of examples for hierarchical clustered graphs such that in every c-planar orthogonal drawing with rectangular clusters, there are at least $O(n)$ edges that bent more than twice. G_n is a sequence

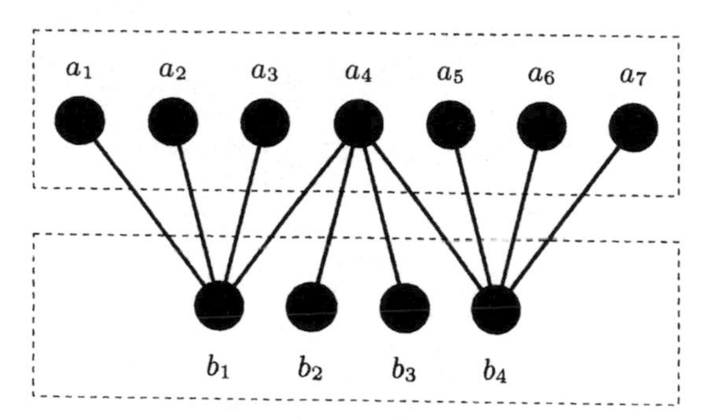

Fig. 8.11. Example for a lot of bends.

of n copies of the graph H shown in Figure 8.11 such that vertex a_7 of a previous copy of H serves as vertex a_1 of the next copy. It is partitioned into two clusters. Cluster A containing the a-vertices and cluster B containing the b-vertices. The embedding is as sketched in the figure. Each copy of H has at least one edge with more than two bends: At least one of the edges $\{a_4, b_1\}$ and $\{a_4, b_4\}$ has two or more bends in the cluster region of A. Suppose it's $\{a_4, b_1\}$. Then at least one of the edges $\{a_4, b_1\}$ and $\{a_1, b_1\}$ has three or more bends. Thus G_n has at least n such edges, but $10n + 1$ vertices.

8.3.3 Multilevel Visualization of Clustered Graphs

Eades and Feng (1996) show a way to represent both the *adjacency* and *inclusion* relations of a *clustered graph* in the same drawing. Here, the inclusion relation is not only just drawn as simple regions containing the drawing of their corresponding vertices, but as a tree structure that also geometrically visualizes this relation. As graphs get larger and larger it is a common strategy to visualize them at multiple abstraction levels. If the graph has a recursive clustering it is a natural approach to take the clustering of the graph as abstraction levels, which provides the possibility to *zoom* in and out within the clustered structure of the graph. The method presented in Eades and Feng (1996) is a three dimensional representation of the clustered graph with each cluster level drawn at a different z-coordinate, and with the inclusion relation drawn as a tree in three dimensions. This kind of representation also keeps track of the abstractions from one level to the next.

Teminology. The *height* of a cluster v, denoted by *height*(v), is defined as the depth of the subtree of T rooted at v.

For a clustered graph, its *view at level* i is a graph $G_i = (V_i, E_i)$ where V_i consists of the set of nodes of height i in T. There is an edge (μ, ν) in E_i if there exists an edge $(u, v) \in E$ where u belongs to cluster μ and v belongs to cluster ν; in other words, edge (μ, ν) is an abstraction of *all* edges between cluster μ and ν.

In a *plane drawing* of a clustered graph, the vertices are drawn as points and edges as curves in the plane as usual. Each cluster $\nu \in T$ is drawn as a simple closed region R that contains the drawing of $G(\nu)$, as defined in Definition 8.14. If a clustered graph has a c-planar representation (Definition 8.15), then it is *c-planar* (Figure 8.12).

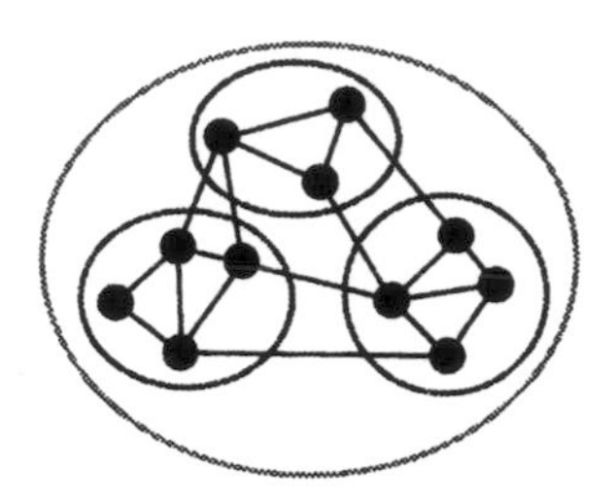

Fig. 8.12. A plane drawing of a clustered graph; the graph shown here is *c-planar*.

Multilevel Drawings. A *multilevel drawing* (Fig. 8.13) of a clustered graph $C = (G, T)$ consists of:

- a sequence of plane drawings of representations from the leaf level (level 0) to the root level of T, where the view of level i is drawn on the plane $z = i$.
- A three dimensional representation of T, with each node $\nu \in T$ of height i drawn as a point on the plane $z = i$, and within the region of ν in the drawing of a view of that level.

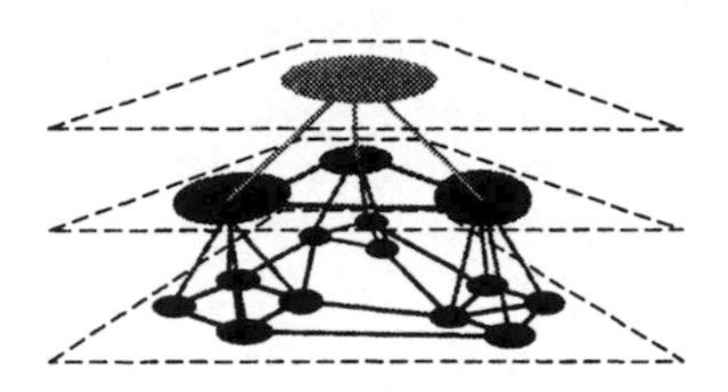

Fig. 8.13. Multilevel drawing.

For the plane drawing of a clustered graph several already presented automatic drawing algorithms can be used, such as the *straight-line convex* or *orthogonal rectangular* drawing methods (Section 8.3.1 and Section 8.3.2, see also Feng (1997)).

Here, the focus is only on the construction of the multilevel drawing which consists of two steps, the construction of view drawings for each level i, $i = 0, \ldots, \mathit{height}(\text{root of } T)$, and the drawing of the inclusion tree T.

1. View drawing for each level:

1. For each level i, $i = 0, \ldots, \mathit{height}(\text{root of } T)$, construct a plane drawing and translate it to the plane $z = i$, starting at the leaf level.
2. An edge (μ, ν) in level $i+1$ is an abstraction for all edges between cluster μ and ν in level i. Choose one of these edges (u, v) between cluster μ and ν as a representative edge, and derive the drawing of edge (μ, ν) in the view of level $i + 1$ from the drawing of edge (u, v) in the view of level i. In the two dimensional plane, cluster μ and ν are drawn as simple closed regions $R(\mu)$ and $R(\nu)$; the drawing of edge (u, v) intersects the boundaries of these regions at points x and y (Figure 8.14). To construct the drawing of edge (μ, ν) in the view of level $i + 1$, use the segment between x and y and translate it to the plane $z = i + 1$.

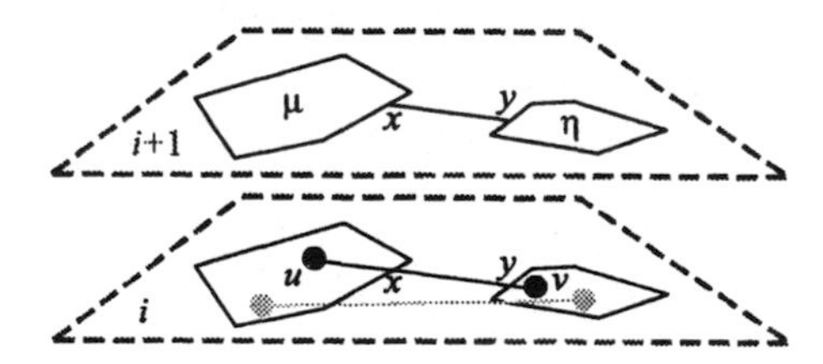

Fig. 8.14. Deriving a drawing for abstraction edges.

2. 3D drawing of the inclusion tree T:

General results on 3D drawings can be found in Chapter 7. To find a three dimensional drawing of the inclusion tree, every node $\mu \in T$ with height i has to be placed on the plane $z = i$ and it also must be positioned in the corresponding region $R(\mu)$. This is achieved as follows:

1. Compute the position of the nodes of the inclusion tree recursively:
   Level $i = 0$*:* (leaf level)
   Take the positions as computed in the plane drawing of level 0;
   Level $i > 0$*:*
   Let the position for node $\mu \in T$ be the average of all xy-coordinates of its children from level $i - 1$;
2. Route the inclusion edges as straight line segments between the corresponding nodes.

8.4 Hierarchical Representation of Compound Graphs

Sugiyama and Misue (1991) introduce an extension to the class of *clustered graphs*, the class of *compound digraphs*, and they also present an algorithm to produce an automatic hierarchical representation of compound digraphs. The main difference between these two graph classes is the use of the *inclusion relation.* A clustered graph is a graph with a partition of its vertex set into clusters. In the representation of a clustered graph, the cluster regions are drawn as simple closed regions that contain the drawing of all the vertices belonging to that cluster; the inclusion relation is restricted to these cluster regions, and there are no edges connecting them. In a compound digraph, the inclusion relation as well as the adjacency relation is defined on the *same* set of vertices.

Definition 8.19 (Compound Graph). *An* inclusion digraph *is a pair $D_c = (V, E)$ where E is a finite set of* inclusion edges *whose element $(u, v) \in E$ means that u includes v (Figure 8.15 (a)).*

An adjacency digraph *is a pair $D_a = (V, I)$ where F is a finite set of* adjacency edges *whose element $(u, v) \in E$ means that u is adjacent to v (Figure 8.15 (b)).*

A compound digraph *is defined as a triple $D = (V, E, I)$ obtained by compounding these two digraphs (Figure 8.15 (c)).*

In Sugiyama and Misue (1991), the inclusion digraph D_c is required to be a rooted tree and is also called the *inclusion tree* of D. The depth of a vertex $v \in V$ is the number of vertices on the path between v and the *root* of D_c and is denoted by $depth(v)$, with $depth(root) = 1$, where *root* denotes the root of D_c. The parent of a vertex $v \in D_c$ is denoted by $Parent(v)$.

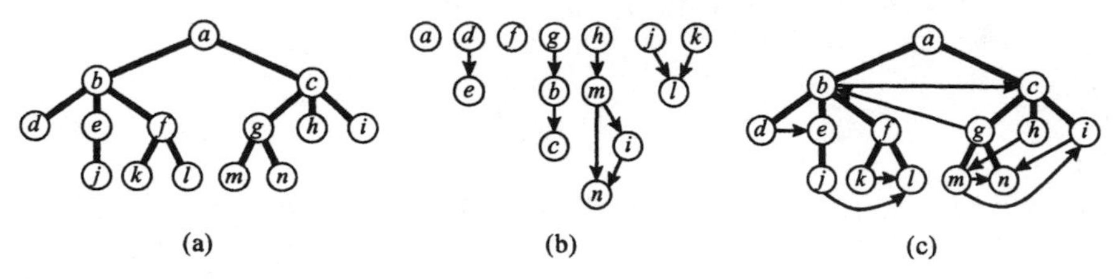

Fig. 8.15. (a) Inclusion tree D_c (b) Adjacency graph D_a (c) The compound digraph D obtained from (a) and (b).

Figure 8.16 shows a representation of the compound digraph and its compund levels of Figure 8.15 (c). The adjacency edges drawn with solid lines have downward orientation and edges drawn with broken lines upward. The vertices of the compound digraph are drawn as rectangles. The inclusion relation $(u, v) \in E$ is realized as the rectangle representing vertex u is inside the rectangle representing vertex v. The conventions for the drawing of compound digraphs are specified more precisely below.

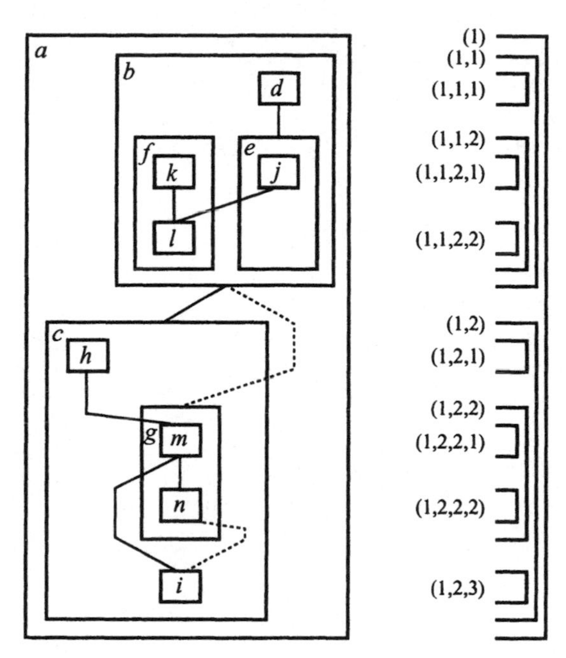

Fig. 8.16. The representation of a compound digraph from Figure 8.15 (c) and its compound levels.

8.4.1 Conventions

Drawing Conventions.

C1 : *Vertex Shape:*
A vertex is drawn as a rectangle with horizontal and vertical sides.

C2 : *Inclusion:*
An inclusion edge (u, v) is drawn in such a way that the rectangle corresponding to u includes geometrically the rectangle corresponding to v.

C3 : *Hierarchy:*
Vertices are laid out hierarchically in terms of both inclusive and adjacent relations on parallel-nested horizontal bands, called *compound levels.*

C4 : *Down-Arrow:*
An adjacency edge (u, v) is drawn as a downward arrow with possible bends, originating from the bottom side of the rectangle corresponding to u and terminating on the top side of the rectangle corresponding to v.

Drawing Rules. To enhance the readability and the aesthetic of the drawing, the following objectives should be satisfied as much as possible.

R1 : *Closeness:*
Connected vertices are laid out as close as possible to each other.

R2 : *Edge Crossings:*
The number of crossings between adjacency edges is reduced as much as possible.

R3 : *Edge-Rectangle Crossings:*
The number of crossings between adjacency edges and the vertex rectangles is reduced as much as possible.

R4 : *Line-Straightness:*
One-span adjacency edges, i.e., edges between adjacent levels, are drawn as straight lines, whereas long span adjacency edges are drawn as polygonal lines with as few bends as possible.

R5 : *Balancing:*
Edges originating from and ending at a vertex rectangle are laid out in a balanced form.

The above rules specify topological and metrical layout properties; their priority is top-down.

8.4.2 The Layout Algorithm

The algorithm consists of the following four steps that are similar to those of an algorithm by Sugiyama et al. (1981) and Sugiyama (1987) for the layered representation of a general digraph. Due to the complex structure of a compound digraph, these steps must be modified and extended, in particular to display the inclusion relations. A brief description of the four steps follows, but the focus will be mainly on the ideas and the modifications that have to be done to the original Sugiyama algorithm (see Chapter 5). A detailed description of this algorithm is given in Sugiyama and Misue (1991).

Step 1: *Hierarchization*

Due to the two kinds of relations existing in a compound graph, this step is different to the one in Sugiyama's original algorithm for general graphs, so it will be explained in more detail.

A. Compound Level Assignment

In this step, the vertices of the compound digraph are assigned to *compound levels* to satisfy the drawing conventions. This level assignment places the vertices on parallel-nested horizontal bands. As shown in Figure 8.16, the compound levels can be expressed by the assignment of a sequence of positive integers to every vertex $v \in V$.

Let $\Sigma = 1, 2, 3, \ldots$ and $\Sigma^+ = \Sigma^1 \cup \Sigma^2 \cup \Sigma^3 \cup \ldots$, and suppose that a lexicographical ordering is introduced for elements of Σ^+, e.g., $(1,1,2) < (1,2) < (1,2,1) < (1,2,2)$. Then the problem of assigning compound levels to the vertices of D is to find a mapping $c\text{-}level : V \rightarrow \Sigma^+$ satisfying the *Inclusion* and *Down-Arrow* conventions (C2, C4).

The *Inclusion* convention (C2) can be expressed as follows:

I1 : $\forall v \in V : c\text{-}level(v) \in \Sigma^{depth(v)}$
I2 : For any inclusion edge $(v, w) \in E : c\text{-}level(w) = append(c\text{-}level(v), s)$; $s \in \Sigma$ and *append* is a function that appends a component to a sequence.

The *Down-Arrow* convention (C4) is more complicated: For any adjacency edge $(v, w) \in F$ there is a unique path P from v to w in the inclusion tree D_c:

$$P : \quad (v = p_m, p_{m-1}, \ldots, p_1, t, q_1, q_2, \cdots, q_{n-1}, q_n = w),$$

where t is the *top vertex*, i.e., t has minimal depth. P originates from the rectangle of v, goes out across $p_{m-1}, \ldots, p_1$, passes t, goes in across $q_1, q_2, \cdots, q_{n-1}$, and terminates on w. To formulate the *Down-Arrow* convention, the order among each pair (p_i, q_i) of vertices that have the same depth for any adjacency edge $(v, w) \in F$ must be specified as follows:

D1 : if $depth(v) > depth(w)$ (or $m > n$),

(a) $c\text{-}level(p_i) \leq c\text{-}level(q_i), \ i = 1, \ldots, n-1$
(b) $c\text{-}level(p_n) < c\text{-}level(w)$

D2 : if $depth(v) \leq depth(w)$ (or $m \leq n$),

(a) $c\text{-}level(p_i) \leq c\text{-}level(q_i), \ i = 1, \ldots, m-1$
(b) $c\text{-}level(v) < c\text{-}level(q_m)$.

For example, in Figure 8.15, the path corresponding to adjacency edge (j, l) is j, e, b, f, l where b is the top vertex. Since $depth(j) = depth(l)$, we have $c\text{-}level(e) \leq c\text{-}level(f)$ and $c\text{-}level(j) < c\text{-}level(l)$ from D2.

A compound digraph has a *compound level assignment* if and only if there exists a mapping $c\text{-}level: V \to \Sigma^+$ satisfying I1, I2, D1, and D2.

B. Hierarchization Algorithm

A hierarchical map of the graph can not always be determined because the digraph might be cyclic. If there are cycles in D, some of the adjacency edges need to be reversed in order to obtain a hierarchization. Because the problem of finding this minimum set of feedback adjacency edges is $\mathcal{NP}$-complete (Garey and Johnson, 1991; Lempel and Cederbaum, 1966), heuristics are introduced for determining the edges that have to be reversed (see also Section 5.2).

In order to meet the requirements D1 and D2, every adjacency edge of D is replaced with one of the two following types of adjacency edges, $\to$ and $\Rightarrow$, which represent the relations $<$ and $\leq$, respectively, in D1 and D2. If edges between the same pair of vertices are duplicated during the replacement, reducing rules such as $\to\, = \,\to\, + \,\to$, $\Rightarrow\, = \,\Rightarrow\, + \,\Rightarrow$, and $\to\, = \,\to\, + \,\Rightarrow$ are applied to determine the resulting edge type. The graph derived by this edge replacement is called the *derived graph* of D. An adjacency edge in the derived graph is called an *original edge*, if $e \in F$, *derived edge* otherwise (Figure 8.17). Note that in the derived graph of D, every adjacency edge connects two vertices with identical depth. The derived graph of $D = (V, E, I)$ is denoted by $DD = (V, E, ID, type)$, where ID is the derived set of adjacency edges and $type : ID \to \{\to, \Rightarrow\}$.

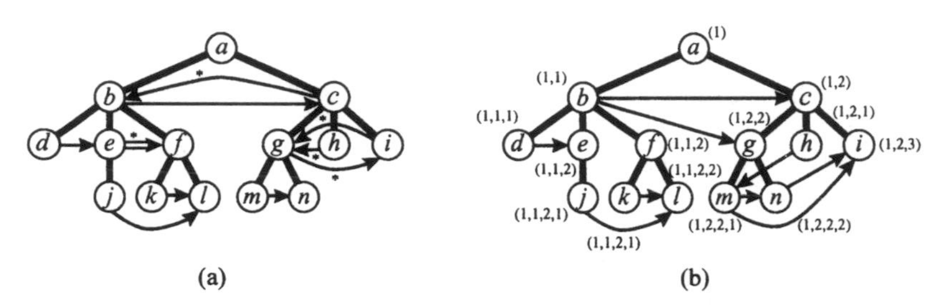

Fig. 8.17. The *derived graph* (a) and the *assigned compound digraph* (b) obtained from the compound graph of Figure 8.15 (c); edges in (a) marked with asterisk ($\star$) are derived edges.

Now, the *Down-Arrow* convention is established and the compound levels can be assigned to the vertices. If DD is not cycle-free, the cycles have to be resolved. To do this, the strongly connected components of DD are investigated, and the cycles are destroyed by either contracting strongly connected components into a so-called *proxy vertex* or by deleting some of the adjacency edges. This procedure leads to a cycle-free, hierarchical graph to which one finally can assign the compound levels. Of course, the deleted edges have to be put back in again, as well as the proxy vertices must be replaced again later on. All vertices of the component of DD, which have been contracted to a proxy vertex, are assigned the same compound level. Each adjacency edge (v, w) of the compound digraph $D = (V, E, I)$ is now checked, whether $c\text{-}level(v) < c\text{-}level(w)$. If this does not hold, the direction of that edge is reversed. The result is an *assigned compound digraph* $DA = (V, E, IA, c\text{-}level)$ (Figure 8.17 (b)).

Step 2: *Normalization*

In an assigned compound digraph DA, an adjacency edge $(v, w) \in IA$ is said to be *proper* if and only if $c\text{-}level(Parent(v)) = c\text{-}level(Parent(w))$ and $tail(c\text{-}level(v)) = tail(c\text{-}level(w)) - 1$, where *tail* is a function, that returns the last number of the *c-level*-string as an integer. The assigned compound digraph DA is now transformed into a *proper* compound digraph by replacing every non-proper adjacency edge with appropriate dummy vertices, dummy inclusion edges and dummy proper adjacency edges (cf. Sugiyama and Misue (1991) for more detail).

Step 3: *Vertex Ordering*

The idea of this step is similar as in Sugiyama's original algorithm for general graphs. The horizontal order of the vertices per level is determined by permuting their order on each level in such a way that the drawing rules *Closeness*, *Edge Crossings*, and *Edge-Rectangle Crossings* (R1 – R3) are satisfied as much as possible. The problem of minimizing edge crossings is $\mathcal{NP}$-complete even for only two levels (Garey and Johnson, 1991). Minimizing of edge-rectangle crossings, which is equivalent to the *linear arrangement* problem, is also $\mathcal{NP}$-complete (Garey and Johnson, 1991). Hence, heuristics, i.e., *barycentric ordering* as in Sugiyama's algorithm, are used to accomplish these tasks. In a compound digraph there exist also *local hierarchies* due to the inclusion relation, so vertex ordering must also be applied to these local hierarchies, i.e., subtrees of D_c. This step leads to an *ordered compound digraph.*

Step 4: *Metrical Layout*

In this last step, the positions of vertices (i.e., horizontal and vertical positions, widths and heights of rectangles) are determined by attaining the *Closeness*, *Line-Straightness*, and *Balancing* rules (R1, R4, R5) as much as possible. This problem can be expressed as a quadratic programming problem; a heuristic called the *priority layout method* is also developed to solve the problem. Once the vertex positions are determined, a routing for the edges can easily be achieved. The orientation of the reversed edges is changed back again, and all inserted dummy vertices and dummy edges are deleted and their corresponding originals are rearranged. This step finally leads to an automatic drawing of the original compound digraph.

8.5 Force-Directed Methods for Clustered Graphs

Force-directed graph drawing methods (cf. Chapter 4) can also be adopted to support and show the structure of a clustered graph. In the following sections we will see different ways of adaptation with different design goals.

One possibility to receive a more structured layout for clustered graphs is to decide between different spring forces. In Sections 8.5.2 and 8.5.3 we show two different approaches for an expanded force model for clustered graphs.

8.5.1 Inserting Dummy Vertices

The easiest way to achieve clustering is to insert *dummy vertices* as follows (Figure 8.18):

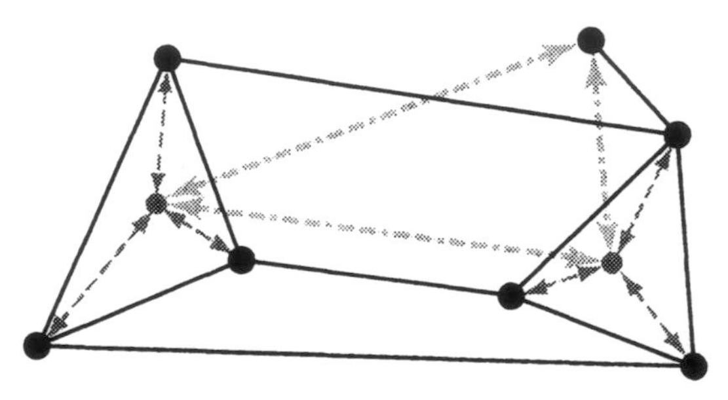

Fig. 8.18. Realizing clustering constraints in a force-directed approach by inserting dummy *attractors* (shaded vertices) in each cluster.

1. Let $G = (V, E)$ be a graph with a partition $(C_1, C_2, \ldots, C_k)$ on the vertex set V. For each C_i, $1 \leq i \leq k$, add a dummy *attractor* vertex c_i to the graph.
2. Add *attractive* forces between an attractor c_i and each vertex of the corresponding cluster C_i.
3. Add *repulsive* forces between pairs of attractors and between attractors and vertices not belonging to any cluster (i.e., if $\bigcup_{i=1}^{k} C_i \subset V$).

In this approach, no new forces have to be added. After inserting these attractors the vertices within a cluster will be closer to each other than before, and the distance between the clusters will grow.

8.5.2 Interactive Clustering

Huang and Eades (1998a) describe an animated interactive system for clustering and navigating huge graphs, called DA-TU, where they use the following expanded force model consisting of three different spring forces (Figure 8.19):

- *internal-spring*
  A spring force between a pair of vertices that belong to the same cluster.
- *external-spring*
  A spring force between a pair of vertices that belong to different clusters.
- *virtual-spring*
  In each cluster there is a virtual vertex (black vertices in Figure 8.19) that is connected to all vertices belonging to the same cluster by virtual edges; this is a similar approach to the concept of *attractors* described above. A virtual spring force exists between a vertex and a virtual vertex along a virtual edge.

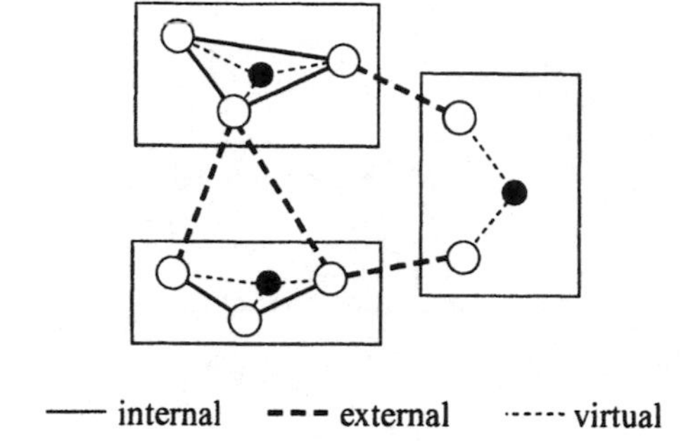

Fig. 8.19. Expanded spring model.

Additionally there is a gravitational repulsion force between each pair of vertices. All forces are applied additively to each vertex.

Some of the features of the DA-TU system are worth to be mentioned here because they show a possible application for clustering:

- The user can interactively change the graph.
- The user can interactively change the clustering of the graph.
- All transitions from one state of the graph to another are animated.
- Clusters can be interactively *contracted* and *expanded*, respectively. This is in particular useful for large graphs that do not fit on the screen or are too large to comprehend, so the clustered structure is used to navigate through the graph. If some of the clusters are contracted, DA-TU draws a so-called *abridgement* of the given graph.

Definition 8.20 (Abridgement, Ancestor Tree). *A clustered graph $C' = (G', T')$ is an* abridgement *of the clustered graph $C = (G, T)$ if T' is an* ancestor tree *of T with respect to a set U of nodes of T and there is an edge between two distinct nodes u and v of G' if and only if there is an edge in G between a descendant of u and a descendant of v.*

In other words, the ancestor tree T' is a subtree of T consisting of all nodes and edges on paths between elements of U and the root. Figure 8.20 shows such an ancestor tree for the set U of black nodes as the shaded area of the original inclusion tree.

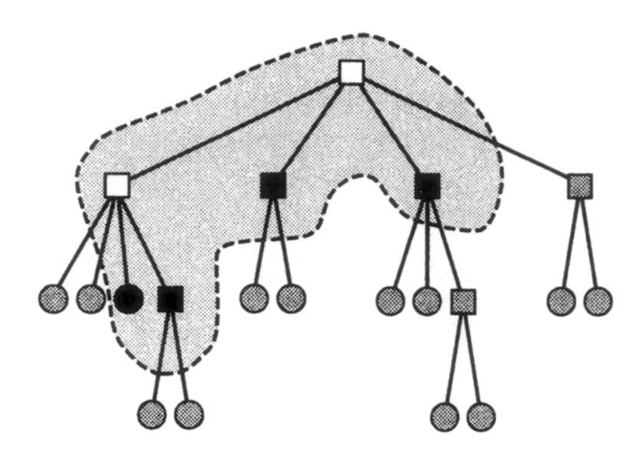

Fig. 8.20. The shaded area is the ancestor tree of the set of black nodes.

8.5.3 Meta Layouts

A similar approach is taken by Wang and Miyamoto (1995). They also use three forces but in a slightly different way. Instead of inserting virtual vertices as attractors in each cluster they use the concept of a *meta-graph*.

First, the edge set of the given graph is divided into *intra-edges*, i.e., edges between vertices belonging to the same cluster, and *inter-edges*, i.e., edges between vertices belonging to different clusters.

The forces in the force-directed placement are also divided into two categories:

- *intra-force*
  A spring force between a pair of vertices that belong to the same cluster.
- *inter-force*
  A spring force between a pair of vertices that belong to different clusters.

A force-directed placement is constructed by applying the intra- and inter-forces. An undirected graph G_{meta} is then constructed by collapsing the clusters of G into *meta-vertices* and transforming inter-edges of G existing between a pair of clusters into one *meta-edge* each. This sounds similar to the concept of a *quotient graph* (Definition 8.3), but goes further than that. A layout for G_{meta} is called *meta-layout* of G and can be obtained from the force-directed placement where the dimensions and center of each meta-vertex are set to the dimensions and center of the underlying subgraph, respectively (Figure 8.21). To calculate the forces between the meta-vertices, an improved force-directed placement is used that takes the different vertex-sizes into account (Wang and Miyamoto, 1995). The net force on a meta-vertex is defined as the *meta-force* on all vertices contained in the cluster represented by that meta-vertex.

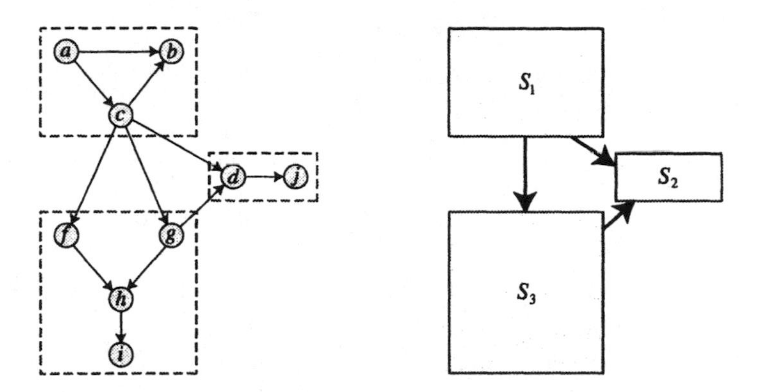

Fig. 8.21. Meta-graph and meta-layout.

In Figure 8.21 a force-directed layout and a partition are given on the left side, the corresponding meta-layout is shown on the right. The forces that are applied to vertex c are:

- The intra-force on c is the sum of forces between c, a and c, b
- The inter-force on c is the sum of forces between c and the vertices of subgraphs S_2 and S_3
- The meta-force on c is the net force on meta-vertex S_1 in the meta layout.

To finally compute a drawing of the clustered graph, Wang and Miyamoto (1995) propose a divide-and-conquer approach.

Divide-and-Conquer Drawing Approach. A divide-and-conquer drawing algorithm would draw a clustered graph in the following three steps:

1. divide the graph into subgraphs (cluster);
2. draw the subgraphs;
3. compose the subgraph layouts together to form the resulting layout.

The problem with this approach is that inter-edges are not taken into account which may result in a drawing with many crossings between inter-edges or long inter-edges.

In Wang and Miyamoto (1995), the last two steps of this divide-and-conquer approach are combined within one force-directed placement algorithm by using the following composite force F_{comp} to position a vertex:

$$F_{comp} = F_{intra} + S(t)F_{inter} + \big(1 - S(t)\big)F_{meta}$$

where F_{intra}, F_{inter}, F_{meta} are the intra-, inter- and meta-force on a vertex, respectively, and $S(t) \in [0,1]$ is a function of layout time t such that $S(t)$ decreases as t increases after a threshold t' and reaches 0 at another threshold $t'' > t'$.

By applying this composite force, the force-directed placement can be divided into three phases:

1. *Between time* 0 *and time* t': $S(t) = 1 \implies F_{comp} = F_{intra} + F_{inter}$.
   The force-directed placement leads to a layout with uniform edge lengths and a small number of edge crossings, as shown in Figure 8.22 (a).
2. *Between time* t' *and time* t'': $S(t)$ decreases:
   The strength of the inter-forces is reduced while the strength of the meta-forces is increased at the same time.
3. *at time* t'': $S(t) = 0 \implies F_{comp} = F_{intra} + F_{meta}$.
   Inter-forces do not count anymore; the intra-forces keep the vertices contained in the same cluster close together while the meta-forces fix the final positions of the clusters and eliminate possible overlaps between clusters. The resulting structured layout is shown in Figure 8.22 (b).

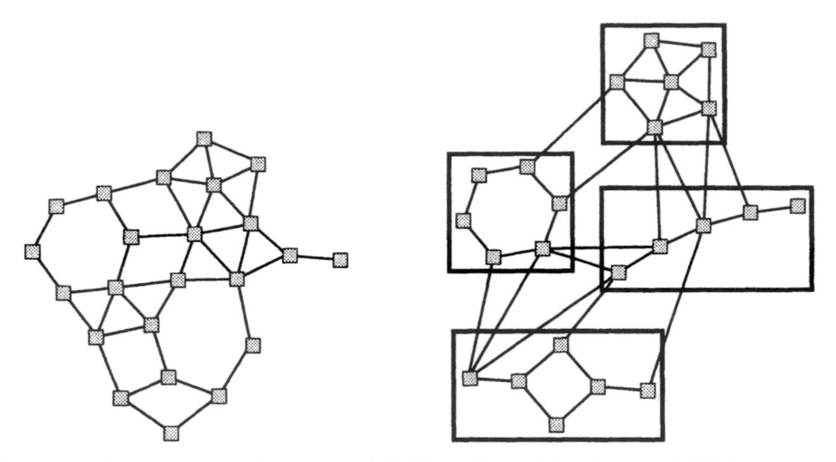

Fig. 8.22. Layout created at time t' (*left*) and resulting layout (*right*).

Wang and Miyamoto (1995) also present a way to add layout constraints to their force-directed placement algorithm by integrating a constraint solver.

Integration of a Constraint Solver. Layout constraints of the three following types can occur:

- *absolute constraints*, to fix an absolute vertex position
- *relative constraints*, to constrain the position of a vertex in relation to other vertices
- *cluster constraints*, to cluster several vertices into a subgraph that can be processed as a whole.

While trying to solve given constraints, some vertices may block others from reaching their optimal positions calculated by the force-directed placement algorithm. This may lead to a poor layout.

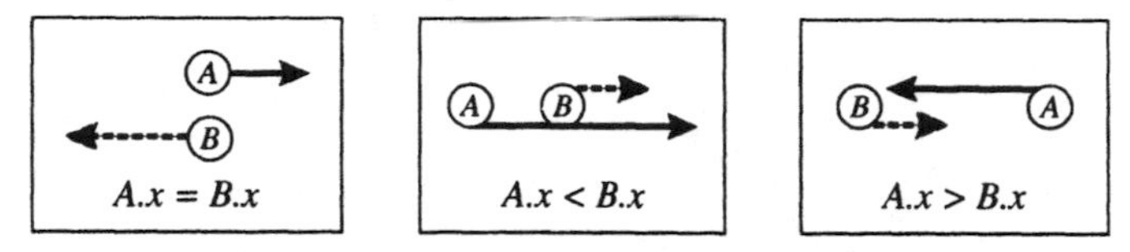

Fig. 8.23. Examples for constraints that become barriers.

If a constraint for two vertices A and B is given that prevents vertex A from reaching its optimal position, then vertex B is called a *barrier* for vertex A. Figure 8.23 shows several examples of barriers; the arrows indicate the direction and the strength of the forces calculated by the layout algorithm, the corresponding constraint shown below would be violated if these movements would be performed. In Figure 8.23 (a) no movement of either vertex is possible without violating the corresponding constraint. In the other two examples, the vertices could at least be partially moved until one of the vertices becomes a barrier for the other.

To avoid barriers while at the same time improving the layout, the vertices could be moved together without changing their relative positions. This is done by introducing *rigid sticks* to represent constraints in the force-directed placement. If vertex v_1 becomes a barrier for vertex v_2, a rigid stick is introduced between them so that they have to move like one rigid object. The movements of v_1 and v_2 are determined by the weighted average of the forces that are working on them:

$$f = \frac{w_1 f_1 + w_2 f_2}{w_1 + w_2}$$

where f is the new resulting force on v_1 and v_2, f_1 and f_2 are the old forces on v_1 and v_2, respectively, and w_1 and w_2 are weights of v_1 and v_2, respectively.

The layout algorithm and the solver cooperate to solve the given constraints as much as possible while at the same time keeping a good layout resolution. This cooperation works in an iteration of the following four steps:

Step 1 : Calculate forces ;

Step 2 : Introduce sticks and distribute forces ;

Step 3 : Calculate new positions ;

Step 4 : Satisfy constraints .

Steps 1 and 3 are performed by the layout algorithm, whereas the other two steps are performed by the solver. Figure 8.24 shows an example of applying constraints to the graph of Figure 8.22.

```
d3.y = d5.y

abs(a4.x - a2.x) <= 120
abs(a1.x - a5.x) <= 70
a5.y = a1.y = a2.y - 64
a4.y = a3.y = a2.y

c2.y = c5.y = c4.y = c1.y
```

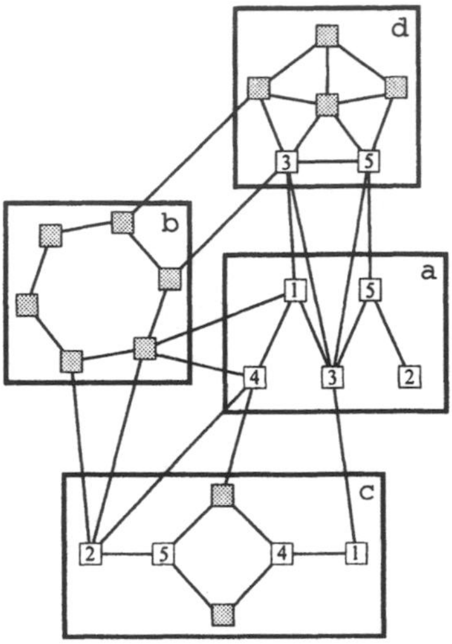

Fig. 8.24. Resulting layout of the graph of Figure 8.22 after adding the constraints on the left side.

8.6 Online Graph Drawing of Huge Graphs – A Case Study

Traditional graph drawing algorithms assume that the given graph can be laid out in a readable and understandable way on the screen or on paper. But there are important situations where this assumption does not hold. Suppose for example the graph displaying parts of the WWW or graphs arising in information retrieval. These graphs can be very large and there is no way to fit them in a readable way on a display medium.

Most graph drawing systems approach the layout of huge graphs in the following way:

1. Layout the graph on a virtual and very large page.
2. Provide a smaller window with scroll bars to show the part of interest, and to allow the user to navigate through the graph.

However, some problems are involved in this approach:

- The whole graph may not be known, e.g., in distributed systems, where a local vertex only knows part of the graph.
- To explore the graph, the user can only move geometrically through the graph by the means of the scroll bars. But the user might want to explore the graph in a logical way, in particular, if the graph contains relational data or hyperlinks. Moreover, long edges that do not fit on the screen are hard to follow. A more user-oriented approach would be better; the user should be able to control the logical content of the display.
- There is no *mental map* (Eades et al., 1991) that helps the user to keep track of his exploration so far. Even worse, the user can not see the whole graph and might get lost in empty areas.
- Besides that, it costs a lot of memory to store and display the large virtual screen.

To deal with these problems, several techniques have been proposed (see Sarkar and Brown (1994); Eades et al. (1991); Nielsen (1990); Mukherjea et al. (1994); Robertson et al. (1993)). For example, in Sarkar and Brown (1994) the *fish-eye* view technique is described where a detailed picture of a subgraph is shown along with the so called context of that subgraph. This kind of view provides the user with more information about the position of the subgraph within the whole graph. Another approach are three-dimensional methods, such as *cone trees* (Robertson et al., 1993) which lead to an increase in density of information on the screen.

These techniques work effectively for graphs of moderately large size, but they can not be applied when the graph is not completely known. Moreover, they still predefine the geometry of the graph.

The aim of *Online Graph Drawing* is the visualization of huge graphs which may be partially unknown. At any time, a tiny but non-empty subgraph called the *logical frame* is known and displayed on the screen. The user can explore the huge graph by changing the logical frame.

The layout of such a logical frame has to satisfy the usual aesthetic criteria for drawing graphs, e.g., minimization of edge crossings and uniform vertex distribution. Additionally, the transition from one picture of a logical frame to the next should preserve the user's *mental map* (Eades et al., 1991), i.e., successive drawings should not differ much, so that the user can easily follow the change in the drawing and does not loose orientation in a completely different layout when changing the focus.

Eades et al. (1997b) describe a model of Online Graph Drawing as well as an instantiation of that model in a system for *Online Force-Directed An-*

imated Visualization (OFDAV) for assisting web navigation (see also Eades et al. (1997a)). An interesting part of that model is a new force-directed drawing algorithm, that can be used to produce a continuous sequence of layouts according to the above mentioned criteria.

In OFDAV, the view of the user is focused on a small subgraph, the logical frame, that is defined by a *focus vertex* v. A force-directed graph drawing algorithm is used to draw this subgraph as well as its logical neighborhood. The user can change focus by selecting another vertex within the displayed frame which then becomes the new focus node, and the view changes according to this selection. Multiple animation steps are used to guide the user through the change of view and to preserve the mental map. A linear *history* is also kept by lining up a certain number of previously visited focus vertices.

The focus vertices together with their neighborhoods form a *clustering* of the graph.

The Online Graph Model. To explore a huge, partially unknown graph $G = (V, E)$, a sequence of *logical frames* $F_1 = (G_1, Q_1), F_2 = (G_2, Q_2), \ldots$ is used (Figure 8.25):

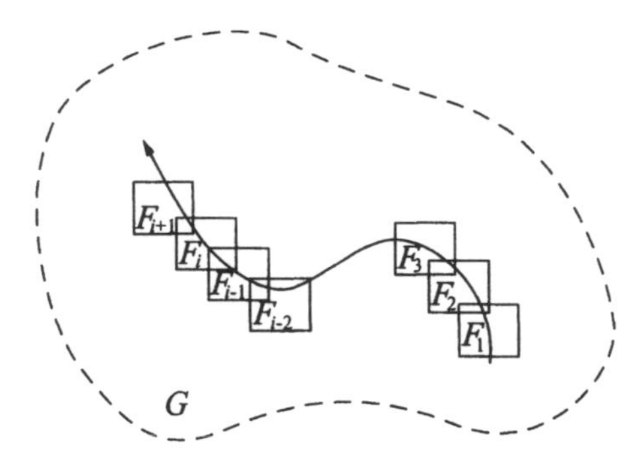

Fig. 8.25. The path of exploration of a huge graph G by a sequence of logical frames.

Each logical frame $F_i = (G_i, Q_i)$ consists of a connected subgraph $G_i = (V_i, E_i)$ of G and a queue Q_i of *focus vertices*. Successive frames differ only by a few vertices. The sequence of logical frames represents the sequence of subgraphs that are viewed by the user of the system and is determined by the interaction of the user who can change the focus and thereby decides which new logical frame has to be displayed. To define the logical frame more precisely, we need to explain the concept of neighborhood.

Suppose that $G = (V, E)$ is a graph, $v \in V$, and d is a non-negative integer. Then, the *distance-d neighborhood* $N_d(v)$ of v is the subgraph of G induced by the set of vertices whose graph-theoretic distance from v is at most d; note that $v \in N_d(v)$. In OFDAV only distance-1 neighborhoods are used, so we write $N(v)$ instead of $N_1(v)$ in the following, and call it the *neighborhood* of v.

Definition 8.21 (Logical Frame, Focus Vertex). *Given a queue $Q = (v_1, v_2, \ldots, v_s)$ of vertices of graph $G = (V, E)$, the subgraph of G induced by the union of $N(v_1), N(v_2), \ldots, N(v_s)$ is called a* logical frame $F = (G', Q)$ *(Figure 8.26 (a)), with $G' = (V', E')$ and*

$$V' = \bigcup_{i=1}^{s} N(v_i) \qquad E' = \{(u,v) \in E \mid u, v \in V'\}.$$

The vertices of the queue Q are the focus vertices *of the logical frame F.*

Clustering. Suppose that $Q = (v_1, v_2, \ldots, v_s)$ is the queue of focus vertices in G. Each neighborhood $N(v)$, for $v \in Q$, can be divided into two parts, the *common part* $C(v)$, and the *local part* $P(v)$, defined as follows:

- $C(v)$ is the part of the neighborhood $N(v)$ that also occurs in the neighborhood of other focus vertices $v' \neq v$, that is:

$$C(v_j) = \bigcup_{i=1, i \neq j}^{s} N(v_j) \cap N(v_i).$$

- $P(v)$ is the part of the neighborhood $N(v)$ that does not occur in the neighborhood of any other focus vertex $v' \neq v$ (Figure 8.26 (b)), that is:

$$P(v_j) = N(v_j) - \bigcup_{i=1, i \neq j}^{s} N(v_i).$$

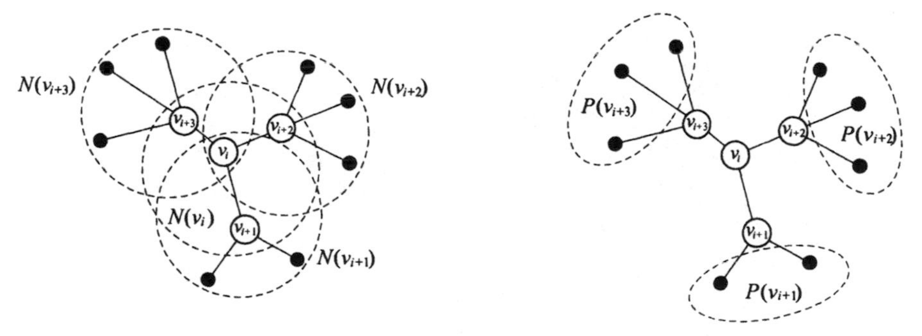

Fig. 8.26. Neighborhood of the focus vertices (*left*) local part neighborhood (*right*).

The user can explore the graph by changing the focus vertex, and this exploration is visualized by a sequence of logical frames. In practice, only a small number of vertices can be displayed on the screen at a time, in particular if the vertices are labeled, e.g., by the names of the html-pages as in OFDAV. Here, a global constant B is introduced as an upper bound for the length of the focus queue Q. For www-graphs, small values of B between 7 and 10 ensure that there are about 20 to 60 vertices on the screen at a time.

The transition from one logical frame to the next is obtained by adding the new focus vertex with its local neighborhood. If the length of Q was already B before, then the least recently used focus vertex and its local neighborhood are deleted (FIFO policy).

The local neighborhoods of the focus vertices in each frame can be viewed as the *clusters* of this frame. This will become clearer in the next section where the force model is explained.

The Force Model. The force model is based on Eades (1984) and consists of a combination of Hooke's law springs and Newtonian gravitational forces. In order to address the specific criteria of this online drawing approach, extra Newtonian gravitational forces among the neighborhoods, $N(v_i), N(v_{i+1}), \ldots, N(v_{i+B})$, of the focus vertices are added. These forces are used to separate the neighborhoods so that the user can visually identify the changes induced by changing the focus. This leads to a clustering of the displayed subgraph.

The total force applied to a vertex v of a logical frame $F_i = (G_i, Q_i)$, with $G_i = (V_i, E_i)$ is

$$f = \sum_{u \in N(v)} f_{uv} + \sum_{u \in V_i} g_{uv} + \sum_{u \in Q_i} h_{uv}$$

where f_{uv} is the force exerted on v by the spring between u and v, and g_{uv} and h_{uv} are the gravitational repulsions exerted on v by one of the other vertices u in F_i (Figure 8.27).

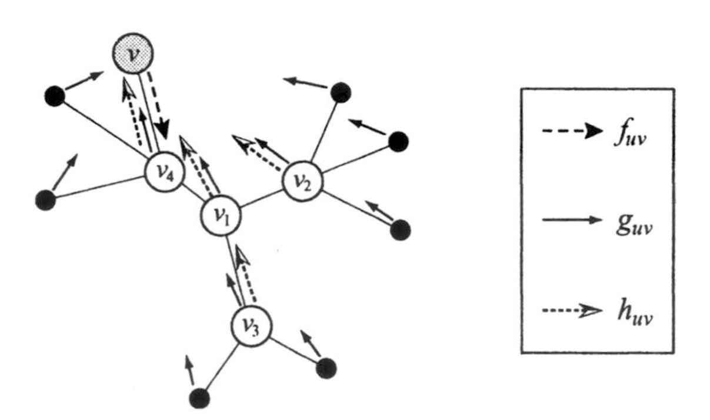

Fig. 8.27. The modified force model applied to the logical frame F_i.

The details of the modified force-directed drawing algorithm are given in Eades et al. (1997b). Here, we will only explain the idea of this approach.

The modified force model aims to satisfy the following four aesthetic criteria:

1. The spring force f_{uv} between adjacent vertices is aimed to ensure that the distance between vertices u and v is approximately equal to the zero energy spring length.
2. The gravitational force g_{uv} ensures that the vertices are not too close together and distributed evenly.
3. The extra gravitational force h_{uv} aims to minimize the overlaps among the (local) neighborhoods within a logical frame. This also ensures that the next vertices that have to disappear are placed close together which makes the identifying of the deleted objects easier for the user.
4. h_{uv} also aims to keep the layout of the queue of focus vertices close to a straight line; new vertices appear in one end of that line while old vertices disappear at the other end. This helps the user in understanding the direction of exploration of the huge graph.

8.7 Summary

For a graph that does not have a natural cluster-structure it is not at first glance clear what a good clustering strategy is. Often, the vertices are gathered together with respect to graph connectivity. Two main heuristics to do this are introduced in Section 8.2. One is to integrate cut size and cluster size balance within a single objective function, like the ratio cut partition, and to optimize them with a suitable procedure. Another one makes use of the eigenvalues of the Laplacian matrix of the graph. Besides connectivity, other graph properties like similarity of neighborhoods can also be of interest.

Once a graph has a cluster-structure, the question arises, how to make this structure visible. If we want to draw a really large graph, it seems to be a good method to draw only the quotient graph. But sometimes one is interested to also see what happens within a cluster. In Section 8.3 and Section 8.4 two methods are presented that draw clusters as shapes which include the corresponding vertices. Additionally, crossings between edges and borders of shapes are avoided. Especially in the planar case in Section 8.3, such a crossing is only allowed if one endpoint of the edge is within the corresponding cluster and the other one is outside of it.

Another way to show the cluster-structure is to draw vertices that belong to the same cluster closer together than such that are in different clusters. Using force-directed methods, one can achieve this by adding a dummy vertex to each cluster or by regarding clusters as big vertices. In this case the force-function must respect the different size of the vertices.

9. Dynamic Graph Drawing

Jürgen Branke

9.1 Introduction

Many graph drawing (GD) scenarios are dynamic inasmuch as they involve a repeated redrawing of the graph after frequently occurring changes to the graph structure and/or some layout properties.

For example, an interactive system might allow the user to manually edit the graph by inserting or deleting vertices and edges, or by setting additional layout constraints (e.g. vertex v_i should be placed above vertex v_j). If the graph is large, it might be necessary to scale parts of the layout, or the user might be allowed to expand or collapse subgraphs by clicking on them. Finally, the represented graph structure may be dynamic, for example the web-sites on a server and their interconnectivity might change over time, and so will a corresponding graphical diagram.

The easiest solution to the above mentioned dynamics would be to consider the drawing of the graph, after each modification, as a completely independent problem, and apply an existing, static GD algorithm from scratch at every step. However, this straightforward approach has two drawbacks: first, it may be inefficient. Since the graph has been modified only slightly, much of the old drawing might be reused to save computation time. Second, and even more important, if the user already familiarized with the drawing, it may mean a significant effort for the user to re-familiarize with the drawing after it changed. The user has build up a so-called "mental map" (Eades et al., 1991) that should be preserved when possible. In other words, in addition to the usual optimization criteria for graph layout, dynamic graph drawing should try to maintain the users mental map, and has to find a good compromise between these two goals.

A number of authors have addressed these problems and devised *dynamic* GD algorithms that provide special treatment of layout adjustment after a graph has changed (as opposed to *static* approaches which assume that the whole graph, and all layout constraints, are known in advance and do not change over time). Depending on the application, dynamic GD algorithms are able to handle the addition/deletion of (groups of) vertices/edges, to accommodate additional layout constraints, or to scale parts of the graph.

This chapter explains the special aspects that should be considered when dealing with dynamic graph drawing, provides a survey of the relevant literature, and suggests new avenues for future research.

The chapter's outline is as follows: Section 9.2 first tries to elucidate the concept of the mental map and surveys different approaches that authors

M. Kaufmann and D. Wagner (Eds.): Drawing Graphs, LNCS 2025, pp. 228-246, 2001.

have suggested to capture it. Then, in Section 9.3, a number of frameworks and algorithms proposed in the literature for dynamic graph drawing are presented.

The chapter concludes with a summary and some remarks on possible future work.

9.2 Maintaining the Mental Map – What Does It Mean?

When a user looks at a drawing, he or she will learn about the drawing's structure, will learn to navigate in the drawing and try to understand its meaning. This effort to become familiar with a drawing has been termed "building a mental map" (Eades et al., 1991). In the case of dynamic graph drawing problems, when the layout changes over time, the user has to repeatedly adjust his/her mental map. Clearly, it would be advantageous to minimize this effort for the user. An example is given in Figure 9.1. If the diagram depicted in (a) is the current layout, and the user adds a new edge from vertex 28 to vertex 22, then rerunning the layout algorithm from scratch might result in (b), which looks, at first glance, quite different from (a).

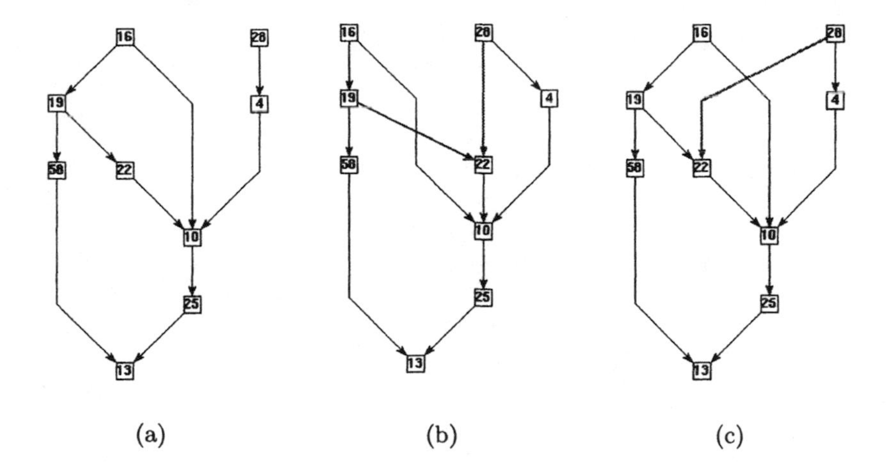

Fig. 9.1. (a) Current layout, (b) New layout after inserting edge 28-22 and re-running static layout algorithm, (c) New layout with preservation of the mental map.

There have been two suggested solutions to this problem:

1. Support the user by animating and highlighting the changes so that the changes can be easily recognized and the transitions are smooth.

2. Minimize the changes such that the effort to regain familiarity is minimized. For the example in Figure 9.1, drawing (c) keeps the position of all vertices of (a), and the graph is immediately recognized as almost the same graph as in (a). Of course this aim is often in conflict with traditional aesthetic criteria (like e.g. minimization of edge crossings, even distribution of vertices on the page etc.), thus a compromise needs to be found.

Obviously, animation and minimization of changes are complementary and can be applied simultaneously for best results, i.e. first a new layout is computed that minimizes the changes to the current drawing, and then the transition from the old to the new layout is animated.

Animation seems to be relatively straightforward and will not be discussed here in more detail. Instead, the focus of this chapter is put on the second issue, change minimization.

This requires to clarify the intuitive but rather fuzzy meaning of "minimizing changes to a layout" in a way that the "mental map" is preserved, an effort that has also been termed "maintaining dynamic stability". So far, numerous models have been suggested in the literature to capture the notion of the "mental map" or "dynamic stability". They basically can be grouped into two categories: either the allowed changes are restricted to a subset of the vertices, or a distance metric is used to measure the change, which allows to trade-off aesthetics with change. These two general approaches will be treated in more detail in the following subsections.

Which of the many suggested models is the most appropriate for which application is still an open issue. Bridgeman and Tamassia (1998) systematically examine and compare a number of difference metrics for orthogonal layouts (most of them can be applied to other layout paradigms as well). They conclude that most metrics behave well in the sense that their result increases as the number of vertices allowed to be moved by a layout adjustment algorithm is increased, at least as long as the changes to the graph structure are not too large. Nevertheless, more extensive comparisons and user studies are needed, maybe similar to Purchase et al. (1996) and Purchase (1997), in which the relevance of aesthetic criteria has been examined.

9.2.1 Restricting Adjustments to Parts of the Layout

A rather stringent concept of preserving the mental map is to not allow any changes to the current drawing i.e. to the current placement of vertices and edges that are not directly affected by a change. Then, the layout algorithm's only decision variables are the placement of new vertices and the routing of new edges, which is done in a way to minimize common static aesthetic criteria. Of course, this approach perfectly maintains the mental map. But under such stringent restrictions, not even consistency, i.e. adherence to the fundamental layout style rules like tree structure or orthogonality (North,

1996), may always be guaranteed, and often the resulting layout will be quite bad according to other aesthetic criteria, because e.g. many edge crossings are usually unavoidable. Algorithms for orthogonal graphs of maximum degree 4 that allow the insertion of new vertices without changing the placement of the other vertices may be found in Fößmeier (1997a) and Papakostas and Tollis (1998). Also, Miriyala et al. (1993) suggest a heuristic to route additional edges in an otherwise fixed orthogonal layout.

The no-change restriction may be weakened by allowing adjustment of vertices in the "vicinity" of a change. In Böhringer and Paulisch (1990), the vicinity of a change has been defined as all vertices directly affected by the change as well as vertices with a distance smaller than a certain edge length (a parameter to be specified). This reflects the idea that a user may tolerate changes in a small portion of the layout around the area where the graph structure changed, but would prefer the remainder of the layout to stay fixed.

Restricting the set of vertices that may be adjusted after a change is particularly useful for large graphs, because it also restricts the time for running the heuristic and for re-drawing the layout.

9.2.2 Distance Metrics

Instead of trying to categorically fix portions of the graph, other authors try to define some measure of similarity (or rather dissimilarity) between layouts in order to capture the effort to rebuilt the mental map after a change. The goal is to measure as precisely as possible how much the look of a drawing changes when certain adjustment operations are performed. An algorithm could then aim to construct a new layout which is a good compromise between aesthetic criteria and similarity to the current layout.

This approach has the advantage that it allows arbitrary changes to the current layout if that were necessary e.g. in order to adhere to the basic layout style rules. On the other hand it may be difficult to find a good trade-off between traditional aesthetic criteria and dynamic stability. Also, it does not yet seem to be clear how to actually measure similarity with respect to the mental map.

In the following subsections, the suggested metrics have been grouped along the general idea they rely on.

Absolute Vertex Positions. A more or less straightforward way to measure similarity is the sum of *Euclidean distances* each point has moved from one drawing to the next (e.g. Bridgeman and Tamassia 1998; Lyons et al. 1998).

An alternative measure would be the *Hausdorff distance*, a standard metric for determining the distance between two point sets (Bridgeman and Tamassia, 1998). It measures the largest distance between a point in the old drawing and its nearest neighbor in the new drawing. But since it does not distinguish between vertices (it doesn't matter which vertex is the closest one), the application to graphs may be questionable. A random shuffle of the vertex positions for example would not be considered harmful (there is still a vertex at every previous vertex position).

A general problem with difference metrics relying on point coordinates is that operations like translation, rotation or scaling will clearly yield large dissimilarity values and indicate a large change in the layout, while the user would easily recognize the old drawing. To alleviate this problem to some extent, in Bridgeman and Tamassia (1998), before computing the metrics, the drawings are first aligned by applying a point set matching algorithm taking into account scaling, translation, and rotation.

Also, most measures consider the vertices as single points without physical extension, i.e. the vertex size and shape is totally ignored. As Bridgeman and Tamassia (1998) argue however, distinctive vertices could serve as a landmark, thus their size does matter. Therefore, when the size of the vertex may be altered by a change, the authors suggest to look at all four corners of a rectangular vertex when computing distance measures.

Orthogonal Ordering / Relative Vertex Positions. One can argue that preserving the relative ordering of vertices is more important than preserving their absolute positions. In Eades et al. (1991) it is suggested that the *orthogonal ordering*, i.e. the ordering of vertices projected on each dimension, has an important influence on preserving the mental map and should thus be maintained. Inspired by that idea, the authors in Bridgeman and Tamassia (1998) suggest to *compare the angles* between straight lines between all pairs of vertices in the old and new drawing. Comparing angles is a more gradual measure than to just consider the ordering, and it additionally reflects the intuition that vertices that are further apart can be allowed larger absolute movements relative to each other (which will still result in the same angular move). Additionally, the paper suggests that a change of the angle is more severe for a user's mental map if the relative positioning is more or less diagonal rather than if the two vertices are on the same level or on top of each other (see Figure 9.2). Thus, as further refinement of the above measure, a weighted version is suggested which takes this into account (for details see Bridgeman and Tamassia (1998)). Astonishingly this is contrary to the orthogonal ordering argument, because changes that influence the orthogonal ordering are considered less severe.

The **λ-Matrix Model** proposed in Lyons et al. (1998) is yet another metric based on relative positions. If the graph has n vertices, a $n \times n$ matrix M is computed that at each entry (i, j) contains the number of points left to the directed line from vertex i to vertex j. The derived difference metric then is the sum of the differences in entries of M before and after a change.

Note that neither the metric based on angles nor the λ-Matrix Model captures e.g. a turn of the complete layout.

Proximity. Clusters are another layout property considered important for the user's mental map. Simply put, items that are close in the old layout,

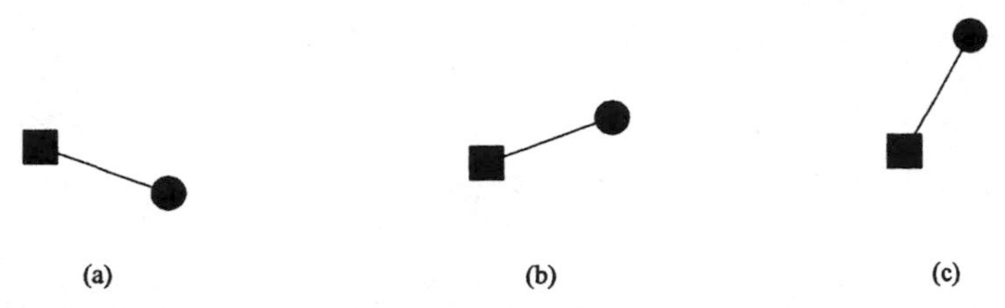

Fig. 9.2. As Bridgeman and Tamassia (1998) argue, the change from (a) to (b) is less severe for the mental map than the change from (b) to (c), although the angular change is the same. Note that from an orthogonal ordering point of view, the change from (a) to (b) would be considered more severe, since it changes the ordering projected to the vertical axis.

should also be close in the new layout. The advantage of measures maintaining clusters is that they capture the intuition that if a subgraph moves (but there are no changes within either subgraph), the distance should be less than if each point in one of the subgraphs moves in a different direction.

Basically all of the many known methods to capture the clustering of a graph might be used to form a metric. The general idea is to use some proximity relation and compare the relationship between the set of vertices of the current graph before and after the layout change.

In the following, let p_i be the position of vertex v_i in the current drawing D, and p'_i be the position of that vertex in the altered drawing D'. Furthermore, let $v_{N(i)}$ be the nearest neighbor to v_i in the old drawing.

Then, the *nearest neighbor* metric determines for how many vertices their nearest neighbor changes, i.e.

$$distance(D, D') = \sum_{i=1}^{n} closer(p'_i, p'_{N(i)})$$

with

$$closer(p'_i, p'_j) = \begin{cases} 0 & : \; d(p'_i, p'_j) \leq d(p'_i, p'_k), \forall k \neq i \\ 1 & : \; otherwise \end{cases}$$

For a weighted variant (proposed in Bridgeman and Tamassia 1998), the number of points closer to p'_i than $p'_{N(i)}$ is considered by using the following alternative interpretation of *closer*:

$$closer(p'_i, p'_j) = \left|\{k | d(p'_i, p'_k) \leq d(p'_i, p'_j)\}\right|$$

For some other metrics, two temporary graphs H and H' are constructed, whose vertices are those of the current graph G and whose edges correspond to relations defined by a proximity relation on D resp. D'. Then, H and H' can be compared by comparing their edge sets (see below).

In particular, the following proximity relations have beens suggested:

- *ϵ-Clustering* (Eades et al., 1991; Bridgeman and Tamassia, 1998): this graph has an edge between every two vertices v_i, v_j with $d(p_i, p_j) < \epsilon$ for some distance measure. As suitable distance, Eades et al. (1991) suggests $\epsilon = \max_{p_i} \min_{p_j \neq p_i} d(p_i, p_j)$.
- *sphere of influence graph* (Eades et al., 1991): this graph has an edge between v_i and v_j whenever

$$d(p_i, p_j) < \min_{k=1,\ldots i-1,i+1,\ldots n} d(p_i, p_k) + \min_{l=1\ldots j-1,j+1,\ldots n} d(p_j, p_l)$$

  i.e. whenever the distance between p_i and p_j is smaller than p_i's distance to its nearest neighbor plus p_j's distance to its nearest neighbor.
- *Delaunay triangulation* (Misue et al., 1995; Lyons, 1992).

If E_H and $E_{H'}$ are the edge sets of graphs H resp. H', then in Bridgeman and Tamassia (1998) the following distance measure is used:

$$distance(D, D') = 1 - \frac{|E_H \cap E_{H'}|}{|E_H|}$$

But since this only measures the *removed* edges, neglecting the *added* edges, maybe the following measure would be more appropriate:

$$distance(D, D') = \frac{|E_H \cup E_{H'}| - |E_H \cap E_{H'}|}{|E_H \cup E_{H'}|}$$

In any case, if the relations/edges between all vertices remain the same, this distance equals zero, if the two graphs have no common edges, distance is one.

Nearest Neighbor Between is yet another metric that has been proposed in Bridgeman and Tamassia (1998). It compares each vertex position with its original position and assumes that a vertex should remain closer to its original position than any other vertex. Then,

$$distance(D, D') = \sum_{i=1}^{n} closer(p_i, p'_i)$$

with

$$closer(p_i, p'_i) = \begin{cases} 0 & : \quad d(p_i, p'_i) \leq d(p_i, p'_k) \\ 1 & : \quad otherwise \end{cases}$$

A weighted variant has also been suggested using the following alternative interpretation of *closer*:

$$closer(p_i, p'_i) = |\{k | d(p_i, p'_k) \leq d(p_i, p'_i)\}|$$

Note that all of those proximity concepts are based on vertex positions only, edges are not considered.

Edge Routing. North (1996) argues that the position of vertices is more important than the routing of the edges because vertices are remembered as locations, while edges are traced "on the fly" to discover connections.

Nevertheless, for the case of orthogonal graph drawing, edge routing has also been used as distance metric: The shape metric in Bridgeman and Tamassia (1998) compares the sequence of directions for each edge and measures the number of edit operations (insert, delete or replace) to transform the old sequence of directions into the new one. Also, in Brandes and Wagner (1998a), a distance metric considering the changes of angles at vertices as well as the number of additions and deletions of bends has been used.

Other Suggestions. Further suggestions to capture the notion of the mental map include to maintain congruence (Eades et al., 1991) which only allows the operations reflection, rotation, translation and scaling, or to maintain topology, i.e. the dual graph of a layout (Misue et al., 1995).

9.2.3 Further Comments

In order to facilitate comparisons of an algorithm on several drawings, Bridgeman and Tamassia (1998) and Lyons et al. (1998) normalize their measures by dividing by the maximum possible value, or the lowest known upper bound if the maximum value is not known, such that all values are between 0 and 1.

So far, the metrics always look at two graphs, and try to minimize change for an isolated, single transition. However, often a sequence of changes are performed, i.e. a history is available. For this case it is argued in North (1996) that a vertex which has recently been moved may be a better candidate for a new move than a vertex that has been at the same location for a long time, i.e. the "age" of a vertex at a certain position should be taken into account when deciding which vertices to move. A similar idea for the relative positioning of vertices has been used in an approach by Brandes and Wagner (1997) and is briefly described in Section 9.3.4.

Sometimes, several changes are known even in advance, e.g. when an animation is built off-line. In those cases, it might be beneficial to minimize the impact on the mental map over the whole sequence of layout changes, rather than considering each change independently.

9.3 Coping with the Dynamics

This section presents the basic ideas of a number of dynamic GD algorithms that can be found in the literature. The approaches have been categorized according to the kind of layout they produce.

9.3.1 General Frameworks

Some papers suggest a general framework on how to treat the dynamics rather than suggesting and examining a specific algorithm for a specific type of application or drawing style.

Among those are the papers by Brandes and Wagner (Brandes and Wagner, 1998a, 1997) who suggest to use a Bayesian perspective to formulate a cost model that represents the trade-off between the usual (static) optimization criteria and the minimization of changes. The main advantage lies in the possibility to formulate the compromise in a generic framework.

With X being the new layout, and Y representing the previous layout, the aim is to search for a new layout X that maximizes

$$P(X = x|Y = y) = \frac{P(Y = y|X = x) \cdot P(X = x)}{P(Y = y)}$$

where $P(X = x)$ basically is the static cost function for the new layout, and $P(Y = y|X = x)$ represents the cost for the difference between the old and the new layout.

The framework is independent of a specific algorithmic approach and also of the distance metric used. However, as the authors demonstrate for the case of the spring embedder algorithm (Brandes and Wagner, 1997) and for Tamassia's grid embedding algorithm (Brandes and Wagner, 1998a), it is often possible to adapt an existing algorithm to reflect the new optimization criterion, at least with respect to specific distance metrics (see also Section 9.3.2 resp. 9.3.4). In any case, a major practical problem might be to set the parameters that adjust the weighting between dynamic stability and conventional layout criteria.

In the paper by Böhringer and Paulisch (1990) the GD problem (all aesthetic criteria as well as criteria preserving the mental map) is transferred into a set of linear constraints which is then solved by constraint propagation. The constraints may be assigned priorities which are used to resolve inconsistencies. Note that this approach has the additional advantage that many user constraints (like vertex A should be above vertex B) can also be formulated as linear constraints and thus easily be integrated. Different dimensions are treated independently. The authors suggest that numerous static layout algorithms could be integrated into their approach by formulating them in terms of constraints. In particular, the paper describes integration of the Sugiyama heuristic for directed, acyclic graphs (see Section 9.3.5).

9.3.2 Orthogonal Drawings in 2D

For the case of simply adding edges to an orthogonal drawing, Miriyala et al. (1993) suggest a heuristic to route additional edges without modifying the existing layout. The heuristic separates the drawing area into regions and then uses a variant of Dijkstra's algorithm to determine the sequence of regions through which the edge should be routed to achieve minimal cost in terms of edge length, crossings, and bends.

Papakostas and Tollis (1998) propose two algorithms for orthogonal planar graphs of maximum degree 4 (i.e. at most one edge at each side of a vertex) for two different scenarios. In both cases, it is tried to minimize the number of bends in the drawing, other common criteria like the area or the number of edge crossings are not optimized directly. The placement of a new vertex is explicitly specified for each combination of their adjacent vertices' free directions (i.e. the directions in which they do not yet have a incident edge).

In the no-change scenario, the current drawing may not be modified when adding new vertices with adjacent edges. The basic approach here is to start with an empty drawing (no vertices or edges) and to add vertices sequentially one by one. New vertices are placed always outside the current drawing area and such that a set of layout properties (invariants) is maintained. It is shown by means of total enumeration of all possible new vertex/edges combinations that these properties can always be maintained.

In the relative coordinates scenario, the existing layout may be modified by introducing a limited number of new rows and/or columns anywhere in the current drawing which, implicitly, maintains the drawing's orthogonal ordering, all bends, and the embedding. Again, the algorithm specifies the placement of new vertices depending on the number of existing neighboring vertices and their free directions. However, since the model does not use any invariants for every step, it can also be used to insert vertices into any existing drawing, produced by any other algorithm.

The derivation of the upper bound on the number of bends when the graph is constructed by introducing vertices one by one according to the suggested algorithms is rather neat and shall be presented here in more detail: Denote by n_i the number of vertices inserted with local degree i, i.e. i connected neighbors at the time of insertion. From the algorithm it follows that the insertion of a vertex with local degree 1 does not introduce any bends, while vertices with local degree 2 (3, 4) introduce at most 3 (5, 8) bends. Clearly, the number of bends is $3n_2 + 5n_3 + 8n_4$, which has to be maximized subject to the following constraints in order to derive an upper bound:

$$\begin{aligned} n_1 + n_2 + n_3 + n_4 &= n - 1 \\ n_1 + 2n_2 + 3n_3 + 4n_4 &\leq 2n \end{aligned}$$

$$n_1 \geq 1$$
$$n_i \geq 0, i = 2 \ldots 4$$

The first constraint ensures that the number of inserted vertices equals n, the second reflects the fact that each edge has to connect to two vertices, thus the number of edges has to be less or equal to $2n$. The above maximization problem can be solved as a linear program and yields the non-integer solution $3n + \frac{3}{2}$, which happens when $n_1 = 1, n_2 = n - \frac{7}{2}, n_3 = 0$, and $n_4 = \frac{3}{2}$. For the integer solution $n_1 = n_4 = 1, n_3 = 0$, and $n_2 = n - 3$, the upper bound on the number of bends is $3n - 1$.

In a similar way, the upper bound on the required area ($\frac{9}{4}n^2$), and for the no-change scenario the number of bends ($\frac{8}{3}n + 2$) and the required area ($(\frac{4}{3}n)^2$) are derived.

Note that according to these theoretical results, the no-change scenario has better worst-case bounds than the relative change scenario. However, an empirical comparison of the two suggested algorithms reported in Papakostas et al. (1996) indicates that the relative-coordinates scenario always outperforms the no-change scenario in practice (i.e. the average case), not only in terms of required area and number of bends, but also in terms of edge crossings, aspect ratio, average edge length, and maximum edge length.

The no-change approach has been further explored by Fößmeier (1997a). First of all, by slightly modifying the algorithm and by refining the linear program, Fößmeier was able to prove a bound of at most $2.5n$ bends, compared to about $2.66n$ by Papakostas and Tollis. As a further improvement, it is noted that many bends are required in particular when a vertex is inserted which has neighbors with their free direction at opposite sides. The suggested strategy following this observation is to attempt to produce vertices with similar free directions. The algorithm does so by preferring vertices with free directions to the bottom, the top and the right (in this order). With this modification and a further refinement of the corresponding LP (now using 3073 variables), the bound on the number of bends can be lowered to about $2.24n$. If vertices with local degree 0 are allowed (i.e. the graph is temporarily unconnected), the obtained bound is about $2.77n$ bends. Another

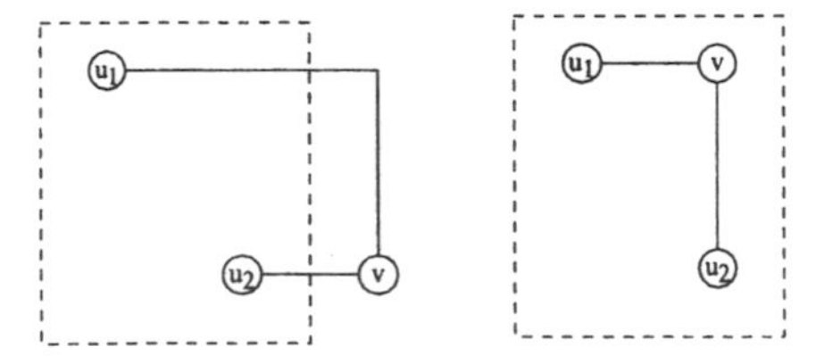

Fig. 9.3. Placing a new vertex v inside the bounding box of the current drawing saves bends and area.

improvement can be obtained when the placement of new vertices inside the current drawing's bounding box is allowed. Fößmeier suggests to allow insertions of vertices with two neighbors at the intersection of free lines of its neighbors (cf. Figure 9.3) which saves not only bends but also area. Whenever one neighbor has free directions to the left and to the right, and the other neighbor has its free directions to the top and to the bottom, there definitely exists such an intersection. Thus, as opposed to the previous approach, now the algorithm should favor semi-critical vertices (vertices with only two free directions) having their free directions at opposite sides whenever possible. The number of bends created by that approach is bounded by approximately $2.22n$, the upper bound on the area needed is $0.937n^2$.

The author tested this algorithm not only against the no-change scenario but also the relative change scenario by Papakostas and Tollis and claims that on the tested examples it worked better in terms of required area and number of bends than any of the two other algorithms. If confirmed on a more extensive test-bed, and for other aesthectic criteria like edge crossings and edge length, this would be remarkable, since as an algorithm under the no-change scenario it perfectly preserves the mental map.

Table 9.1 compares the approaches by Papakostas/Tollis and Fößmeier. The computation time to insert one vertex is constant for all algorithms.

Table 9.1. Upper bounds on no. of bends and area for different approaches.

	Bends	Area
Papakostas/Tollis relative change	$3n$	$2.25n$
Papakostas/Tollis no-change	$2.66n$	$1.77n$
Fößmeier basic no change	$2.5n$	$1.44n$
Fößmeier improved	$2.24n$	$0.937n$
Fößmeier improved, possibly disconnected	$2.77n$	$1.057n$
Fößmeier inside bounding box	$2.22n$	$0.937n$

Brandes and Wagner (1998a) demonstrate how the minimum cost flow approach suggested in Tamassia (1987) (cf. Chapter 6) can be extended to account for their Bayesian framework (cf. Section 9.3.1) and to minimize the changes of angles at vertices as well as of bends in the edges. The basic idea of that approach is to modify the flow network of the old drawing by adding some "residual arcs" in opposite direction to the current flows and by changing the cost and capacity constraints of these arcs to reflect the additional cost of changing a flow. Note that since this approach relies on Tamassia's algorithm, it assumes the embedding to be given and fixed.

The Three-Phase method (Biedl et al., 1997a; Biedl and Kaufmann, 1997) uses a slightly different orthogonal drawing convention that is not restricted to a maximum degree 4 but instead uses "stretched" vertices that may have more than one edge incident at each side (for an example, see Figure 9.4). In such a setting, it is possible (and done in Biedl and Kaufmann (1997)) to draw the graph with exactly one bend in every edge. In the first phase, vertices are considered as points (not boxes) and are placed in the drawing such that no two vertices share the same row or column. Now, edges can be routed with exactly one bend per edge. To obtain a feasible drawing, one finally has to decide on the port assignment (i.e. the place where an edge connects to a vertex) and to adjust the dimensions of the vertices accordingly.

In the dynamic scenario described in Biedl and Kaufmann (1997), the vertices are added sequentially one at a time starting form an empty drawing, and the only modifications allowed to the current drawing is the insertion of a limited number of rows and columns which again preserves the orthogonal ordering. Although in general, the algorithm would allow new vertices to be placed in the middle of the new drawing, in the version described in the paper a vertex is simply placed at the median of rows of the already placed vertices connected to it, and at the extreme right or left of the drawing depending on a greedy heuristic trying to balance the number of edges to the right and to the left at every vertex. The algorithm uses an area of at most $(\frac{m}{2}+n)\times(\frac{2}{3}m+n)$.

The variant described in Biedl et al. (1997a) only achieves an upper bound on the area of $(m+n)\times(m+n)$, but it allows user-specified placements of the vertices, and also moving a vertex from one place to another.

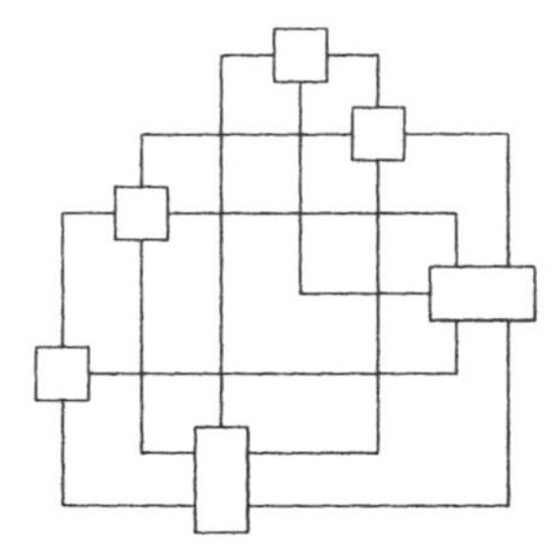

Fig. 9.4. An example for an orthogonal graph with "stretched" vertices.

InteractiveGiotto (Bridgeman et al., 1997) is an interactive variant of the Giotto tool for producing orthogonal drawings. It requires the user to specify the placement of new vertices and to indicate the desired routing. The tool then transforms the current layout into a planar one by replacing each bend and each crossing with a dummy vertex. The embedding and the edge crossings are preserved in this step. The resulting graph is then optimized by a variant of Tamassia's minimum cost flow approach (cf. Chapter 6). The edge bends, the type of the 90° bends and the number of corners are preserved by setting a target value for the flow in some arcs of the network. Note that new bends may be introduced by the algorithm, if needed.

9.3.3 Orthogonal Drawings in 3D

There have also been two algorithms suggested for orthogonal graph drawing in three dimensions (Papakostas and Tollis, 1997a,b). Similar to the two-dimensional relative-coordinates approach from Papakostas and Tollis (1998), the first algorithm assumes graphs of maximum degree six (i.e. at most one incident edge at each side of a vertex), restricts changes to the current drawing to the insertion of a limited number of planes, and bases the decision about where a new vertex will be placed and how its incident edges will be routed entirely on the free directions around the adjacent vertices. The insertion of a vertex needs constant time and the volume of the drawing is at most $4.66n^3$ as has been shown by solving an LP in a similar way as described above. The second algorithm assumes a setting similar to that in Biedl and Kaufmann (1997) described above, i.e. vertices are represented using 3-dimensional boxes with volume of at least one cubic unit, which can be stretched in each dimension to accommodate an arbitrary number of incident edges at each side. Just as in Biedl and Kaufmann (1997), the algorithm maintains the property that no two vertices have overlapping x-, y- or z-coordinates, and produces drawings without any crossings and exactly two bends per edge. The basic idea is to route all new edges from old vertices to the new vertex straight to a plane outside the current drawing where they are going to have their first bend. By this, the problem has been converted to a two-dimensional one, and the new vertex can be inserted by an algorithm similar but different to Biedl and Kaufmann (1997), introducing exactly one additional bend per edge without producing any crossings. A more detailed description of these two algorithms for 3-dimensional orthogonal drawings can be found in Chapter 7.

9.3.4 Force-Directed Methods

Many popular methods for drawing general undirected graphs are some kind of force-based spring model (cf. Chapter 4) with attracting forces between connected vertices and repelling forces between all pairs of vertices. A proper layout according to that model corresponds to an equilibrium of the forces. The model lends itself to dynamic graph drawing, since one might simply make the changes to the graph structure, and let the forces act to find a new equilibrium. The movement of the vertices according to the forces can be easily animated to make the changes more gradual (see for example Eades et al. (1997a)). However, even a small change to the graph may lead to a quite different equilibrium state and may thus destroy the mental map.

Brandes and Wagner (1997) present two ways to adapt the force-directed model to adhere to stability criteria, using their Bayesian framework from Section 9.3.1. In one of them, the change of absolute vertex positions is considered as distance criterion between drawings, which translates nicely into introducing additional forces keeping the vertices at their previous location (springs with natural length zero). In the other approach, the stability criterion is the relative rather than the absolute vertex positions. This is taken into account basically by emphasizing the forces between vertices that are present in the old as well as the new layout. In other words, the unchanged parts of the graph are connected by a stiffer structure than new or altered ones. As the authors note, this approach may also take into account that for consecutive changes, the history plays a role. By cumulating the stiffening effect, the longer a relation existed, the less will it be changed.

9.3.5 Layered Graphs

For acyclic, layered digraphs, the Sugiyama heuristic (cf. Chapter 5) or a variant thereof is quite popular. Böhringer and Paulisch (1990) demonstrate in their paper how this heuristic may be modeled in terms of constraints, and how it can then be transformed to preserve dynamic stability. In the layering step, for each edge (i, j) a constraint is introduced saying that vertex i should be placed above vertex j, and the layering is decided. Then, the barycenter ordering is used to derive constraints determining whether one vertex should be left or right of another vertex.

When the graph is modified, stability constraints are derived from the old layout determining:

1. the ordering of the vertices in each layer
2. that vertices which have been on the same layer in the old layout should also be on the same layer in the next layout.

Only vertices close to the change in the graph (i.e. vertices in the vicinity of the change, see Section 9.2) are exempt from these constraints and allowed to move freely. By setting the size of the vicinity, the user may influence the emphasis on dynamic stability. Given the total set of constraints (Sugiyama plus stability constraints), constraint propagation is used to find a feasible layout. Inconsistencies are resolved by dropping some of the constraints, depending on priorities assigned to them.

The DynaDAG system (North, 1996) is another adaption of the Sugiyama heuristic that allows interactive changes to the graph structure. DynaDAG allows to insert, optimize or delete single vertex or edges. Vertices are originally placed on the highest possible layer and may be moved down when this becomes necessary by an insertion of another vertex or edge. Vertices are moved down layer by layer, shifted in each layer to its median position (w.r.t. its adjacent vertices). When the vertices are in their final position, the adjacent edges are adjusted (shrunk, moved, or stretched). New edges are routed heuristically. The final vertex coordinates are calculated by a linear program, with a linear penalty for moving a vertex from its old assignment. A nice feature is that for positioning a new vertex, the placement by the user is taken into account.

9.3.6 Trees, Series-Parallel Digraphs, and ST-Digraphs

For the special case of drawing trees, Moen (1990) suggests a dynamic GD method based on merging contours of subgraphs. First, an outline is calculated for each vertex in the tree. The algorithm operates by recursively calculating an outline around each subtree from the leaves to the root of the tree. As the algorithm moves toward the root, the contours of the vertex' children are placed as close together as possible, then the children's and the parent vertex contours are joined into one large polygon. When the graph is changed, the contour is disassembled again as necessary and the changed parts are merged anew.

Besides drawing trees, Cohen et al. (1995, 1992) additionally consider series-parallel digraphs, and planar ST-digraphs, and suggest a set of representations and operations that allow an efficient handling of update operations guaranteeing that unaffected components (e.g. subtrees) change only by a translation. This approach belongs more to the area of dynamic data structures and shall thus not be treated here in more detail.

9.3.7 Separating Overlapping Vertices

All of the models described so far assume some sort of incremental scenario, where repeatedly vertices and edges are added to the graph.

A slightly different perspective has been taken on in the approaches described in this section. Basically the graph is considered static, but somehow the layout has overlapping vertices that should be distributed more evenly over the page. This might be the case for example after new vertices have been inserted into the drawing, after a vertex has been expanded into a subgraph, or after some area of the drawing has been expanded to take a closer look at the details. In neither of the approaches described below, the edge routing is considered explicitly.

In Lyons (1992) this problem of dissolving clusters of vertices has also been termed "cluster busting". Four heuristics are presented in that paper to distribute vertices more evenly while retaining similarity to the old drawing. Two of the approaches have been described in more detail in Lyons et al. (1998). The key idea there is to restrict the movement of each vertex to their Voronoi region i.e. the set of all points in the plane that are closer to the specific vertex than to any other vertex of the diagram. Clearly, this guarantees that for every vertex, its new position is closer to its old position than to the old position of any other vertex, which somehow corresponds to the "nearest-neighbor-between" criterion from Section 9.2. However, since often the Voronoi regions are too small to allow the desired adjustments, the authors suggest to iterate the process of determining the Voronoi regions and moving the vertices within these regions. The two heuristics they suggest differ in the way the vertices are moved: the first one, called Voronoi Diagram Cluster Buster Algorithm (VDCB), moves each vertex to the centroid of its Voronoi region. The GeoForce algorithm has been based on the idea of the well known spring algorithm, with repelling forces between vertices and attracting forces between each vertex and its previous position. The vertexes maximum step size is limited to a fraction of its distance to its Voronoi edges. From the results reported in that paper, it does not seem entirely clear which approach is superior.

Eades et al. (1991) and Misue et al. (1995) suggest to use the "push force-scan algorithm" (PFS) for cluster busting which pushes overlapping vertices apart by calculating desired repelling forces f_{ij} between the midpoints of every pair of vertices (i, j) (cf. Figure 9.5), decomposing these forces into their portions parallel to the x- resp. y-coordinate (denoted f_{ij}^x and f_{ij}^y), and using these to shift vertices first along one, then along the other coordinate. The horizontal and vertical scan work analogous, therefore only the horizontal scan is described here: Assume that $x_1 \leq x_2 \leq \cdots \leq x_n$, then the algorithm fixes x-coordinates in the order $v_1, \cdots, v_n$ by moving in the ith step $v_{i+1}, \cdots, v_n$ by $\max_{i<j\leq n} f_{ij}^x$ (vertices with the same initial x-coordinate are decided at the same time). Clearly, the algorithm runs in $O(n^2)$ and preserves the orthogonal ordering.

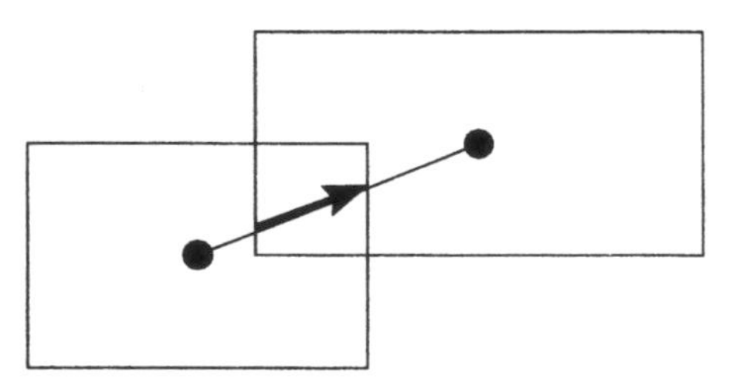

Fig. 9.5. Two overlapping vertices and the shift vector as used by the force scan algorithm.

As has been shown by Hayashi et al. (1998), finding a minimum area layout of a given set of rectangles on a plane, preserving the orthogonal ordering, is $\mathcal{NP}$-complete. As heuristic, the authors suggest to use an improved version of the PFS algorithm which usually results in a smaller total area than standard PFS. Instead of moving all subsequent vertices based on the maximum force between vertex i and all subsequent vertices, the improved PFS moves each vertex i depending on the placement of all vertices $v_1, \ldots, v_{i-1}$ that have already been decided on (except for a special case that will not be discussed here). Again, only the horizontal scan is described in more detail: Let γ_j be the distance by which vertex j has been moved in x-direction. Then, for $i = 1 \ldots n$ vertex v_i will be shifted by $\max_{1\leq j<i}(\gamma_j + f_{ji}^x)$.

In Misue et al. (1995) a variant of the force scan algorithm is suggested that additionally allows contraction operations. However this variant does not guarantee disjoint vertices.

9.3.8 Nonlinear Magnification

For large graphs, it is usually impossible to display the whole drawing on the screen in reasonable resolution. But if only a part of the drawing is displayed, the overall structure of the graph is hidden, which makes it much harder for the user to build up and maintain a mental map, and to navigate in the drawing.

One suggested solution to that problem is to magnify some important parts of the drawing, while demagnifying the other parts correspondingly. This allows for enhanced resolution in some areas of interest, without sacrificing the global view of the entire graph. In an interactive setting, the user may be allowed to select the areas for magnification, while the global context of the surrounding structures is maintained.

There are many variants of nonlinear magnification. Since a detailed analysis lies outside the scope of this paper, the interested reader is referred to e.g. Misue et al. (1995) or Keahey and Robertson (1996).

9.3.9 Deleting Vertices and Edges

Many of the above described approaches mainly address the insertion of vertices. Some argue that deleting vertices or edges while maintaining the mental map is much easier than inserting vertices or edges - simply remove them without changing anything else - and from time to time a compaction algorithm could be run to reduce the empty space created by deletions (Papakostas and Tollis, 1998). But each deletion may open some new ways to improve the aesthetics of the drawing by rearrangement. For the case of a larger number of deletions, the fortified chances to improve the aesthetics by an adjustment after a deletion accumulate. The resulting drawing may then be quite far from the optimum in terms of aesthetic criteria. In such cases, an approach using a cost model, like in the Bayesian framework (cf. Section 9.3.1), might be advantageous since it can be applied after insertions and deletions alike.

9.4 Conclusion and Future Work

Dynamic and interactive graph drawing has many applications but only recently got into the focus of researchers and is thus still in a very early stage.

The basic difference between static and dynamic graph drawing algorithms is that in the dynamic case, in addition to producing an aesthetic layout, the algorithm has to minimize changes to the user's mental map. So far, this has been tried in a variety of ways, basically by restricting the changes allowed to the layout, or by defining a cost function reflecting the severity of changes and then trying to find a good trade-off between aesthetic criteria and dynamic stability. The field however, still seems to lack a basic understanding as to what actually influences the mental map. Because of this, and because there are so many different GD applications, almost all papers published so far are more or less unique and difficult to compare.

There are many areas for future work. First and foremost, it seems to be important to get a clearer concept of how changes in the drawing influence a user's mental map. This should be examined e.g. by user studies to learn about the importance of the many different criteria suggested in the literature so far. Then, the aspect of deleting vertices seems to deserve more attention. Optimizing the layout adjustment for a sequence of graph changes, e.g. for an off-line animation, is still an open yet very challenging area of research. Besides, there is a wealth of static layout algorithms (Di Battista et al., 1999), and for most of them it would be worthwhile to develop a dynamic version.

10. Map Labeling with Application to Graph Drawing

Gabriele Neyer*

When visualizing information, it is often essential to display data with a graphical object. This means that text labels have to be associated with graphical features. Until now, the placement of labels is primarily performed manually, particularly in map production. For example, in the area of Cartography, Geographic Information Systems (GIS), and Graph Drawing map labeling usually has to be performed efficiently. Therefore, it is highly desirable to use automatic map labeling algorithms. The ACM Computational Geometry Task Force Force (1996) has identified label placement as an important area of research.

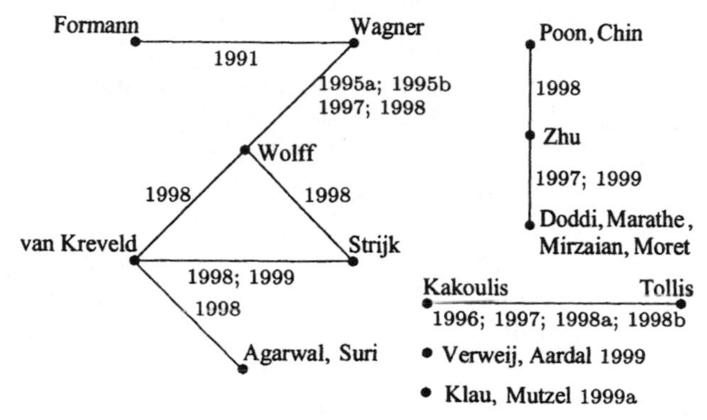

Fig. 10.1. Labeled graph of some map labelers and their articles discussed in this survey.

Often, the solution to a graph labeling problem involves drawing *and* labeling the graph. However, almost all known algorithms for graph labeling start from a given graph drawing. Thus, the graph labeling problem is artificially subdivided into two problems. This can have disadvantages in case that the drawn graph is too dense to label. Due to the fact that there is only one algorithm published so far that *simultaneously* draws and labels a graph, we mainly describe labeling algorithms that start at a drawn graph. Therefore, this chapter is a survey on map labeling algorithms. Nonetheless, since we also survey graph labeling algorithms and most map labeling algorithms

* This work was partially supported by grants from the Swiss Federal Office for Education and Science (Projects ESPRIT IV LTR No. 21957 CGAL and No. 28155 GALIA), and by the Swiss National Science Foundation (grant "Combinatorics and Geometry").

M. Kaufmann and D. Wagner (Eds.): Drawing Graphs, LNCS 2025, pp. 247-273, 2001.

also apply to graph labeling (without drawing), this chapter is also a survey on graph labeling algorithms. For simplicity, we will call an algorithm that labels a drawn graph a *graph labeling* algorithm. In case that an algorithm draws and labels a graph, we point that out explicitly.

10.1 Formal Background

We distinguish between three kinds of graphical features according to their dimension.

Point Features. Cities, summits, area features on small scale maps, and vertices of graphs or diagrams.

Line Features. Rivers, boarders, streets, straight edges, polygonal lines, and edges or arcs of graphs or diagrams.

Area Features. Mountains, islands, countries, and lakes.

Point and line feature labels are arranged next to the object and area feature labels are usually placed within the boundary of the feature to be labeled.

In the last ten years, the amount of research in automatic map making has increased significantly, as the number of published articles illustrates[1]. A detailed and up-to-date map labeling bibliography can be found at `http://www.inf.fu-berlin.de/map-labeling/bibliography/`.

Although most variants of map labeling are $\mathcal{NP}$-complete, many good labeling approximations and heuristics, especially for point labeling, have been suggested. In this article we want to give an overview of the most important map labeling algorithms that apply to graph labeling. See Figure 10.1 as an example of a labeled graph. The majority of map labeling algorithms is easily applicable for graph labeling. A point labeling algorithm can be applied for the labeling of the nodes of a graph. If the point labels have to be placed without overlaps with other graphical features, e.g. edges, the number of applicable algorithms decreases. This case is barely considered in literature. We discuss the applicability of the point labeling algorithms for labeling the point feature of a graph in the respective sections. A line feature labeling algorithm can be used for labeling edges of a graph and a general graphical feature labeling algorithm can be applied to labeling nodes, edges, and faces of a graph.

We give a survey of labeling algorithms that are intended to label graphs, and of the most important map labeling algorithms that were intended to label geographical maps. Our intention is to keep the description of the algorithms as general as possible. We do not present algorithms that label line features with curved labels, since they are usually not applied to graph labeling. For the latter problem see (Edmondson et al., 1997; Knipping, 1998; Wolff et al., 1999).

---
[1] `http://liinwww.ira.uka.de/bibliography/Theory/map.labeling.html`

Extensive effort has been spent by cartographers like Imhof (1962, 1975) and Yoeli (1972) to devise rules that measure the semantic clarity of a labeling assignment. We state three concepts that are widely accepted as the basic rules for accurate map labeling.

Readability. The labels are of legible size.

Unambiguity. Each label can be easily identified with exactly one graphical feature of the layout.

Avoidance of Overlaps. Labels should not overlap with other labels or other graphical features of the layout.

We denote the possible label positions of a feature as its *label candidates.* Sometimes, a *cost* is assigned to an individual label candidate which reflects the quality of this label in terms of unambiguity, overlap with graphical features, and preferences between the label candidates.

How features are labeled depends on the specific labeling model. The most important models are:

Fixed Position Model. Each feature has a finite set of label candidates. For point labeling, typical examples are the 2- or 4-position model as shown in Figure 10.2(a).

Fixed Position Model with Scalable Labels. Each feature has a finite set of label candidates, where the size of all labels can be scaled.

Slider Model. Each feature has a fixed label that can be placed at any position that touches the feature. Figure 10.2(b) shows the 1-, 2-, and 4-slider model for point features, where the labels can be shifted continuously as indicated by the arrows. Figure 10.2(c) shows the point-adjacent slider model, where the label is adjacent to its point feature but can be arbitrarily rotated.

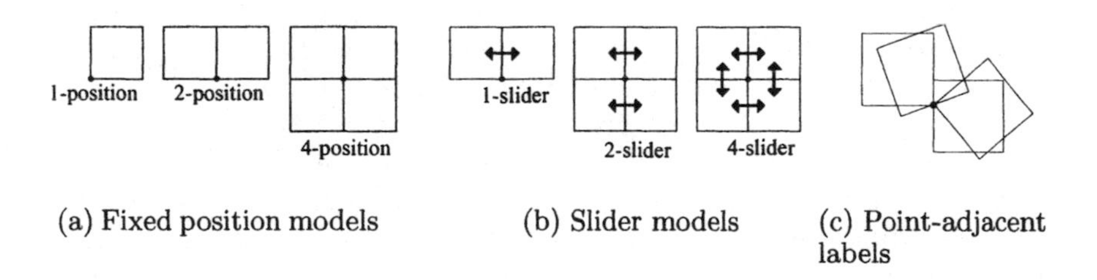

(a) Fixed position models (b) Slider models (c) Point-adjacent labels

Fig. 10.2. Labeling models.

Subject to these basic constraints, the most common problems are:

Decision Problem. Does there exist a label assignment, such that each feature is labeled with a label of its candidate set, and no two labels overlap?

Label Problem. In case the Decision Problem yields a *yes* answer – Find a label assignment, such that each feature is labeled with a label of its candidate set, and no two labels overlap.

Number Maximization Problem. Assign as many labels as possible, such that each feature is labeled with at most one label of its candidate set, and no two labels overlap.

Size Maximization Problem. Find a maximum scaling factor s and a corresponding label assignment, such that each feature is labeled with a label of its candidate set, scaled with s, and no two labels overlap.

Note that the Label Problem is of prime importance for graph labeling: Often, the coordinates in a graph are adapted until a graph labeling exists.

Let P be a label problem and A be an algorithm for P. Then it is clear that a binary search on all label sizes combined with algorithm A solves the size maximization problem. Similarly, an algorithm that solves the size maximization problem also solves the label problem and the decision problem. Thus, the label problem and the size maximization problem are in the same complexity class and at least as hard as the decision problem.

Furthermore, an optimal algorithm for the number maximization problem solves the label problem. Thus, the number maximization problem is at least as hard as the label problem.

Since the point feature label problem or the line feature label problem are special cases of the graphical feature label problem it is clear that the graphical feature label problem is at least as hard as the point or line feature label problem.

For convenience, we recall the definitions of some complexity terms used in this chapter in Section 10.1.1. In Section 10.2 we give three tables that give an overview about the label problems discussed in this chapter, with their complexity. We divide the label problems into point feature label placement, line feature label placement, and graphical feature label placement. Consequently, point labeling is discussed in Section 10.3, line labeling in Section 10.4 and graphical feature labeling in Section 10.5. We discuss the graphical feature labeling approach more detailed in Section 10.5.2, as the tutorial example. In Section 10.6 we shortly present general optimization strategies such as simulated annealing, gradient descent, and zero-one integer programming that are widely used to solve map labeling problems. The only known graph labeling algorithm that simultaneously draws and labels a graph so far is the combined labeling and compaction approach from Klau and Mutzel (1999a). For a given orthogonal representation, an orthogonal labeled graph drawing with small total edge length is computed. This algorithm is presented in Section 10.3.5.

10.1.1 Complexity Dictionary

Many map labeling problems are $\mathcal{NP}$-hard. That means not only that there is no known efficient (polynomial time) algorithm for solving the problem, but also that it is quite unlikely that one exists. If the optimal solution is unattainable we settle for feasible approximative solutions that are "close" to the optimum. In order to evaluate the limits of approximability we give the following short dictionary, partly taken from Garey and Johnson (1991) and Hochbaum (1995):

δ-approximation. A polynomial algorithm is said to be a *δ-approximation* algorithm ($\delta \geq 1$) for a minimization problem P, if for every problem instance I with an optimal solution value $OPT(I)$, it delivers a solution that is at most δ times the optimum. Similarly, for maximization problems a δ-approximation algorithm ($\delta \leq 1$) delivers for every instance I a solution that is at least δ times the optimum.

Best achievable performance ratio. An optimization problem P has *best achievable performance ratio* α if there exists an α-approximation for P and no δ-approximation algorithm for P exceeding α exists (unless $\mathcal{P}$=$\mathcal{NP}$).

PAS. A family of approximation algorithms for a problem P, $\{A_\epsilon\}_\epsilon$, is called a *polynomial approximation scheme* or PAS if algorithm A_ϵ is a $(1+\epsilon)$-approximation algorithm and its running time is polynomial in the size of the input for a fixed ϵ.

10.2 Contents and Complexity Overview

Tables 10.1, 10.2, and 10.3 give an overview about the different label problems discussed in this chapter. The currently best known complexity results are given. The tables also serve as an index to this chapter.

10.3 Point Feature Label Placement

10.3.1 Map Labeling Related to SAT

Formann and Wagner (1991) have studied the point labeling problem in the 4-position model (see Figure 10.2(a)). More precisely, for a given set of points in the plane the aim is to label each point with an axis parallel rectangle such that every point coincides with one of the four corners of its label and no two rectangles overlap. The restriction that each point coincides with a corner of its label rectangle implies that each rectangular label can be placed in exactly four positions.

Formann and Wagner (1991) have proved that it is $\mathcal{NP}$-complete to decide this problem, even if all labels are equally sized squares. Independently, Kato and Imai (1988) and Marks and Shieber (1991) also have proved the $\mathcal{NP}$-completeness of this label problem. For brevity, we only sketch the idea of the $\mathcal{NP}$-completeness proof of Formann and Wagner. In their proof an instance

Table 10.1. Complexity results for map labeling decision problems.

Labeling Model		Decision Problem	References
point features	4-position model	$\mathcal{NP}$-complete	Section 10.3.1, Formann and Wagner (1991)
	2-position model	$\in \mathcal{P}$	Section 10.3.1, Formann and Wagner (1991)
	finite candidate set	$\mathcal{NP}$-complete	follows from Formann and Wagner (1991)
	1-, 2-, 4-slider model	$\mathcal{NP}$-complete	Section 10.3.3, Iturriaga and Lubiw (1997)
	point-adjacent model	$\mathcal{NP}$-complete	follows from Iturriaga and Lubiw (1997)
edge features	finite candidate set	$\mathcal{NP}$-complete	Section 10.4.2, Kakoulis and Tollis (1996)
	3-position model	$\in \mathcal{P}$	Section 10.4.1, Doddi et al. (1997); Poon et al. (1998); Strijk and van Kreveld (1999)
graphical features	finite candidate set	$\mathcal{NP}$-complete	follows from Formann and Wagner (1991)
	slider model	$\mathcal{NP}$-complete	followrs from Iturriaga and Lubiw (1997); Kakoulis and Tollis (1996)

of 3SAT is reduced to this point labeling problem. Gadgets for variables and clauses are constructed consisting of overlapping label candidates of point features. The variable gadgets are connected with clause gadgets according to the 3SAT formula by pipe gadgets that conserve encoded variable settings. The resulting map labeling problem has an overlap free solution if and only if the 3SAT formula is satisfiable.

Furthermore they have shown that the decision problem for the 2-position model (see Figure 10.2(a)) can be solved in time $\mathcal{O}(n \log^2 n)$ time by reduction to a 2SAT problem (Imai and Asano, 1986). Any point in the labeling instance is identified with a Boolean variable. The two label positions of the ith point are denominated with the variable setting x_i and $\overline{x_i}$. Let the two label candidates x_i and $\overline{x_j}$ overlap. Then x_i and $\overline{x_j}$ can not simultaneously appear in a solution, which is ensured by a clause of the form: $\overline{(x_i \wedge \overline{x_j})} = (\overline{x_i} \vee x_j)$. A satisfying truth assignment for the set of clauses yields a solution to the point labeling problem. If no such assignment exists it follows there is no overlap free solution to the point labeling problem.

Based on the solution method of the 2-position model they have further investigated the problem of finding a labeling for the 4-position model in which the labels have maximum size. The $\mathcal{NP}$-completeness proof implies that no polynomial approximation algorithm with an approximation factor

Table 10.2. Complexity results for map labeling number maximization problems.

Labeling Model		**Number Maximization**	References
point features	2-position model	unknown complexity	
	finite candidate set	$(1/log\, n)$-approximation algorithm, PAS	Section 10.3.2, Agarwal et al. (1998)
	1-, 2-, 4-slider model	$(1/2)$-approximation algorithm, PAS	Section 10.3.3, van Kreveld et al. (1998)
	point-adjacent model	polynomial approximation algorithm that places at least $(1-\epsilon)n$ squares of size at least $\frac{1}{1+\epsilon}$ OPT	Section 10.3.4, Doddi et al. (1997)
edge features	finite candidate set	heuristics	Section 10.4.2, Kakoulis and Tollis (1996, 1997)
	3-position model	unknown complexity	
graphical features	finite candidate set	heuristics	Section 10.5.1, 10.5.2, Kakoulis and Tollis (1998a,b); Wagner and Wolff (1998)
	slider model	heuristics	Section 10.6, Christensen et al. (1993, 1995); Edmondson et al. (1997); Zoraster (1986, 1990)

Table 10.3. Complexity results for map labeling size maximization problems.

Labeling Model		**Size Maximization**	References
point features	4-position model	best achievable performance ratio $\frac{1}{2}$, $\frac{1}{2}$-approximation algorithm	Section 10.3.1, Formann and Wagner (1991); Wagner (1994); Wagner and Wolff (1995a,b, 1997)
	point-adjacent model	$\frac{\sin(\pi/10)}{8\sqrt{2}}$-approximation algorithm	Section 10.3.4, Doddi et al. (1997)
edge feat.	3-position model	exact algorithm	Section 10.4.1, Doddi et al. (1997); Poon et al. (1998); Strijk and van Kreveld (1999)

exceeding $\frac{1}{2}$ exists, unless $\mathcal{P}=\mathcal{NP}$. Furthermore, Wagner has proved that an approximation algorithm that achieves this bound must take $\Omega(n \log n)$ time (for a detailed proof see (Wagner, 1994)). Formann and Wagner have introduced a $\frac{1}{2}$-approximation algorithm that increases stepwise the size of the squares. In each step label candidates are deleted permanently and then two label candidates are chosen for each point. For these candidates the algorithm for the 2-position model is applied. This procedure is repeated until there is no solution to the corresponding 2SAT problem. The resulting algorithm achieves the $\frac{1}{2}$-approximation bound and the time bound $\mathcal{O}(n \log n)$. Although optimal in a theoretical sense, this result is not useful for practical purposes because the solutions are usually too far off the optimal size. This approximation algorithm only works for equal sized square labels and not for arbitrary rectangles.

Because of this lack of practical results, Wagner and Wolff (1997, 1995a,b) have spent more time on the problem and published several papers that are based on the initial approach of Formann and Wagner discussed above. In their approaches, the important sizes of the labels, where conflicts are appearing, are computed. The maximum label size is determined with binary search on the list of important label sizes. Furthermore, label candidates are eliminated only temporarily for the current size of the binary search and then the 2SAT problem is used to decide solvability. All algorithms have the $\frac{1}{2}$-approximation bound and $\mathcal{O}(n \log n)$ runtime. To test their algorithms, Wagner and Wolff have implemented an example generator that creates random examples under several distributions. Christensen et al. (1997) have tested the latest algorithm of Wagner and Wolff (1997) against simulated annealing. Both algorithms have had about the same performance but the algorithm of Wagner and Wolff has been about three times faster.

Since each point feature has only four label candidates, we mention that these algorithms are best applicable to label problems where the labels are restricted to the four positions of the 4-position model. When labeling the nodes of a graph that contains other graphical features like edges, these algorithms have to be modified not to place labels that overlap them. It appears that overlaps with other graphical features can be avoided by simultaneously conserving the $\frac{1}{2}$-approximation bound. Nevertheless, the time needed to compute intersections of labels with other graphical features influences the running time of the algorithm.

10.3.2 Label Placement by Maximum Independent Set in Rectangles

Agarwal, van Kreveld, and Suri (1998) have formulated the point labeling problem as a maximum independent set problem in rectangles. The aim here is to label a maximum number of point features. For each point feature p_i a label r_i is given, that is, an axis-parallel rectangle of fixed size and a set of label positions such that each label position coincides with p_i. A feasible labeling is a subset of all label positions R such that the labels are pairwise disjoint and each point feature is labeled at most once. The labeling problem is to find the largest feasible labeling. Since all label candidates of a point p_i overlap at point p_i, it follows that at most one label candidate of a point is chosen. Thus, the labeling problem corresponds to finding a largest subset of pairwise disjoint rectangles in R. This corresponds to finding a maximum independent set of rectangles in R.

Since the computation of a maximum independent set of rectangles is known to be $\mathcal{NP}$-hard (Fowler et al., 1981; Imai and Asano, 1983), the authors have presented a $(\frac{1}{\log n})$-approximation algorithm based on the divide and conquer paradigm, that runs in polynomial time. The rectangles are divided according to the median x-coordinate of all rectangles into three sets: the rectangles that lie left of the median, lie right of the median, and intersect the median. They have computed the maximum independent set of the set of rectangles that intersect the median and recursively compute the approximate maximum independent set of the two other sets. In the merge step of the algorithm they choose the set with maximum cardinality of the following two sets: (1) the union of the approximative independent set of the rectangles left of the median and right of the median and (2) the maximum independent set of the rectangles that intersect the median.

For the case that all rectangles in R have unit height, an $(1 - \frac{1}{k+1})$-approximation algorithm is described that runs in $\mathcal{O}(n \log n + n^{2k-1})$ time, for any $k \geq 1$. The algorithm uses dynamic programming combined with the shifting technique of Hochbaum and Maass (1985). The rectangles are partitioned by a set of horizontal lines $l_1, \ldots, l_m$ such that the separation between two lines is strictly more than one, each line intersects at least one rectangle, and each rectangle is intersected by exactly one line. Let R_i be the set of rectangles in R that intersect l_i. They have defined subgroups $R_i^k = \bigcup_{j=0}^{k-1} R_{i+j}$ and sets $G_j = \bigcup_{i \geq 0} R_{i(k+1)+j}^k = R \setminus \bigcup_{i \geq 0} R_{i(k+1)+j}$ (see Figure 10.3). G_j is obtained from R by deleting the rectangles that intersect every $(k+1)$st line, starting with the jth line. Note that two subsets $R_{i_1(k+1)+j+1}^k$ and $R_{i_2(k+1)+j+1}^k$ of G_j are disjoint if $i_1 \neq i_2$. For example, line $l_{(i)(k+1)+j}$ and line $l_{(i+1)(k+1)+j}$ separates $R_{i(k+1)+j+1}^k$ from the other subsets. The union of the independent sets of the subsets of a set G_j yields an independent set for all rectangles in G_j.

The rectangles in $R \setminus G_j$ are intersected by at most $\lfloor m/(k+1) \rfloor$ lines. Therefore, computing a maximum independent set for each G_j and choosing the largest one yields a $(1 - \frac{1}{k+1})$-approximation. The maximum independent set of a subgroup of G_j is computed by dynamic programming by means of a 2-dimensional invariant. One step in the dynamic program takes $\mathcal{O}(n^{2k-1})$ time. The runtime of $\mathcal{O}(n \log n)$ can only be achieved for the special case $k \leq 2$.

This approach works only for labels with unit height, but varying widths, and the authors leave the solution for arbitrary rectangles with a constant

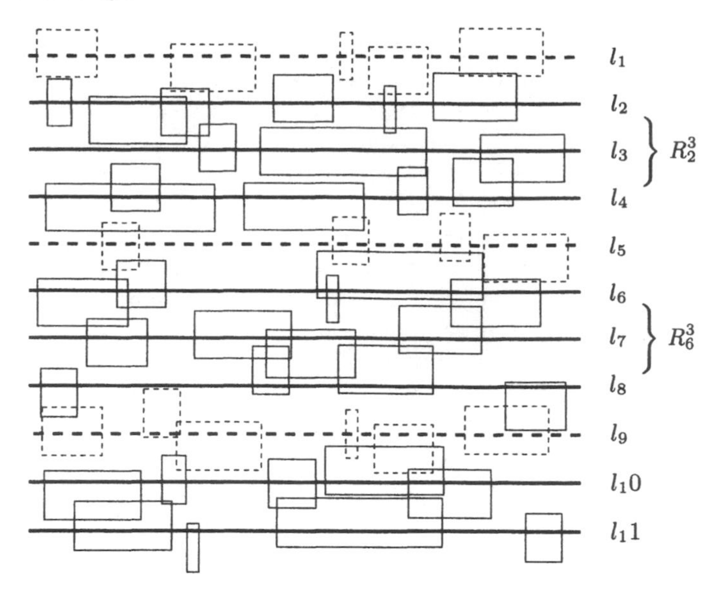

Fig. 10.3. The set G_1 for $k = 3$.

factor approximation scheme as an open problem. The label candidate set for any point may be arbitrarily large, as long as all label candidates coincide with their point feature. This algorithm can be easily modified to label the nodes of a graph overlap free with other graphical features. One simply has to eliminate those label candidates that overlap graphical features. The bounds stay valid in respect to the reduced label candidate set.

10.3.3 Point Set Labeling with Sliding Labels

Van Kreveld, Strijk, and Wolff (1998) have discussed algorithms that aim at labeling a maximum number of point features, where the label positions are not restricted to a finite number of positions. They have defined three models, the 1-slider, 2-slider, and 4-slider model, in which the labels can move continuously (see Figure 10.2(b)). Iturriaga and Lubiw (1997) have shown that the decision problem whether a solution to a point labeling problem in the 1-slider model exists is $\mathcal{NP}$-complete.

Before providing a $\frac{1}{2}$-approximation algorithm for the three slider models, van Kreveld et al. have compared the three slider models with the fixed position models that are shown in Figure 10.2(a). They have been concerned with the question: "How many more points can be labeled in one model than another, in any point set?". In order to quantify this question they have introduced the *ratio of two models*. Let P be a set of n points in the plane. Let M_1 and M_2 be two models for labeling P, and let $\mathrm{opt}_{M_1}(P)$ and $\mathrm{opt}_{M_2}(P)$ be the maximum number of points of P that receive a label in the models M_1

and M_2, respectively. The *ratio of the models M_1 and M_2* is the supremum of the ratio $\text{opt}_{M_1}(P)/\text{opt}_{M_2}(P)$ for $n \to \infty$ and maximized over all point sets P with n points.

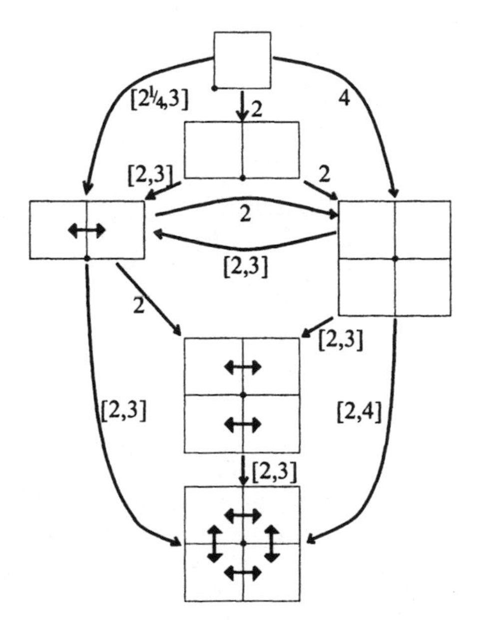

Fig. 10.4. Upper and lower bounds for the ratios of two models.

They have shown lower and upper bounds for the ratios of all pairs of models. These ratios are depicted in Figure 10.4. For example the maximum number of points labeled in the 2-position model can be at most twice the maximum number in the 1-position model. Thus, the lower bound and upper bound for the ratio of the 2-position and the 1-position model is 2. The arcs indicate the partial order on the models.

Furthermore the authors have provided a $\frac{1}{2}$-approximation algorithm for the 4-slider model that runs in $\mathcal{O}(n \log n)$ time. The algorithm can be applied with minor changes to the 2-slider and 1-slider model. Given a set of points, a set of labels that have already been placed, and a set of points that are unlabeled, the *leftmost label* is defined to be the label whose right edge is leftmost among all *possible* labels of unlabeled points. A label is possible if it does not overlap any placed label. The algorithm is a greedy strategy that repeatedly chooses the leftmost label. It labels at least half the number of points labeled in an optimal solution. In order to compute the leftmost label efficiently, the *right envelope* of all placed labels l_i and their copies l_i', precisely one unit below l_i is stored. The right envelope is a function in y, where $f(y) = \max\{\max\{x|(x,y) \text{ is occupied by } l_i \text{ or } l_i'\}, -\infty\}$ (see Figure 10.5). A reference point of a label is its lower left vertex. Therefore, the right envelope

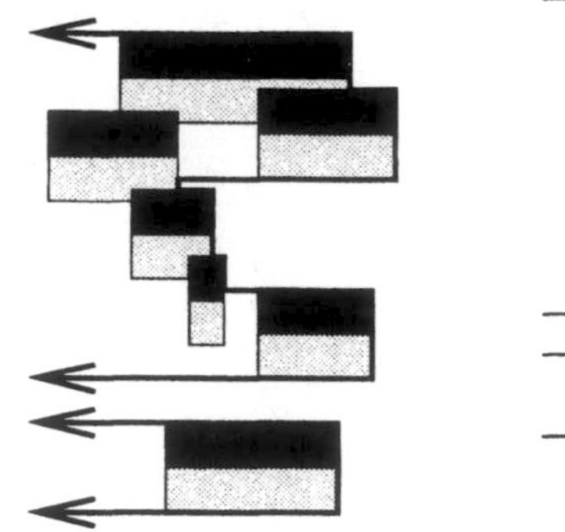

Fig. 10.5. Right envelope.

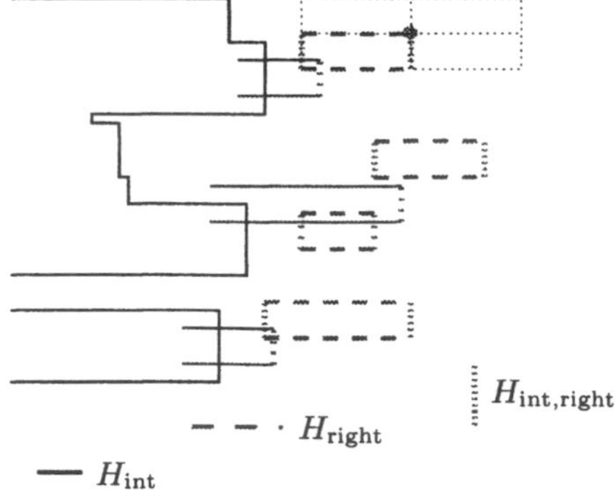

Fig. 10.6. The heaps storing the reference segments.

incloses exactly all reference point positions that are impossible. Observe that the union of the reference points of all possible label positions for one point is the boundary of the lower left label position (see Figure 10.6). These reference points are subdivided into the two horizontal segments and the two vertical segments. These segments are stored in three different heaps: H_{right} contains all horizontal segments that lie right of the envelope, H_{right} contains all horizontal segments that intersect the envelope and $V_{\text{int,right}}$ contains all segments that intersect or lie right of the envelope. These heaps allow to query for the leftmost possible label. By applying common geometric data structures the authors avoided using a brute force method and thus reduced the runtime of the algorithm.

The $\frac{1}{2}$-approximation algorithm is only valid for rectangles with equal height but variable width. The algorithm is applicable for arbitrary rectangles but the authors cannot guarantee the approximation factor any more. In their experiments on real world data the 4-slider model performs 10-15% better than the 4-position model.

Furthermore, they have shown that for each of the slider models and for any constant $\epsilon > 0$, there is a polynomial time algorithm that labels at least $1 - \epsilon$ times the maximum number of points that can be labeled. The concept of the approximation algorithm is similar to the $(1 - \frac{1}{k+1})$-approximation algorithm of Agarwal et al. (1998) described in Section 10.3.2. Due to the higher running time of $\mathcal{O}(n \log n + n^{2k-1})$ time, the algorithm is primarily of theoretical interest.

Strijk and van Kreveld have suggested a modification of the $\frac{1}{2}$-approximation algorithm for the 4-slider model, when other line segments are in the scene which may not be intersected.[2] This could be of interest for labeling graphs.

[2] Marc van Kreveld, personal communication (1999).

10.3.4 Label Placement with Point-Adjacent Labels

Doddi, Marathe, Mirzaian, Moret, and Zhu 1999 have studied the point labeling problem for a set of n points S such that all labels are squares of the same size, the feature may lie anywhere on the boundary of its label region and the label size is maximized. Differently to all other approaches the labels are allowed to rotate around their labeling point (see Figure 10.2(c)).

They have given a $\frac{\sin(\pi/10)}{8\sqrt{2}}$-approximation algorithm for this problem that runs in $\mathcal{O}(n \log n)$ time. In order to achieve this factor, they have derived an upper bound for the maximum size of the labels by means of the well studied *minimum k-diameter* theory (Datta et al., 1993; Eppstein and Erickson, 1994). The *diameter* of a set S is the maximum distance between any two points in the set. The *minimum k-diameter* of S, denoted $D_k(S)$, is the smallest diameter among all subsets of S of size k. It is easy to see that a circle of radius $\frac{D_k(S)}{2}$ centered at a point $p_i \in S$ contains at most $k-1$ points. This led to an upper bound for the maximum label size of at most $\frac{D_5(S)}{\sin(\pi/10)}$. The $\frac{\sin(\pi/10)}{8\sqrt{2}}$-approximation algorithm labels each point with a square of size at least $\frac{D_5(S)}{8\sqrt{2}}$. The algorithm randomly selects a point p_i and labels this and all points that have distance at most $\frac{D_5(S)}{\sin(\pi/10)}$ to p_i according to a case distinction of their number (which is at most 4) and their positions. This procedure is repeated recursively until every point is labeled. The calculation of the minimum 5-diameter dominates the runtime with $\mathcal{O}(n \log n)$ time. The labeling process can be done in $\mathcal{O}(n)$ time.

In case the point features are to be labeled with circles of maximum radius the authors can apply their techniques for regular squares and achieve the following results: Firstly, the labeling circles have radius at most $(4 + 2\sqrt{3})D_3(S)$. Secondly, there exists a $\frac{1}{8(2+\sqrt{3})}$-approximation algorithm that runs in $\mathcal{O}(n \log n)$ time, for the same reasons as in the squared case. This approximation factor has recently been improved by Strijk and Wolf (1999).

Furthermore, they have given a bicriteria approximation algorithm for the variant of the problem in which some point features are allowed to remain unlabeled and each square must be placed adjacent to its point feature such that its sides are parallel to the axes. (This complies to the 4-slider model). For any $\epsilon > 0$ their approximation algorithm finds a placement for at least $(1-\epsilon)n$ squares of size at least $\frac{1}{1+\epsilon}$ OPT, where OPT denotes the size of the squares in an optimal solution. It needs $\mathcal{O}(n \log D_5(S))$ time. The algorithm can be modified such that it applies to other regular label shapes like regular polygons.

The disadvantages of these approaches are obvious. They only work for equal sized and regular shapes of labels. The approximation guarantee for the first two algorithms is very small and far off the optimal size of a labeling. Furthermore, the point-adjacent label model seems to be only of theoretical interest and the algorithm can not be modified easily to label point features of a graph. The algorithms do not provide the possibility of excluding label positions that overlap other graphical features. Note that the proceedings version (Doddi et al., 1997) contains errors.

10.3.5 Combining Graph Labeling with Compaction

In this section we investigate in the Compaction and Labeling (COLA) problem which is an orthogonal graph drawing and labeling problem in which the graph drawing and labeling is solved simultaneously (Klau and Mutzel, 1999a). A graph drawing is called orthogonal, when each edge segment is either horizontal or vertical. See Chapter 6 for a summary on orthogonal graph drawing.

In the COLA problem we are given an *orthogonal representation* of the graph. This orthogonal representation is a description of a planar and orthogonal embedding of the graph. Additionally to the planar topology it describes the shape of the drawing, i.e., for each edge the order of the bends and the angle formed with the following edge in the appropriate face are given. In order to draw the graph, coordinates have to be assigned to the vertices.

For each vertex we are given a set of labels, each of which is fixed in size. The authors extend the compaction techniques known from orthogonal graph drawing (see Chapter 6) to include conditions on the labels and their positions. Analogously to the usual compaction problem, necessary conditions for an orthogonal graph drawing of minimum total edge length are encoded in a special pair of graphs, the so called *shape graphs*. These conditions mainly are relative positioning constraints and minimum distance relations between graph components. The pair of shape graphs is extended with label related conditions. These conditions are that a label has to touch its vertex and that a label may not overlap other labels or objects. Thus, the labeling model is the slider model. After that, an integer linear program (ILP) is constructed from the pair of *labeled shape graphs*. The authors implemented a branch and cut algorithm which solves this kind of ILPs. The solution gives an orthogonal labeled drawing of the graph with small total edge length. We now describe these steps more in detail, starting with the construction of the pair of shape graphs.

The construction of an orthogonal embedding of a graph from its orthogonal representations was shown to be $\mathcal{NP}$-hard (Patrignani, 1999b). For the construction of the pair of shape graphs, Klau and Mutzel transform the orthogonal representation in a *simple* orthogonal representation by replacing each bend with a vertex. The simple orthogonal representation partitions the set of edges in a set of horizontal edges E_h and a set of vertical edges E_v. A horizontal (resp. vertical) *subsegment* in the simple orthogonal representation is a connected component in (V, E_h) (resp. (V, E_v)). A maximally connected component is a maximal set of consecutive edges of one direction and is called a *segment*. Furthermore, each edge is a subsegment and each vertex v is incident to exactly one horizontal and one vertical segment which are denoted by vert(v) and hor(v). Let v_l, v_r, v_t, and v_b be the leftmost, rightmost, topmost, and bottommost vertex on a segment s. Then, the *left, right, top,* and *bottom limits* of s are $l(s) = \text{ver}(v_l)$, $r(s) = \text{ver}(v_r)$, $t(s) = \text{hor}(v_t)$, and $b(s) = \text{hor}(v_b)$.

A pair of shape graphs $\langle(S_v, A_h), (S_h, A_v)\rangle$ is defined as follows. Each horizontal (resp. vertical) segment is represented by a vertex in the set S_h (resp. S_v). Weighted arcs between the segments characterize relative positioning relations between the segments. All arcs (s_i, s_j) in a shape description have weight 1, indicating that the coordinate difference of segments s_i and s_j must be at least 1. Arcs between horizontal segments are in set A_v and arcs between vertical segments are in set A_h. For a detailed description of the construction and properties of the pair of shape graphs see (Klau and Mutzel, 1999a). The compaction problem is discussed in more detail in the Chapter 6.

The resulting pair of shape graphs is extended with a description of the constraints on the labels. Each label is modeled by a rectangle bounded by the segments $l_\lambda, r_\lambda, t_\lambda$, and b_λ (see Figure 10.7). For the segments l_λ and r_λ a vertex is added to S_v, and for the segments t_λ and b_λ a vertex is added to S_h. The set of arcs A_h is enlarged by arcs of the type (l_λ, r_λ) of weight $w(\lambda)$ which corresponds to the width of label λ. Analogously, the set of arcs A_v is enlarged by arcs of the type (l_λ, r_λ) of weight $h(\lambda)$ which corresponds to the height of label λ. Let $v = a(\lambda)$ be the vertex label λ is associated with. In case that a vertex is represented by a box, the vertex is bounded by the segments l_v, r_v, t_v, and b_v as illustrated in Figure 10.8, otherwise $l_v = r_v = \text{ver}(v)$ and $b_v = t_v = \text{hor}(v)$. A feasible position of a label λ relative to it vertex $a(\lambda)$ is given if and only if:

1. The left side of λ does not lie to the right side of $a(\lambda)$.
2. The right side of λ does not lie to the left side of $a(\lambda)$.
3. The bottom side of λ does not lie above the top side of $a(\lambda)$.
4. The top side of λ does not lie below the bottom side of $a(\lambda)$.

These conditions are realized in the pair of shape graphs by adding the arcs $(l_\lambda, r_{a(\lambda)})$ and $(l_{a(\lambda)}, r_\lambda)$ to A_h and adding the arcs $(b_\lambda, t_{a(\lambda)})$ and $(b_{a(\lambda)}, t_\lambda)$ to A_v for each label λ. These arcs have weight 0. For an illustration see Figure 10.8.

A *complete* pair of labeled or unlabeled shape graphs defines a unique labeled or unlabeled orthogonal drawing of the graph. A pair of labeled shape graphs is *complete*, if and only if both arc sets do not contain non negative cycles and for every pair of segments $(s_i, s_j) \in (S_v \cup S_h) \times (S_v \cup S_h)$ one of the four conditions holds:

$$(1)\ r(s_i) \xrightarrow{+} l(s_j) \qquad (3)\ t(s_j) \xrightarrow{+} b(s_i)$$

$$(2)\ r(s_j) \xrightarrow{+} l(s_i) \qquad (4)\ t(s_i) \xrightarrow{+} b(s_j),$$

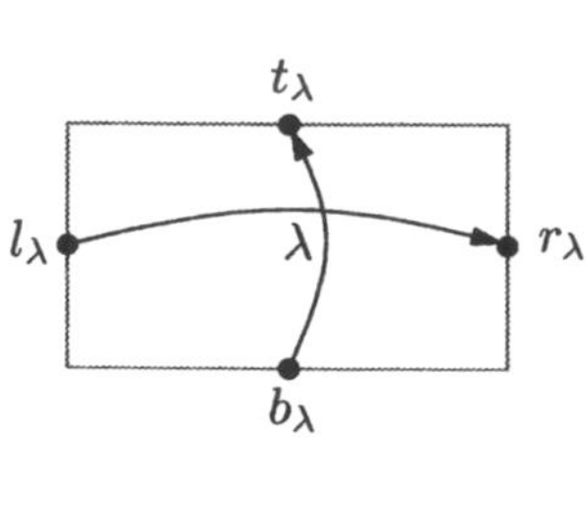

Fig. 10.7. Four segments

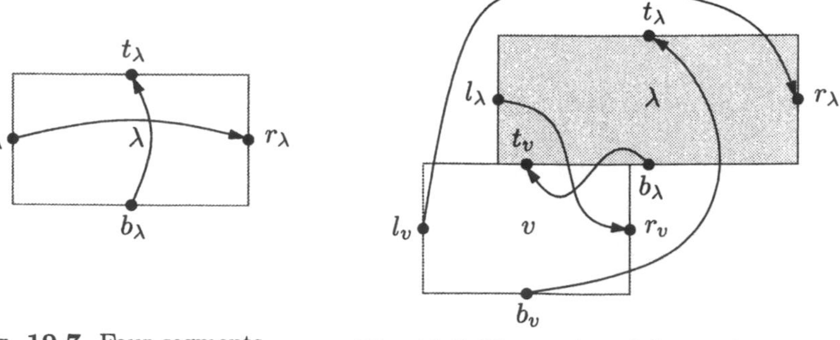

Fig. 10.8. Illustration of the arcs between a vertex and its label.

where $s_i \xrightarrow{+} s_j$ denotes the existence of a path of non negative weight between s_i and s_j. Usually a pair of labeled shape graphs is not complete and there are many possibilities for extending the labeled shape graphs to a complete pair of labeled shape graphs. Klau and Mutzel showed that there exists a simple labeled orthogonal drawing for a pair of labeled shape graphs σ_L if and only if there exists a complete labeled extension τ_L of σ_L. Thus, the problem is to find a suitable extension of the pair of labeled shape graphs. Therefore, the pair of labeled shape graphs is transformed into an ILP.

For the description of the ILP we have to define a set O which denotes the set of objects that should be compacted. It consists of the segments corresponding to consecutive edges, labels, and vertices whose images are boxes of non zero size.

Starting at a given labeled pair of shape graphs $\sigma_L = \langle (S_v, A_h), (S_h, A_v) \rangle$ the authors formulate an ILP to solve the COLA problem. For each potential additional arc (s_i, s_j) which might be in some complete extension, they introduce a variable x_{ij} which is one if arc (s_i, s_j) is contained in the extension, otherwise zero. We refer by X to the set of binary variables. Additionally, for each segment $s \in \sigma_L$ they introduce a variable $c_s \in \mathbb{Q}$ denoting the coordinates of s. The ILP is defined as follows:

$$\min \sum_{e \in E_h} c_{r(e)} - c_{l(e)} + \sum_{e \in E_v} c_{t(e)} - c_{b(e)} \qquad \text{subject to} \tag{10.1}$$

$$x_{r_o,l_p} + x_{r_p,l_o} + x_{t_o,b_p} + x_{t_p,b_o} \geq 1 \qquad \forall (o,p) \in O \times O, o \neq p \tag{10.2}$$

$$c_j - c_i \geq w_{ij} \qquad \forall (s_i, s_j) \in A_h \cup A_v \tag{10.3}$$

$$c_j - c_i - (M + w_{ij})x_{ij} \geq -M \qquad \forall x_{i,j} \in X \tag{10.4}$$

$$x_{ij} \in \{0,1\} \qquad \forall x_{ij} \in X \tag{10.5}$$

Inequalities 10.2 model the characterization of separation, i.e., the existence of necessary paths between distinct objects in an extension. Inequalities 10.3 force the coordinates to obey the distance rules. The coordinates of two segments s_i and s_j that are connected by an arc of weight w_{ij} are forced to have distance at least w_{ij}. Inequalities 10.4 force the segments of potential additional arcs to have distance at least w_{ij} in help of a big constant.

A solution of the ILP can be computed with a branch and bound algorithm. Klau and Mutzel showed that each feasible solution of the ILP defines a labeled orthogonal embedding with appropriate shape.

10.3.6 Optimization Algorithm for Point Set Labeling

Verweij and Aardal (1999) have given an optimization algorithm for point set labeling in the 4-position model in which the number of labels is to be maximized. They have formulated the labeling problem as an 0-1 integer linear program. In contrast to several other approaches using mathematical programming methods, they have used optimization methods that are specific for point labeling problems. For an introduction to linear programming we recommend (Chvátal, 1983a), for an overview of mathematical programming methods used for map labeling see (Zoraster, 1986, 1990).

The authors make use of the observation that the four label candidates of a point p_i overlap at p_i. This implies that at most one label candidate of each point can be in a maximum set of non overlapping label rectangles (see Section 10.3.2). Thus, the label problem corresponds to finding a maximum independent set of label rectangles.

The algorithm is a specialized branch-and-cut algorithm for finding a maximum independent set of rectangles. It can be used to find provably optimal solutions for map labeling instances up to 950 cities within modest computation time.

For a fixed label size σ the authors have formulated the maximum independent set of rectangles problem as an integer linear program with 0/1 variables Π. Kucera et al. (1993) have shown that there are only $\mathcal{O}(n^2)$ sizes the optimal label size can take, where n is the number of label candidates. Optimizing over those can be done by solving $\mathcal{O}(\log n)$ maximum independent set of rectangle problems.

The integer linear program has the following form: For each label rectangle a 0/1 variable x_i is defined. The objective function is to maximize the sum over all x_i. Additionally, there is an inequality $x_i + x_j \leq 1$ if the label rectangle of x_i overlaps the label rectangle of x_j.

A branch-and-cut algorithm is a branch-and-bound algorithm where cutting planes might be added to the set of restrictions. The root of the branch-and-bound search tree consists of the LP relaxation $\overline{\Pi}$ of Π. At level k of the tree we have a collection of problems $\overline{\Pi}_1^k, \ldots, \overline{\Pi}_l^k$ such that all integral solutions of Π are integral solutions of $\overline{\Pi}_1^k \cup \cdots \cup \overline{\Pi}_l^k$.

The branch-and-cut algorithm maintains the best known value of an integer solution (which also is a lower bound) as well as the best upper bound for an integral solution. At iteration i an open problem $\overline{\Pi}_j^i$ is selected and solved. If $\overline{\Pi}_j^i$ is infeasible, then $\overline{\Pi}_j^i$ is removed from the list of open problems and the algorithm proceeds with the next iteration. Otherwise, if the optimal solution $\overline{x}^i$ to $\overline{\Pi}_j^i$ is integral, $\overline{\Pi}_j^i$ is removed from the list of open problems and the best known value is updated if $\overline{x}^i$ is greater than the best known value. Otherwise, a component r of the solution vector that is not integral is chosen and $\overline{\Pi}_j^i$ is substituted by two new open problems that are formulated by adding the constraint $x_r = 0$ or $x_r = 1$, respectively. Since it is desirable to restrict the number of open problems, Verweij and Aardal have first invoked cutting plane algorithms and variable setting algorithms before the generation of the new open problems.

The generation of good lower and upper bounds is crucial for pruning the search tree. In order to obtain good upper bounds the LP-relaxation $\overline{\Pi}$ can be strengthened by adding valid inequalities. The goal of these inequations is to cut off the current non-integral optimal solution of a polyhedra $\overline{\Pi}_j^k$. Therefore, before involving the variable setting algorithms two families of valid inequalities are used: clique inequalities and odd hole inequalities. For the clique inequalities refer to (Verweij and Aardal, 1999; Nemhauser and Sigismondi, 1992). The idea of the odd hole inequalities is as follows. The conflict graph of the set of rectangles consists of a vertex for each rectangle and an edge for each pair of vertices where the corresponding rectangles of the vertices overlap. In case that this conflict graph contains an odd circle of length l, it follows that at most $\frac{l-1}{2}$ rectangles out of the rectangles corresponding to the vertices in the circle can be placed. These circles lead to inequalities that can be used as cutting planes.

Verweij and Aardal have used three algorithms to reduce the number of new open problems. The first algorithm is called variable setting by reduced costs. If for some index v of the solution vector $\overline{x}^i$ we have that $\overline{x}_v^i = 0$ (or $\overline{x}_v^i = 1$) and if the reduced costs of v are smaller than the gap between the best lower bound and $\overline{x}^i$, then we can set $\overline{x}_v^j = 0$ (or $\overline{x}_v^j = 1$) for all j in the subtree of the tree rooted at i.

The second algorithm is called logical implication algorithm. It makes use of the following observation: Let $\overline{x}_v^i = 1$, then it follows that $\overline{x}_w^i = 0$ if the corresponding rectangles of v and w overlap. This situation is searched in the logical implication algorithm.

The third method is the variable setting by recursion and substitution. It is based on the following observation: Let G_1 and G_2 be a partition of the label rectangles where the rectangles of G_1 do not overlap the label rectangles of G_2. We say G_1 and G_2 are independent. Then, the maximum independent set of rectangles of $G_1 \cup G_2$ is the maximum independent set of rectangles of G_1 unified with the the maximum independent set of rectangles of G_2. Focus on a node i in the branch-and-cut tree. We call all variables that are not set to 1 or 0 *free*. In the third variable setting algorithm they exploit the fact that finding a maximum independent set of a small component is easy. Therefore, they search for independent partitions of rectangles corresponding to free variables. Then, they determine the maximum independent set of rectangles for these partitions and substitute the partial solutions back.

By incorporating a local search algorithm feasible solutions can be found and thereby lower bounds for the optimal value. The local search algorithm consists of the generation of an integral solution through rounding and several local optimization steps.

10.4 Line Feature Label Placement

10.4.1 Labeling a Rectilinear Map

A rectilinear map consists of n disjoint horizontal and vertical line segments. The problem is to label each line segment with a rectangle of height B and length the same as the segment. The three positions that are allowed for the label are depicted in Figure 10.9(a). The corresponding decision problem is called Three Position Rectilinear Segment Labeling (3RSL) problem. This problem can be solved in polynomial time. It has been first studied by Doddi et al. (1997) and then by Poon et al. (1998). Recently, Strijk and van Kreveld (1999) improved the solution of Doddi et al. and Poon et al..

In Figure 10.9(b) the rectilinear area around each line segment i is divided into four regions. A label on the line segment will occupy region r_{i2} and r'_{i1}. We introduce two boolean variables x_{i1} and x_{i2} with values 0 or 1. $x_{i1} = 1$ means that r_{i1} is part of the label placement and r'_{i1} is not. Analogously, $x_{i2} = 1$ means that r_{i2} is part of the label placement and r'_{i2} is not. The expression $(\overline{x_{i1}} \vee x_{i2})$ encodes that setting x_{i1} implies the setting of x_{i2} (also denoted as $(x_{i1} \Rightarrow x_{i2})$). Thus, n clauses are needed to force the corresponding behavior for all segments. For two overlapping regions, e.g. r_{i1} and r'_{j2} the clause $\overline{(x_{i1} \wedge \overline{x_{j2}})} = (\overline{x_{i1}} \vee x_{j2})$ enforces that at most one of these regions is labeled.

Poon et al. as well as Strijk and van Kreveld have shown how to find all pairwise intersections of label regions in $\mathcal{O}(n \log n)$ time by computing the horizontal and vertical decomposition and using a sweep line algorithm. In contrast to Poon et al., Strijk and van Kreveld have calculated the intersections implicitly only when needed. This reduces the number of clauses for 2SAT and thus the runtime. The resulting 2SAT instance is solved by a combination of a graph algorithm of Masuda et al. (1983) and the rectilinear segment algorithm of Imai and Asano (1986).

In total they have improved the running time for the 3RSL problem for n line segments from $\mathcal{O}(n^2)$ (Poon et al., 1998) to $\mathcal{O}(n \log n)$ time. Doddi et al. (1997) have shown that there exists a lower bound of $\Omega(n \log n)$ for

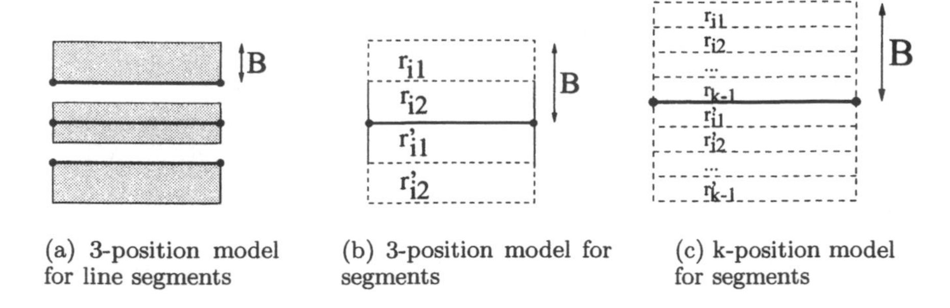

(a) 3-position model for line segments

(b) 3-position model for segments

(c) k-position model for segments

Fig. 10.9. Position models.

the runtime of this problem. The version of Strijk and van Kreveld holds this lower bound.

Given the rectilinear line segments, the maximum height rectilinear segment labeling problem is to find the maximum B such that there exists a placement for n non-overlapping rectangular labels, each attached to a distinct line segment, of height B and of length equal to the segment length. Strijk and van Kreveld have solved this problem by performing a binary search on a list of possible values of maximum label height. Strijk and van Kreveld have computed this list only implicitly and therefore improved the running time of this problem from $\mathcal{O}(n^2 \log n)$ (Poon et al., 1998) to $\mathcal{O}(n \log^2 n)$ time.

In case the axis parallel constraint on the segments is dropped, the computation of the pairwise intersection of labeling regions has to be adapted. This is done using the partition/cutting tree data structure for segment/segment intersection searching of Agarwal and Erickson (1997), which leads to an $\mathcal{O}(n^{4/3}\text{polylog}(n))$ time algorithm for the corresponding decision problem.

Note that the 3RSL can be generalized to the k-position rectilinear segment labeling problem as shown in Figure 10.9(c).

10.4.2 On the Edge Label Placement Problem

Kakoulis and Tollis (1996, 1997) have written several papers on the edge labeling problem with extensions to graphical features. They have defined the edge labeling problem as an integer linear program: Each label position is associated with a cost that reflects the ranking of that label in terms of *unambiguity* and the number of *overlaps* with other graphical features and labels. The objective function is the sum of the costs of all assigned labels. The constraints of the integer program guarantee that any edge will have exactly one label assigned to it. They have shown that the edge labeling problem is $\mathcal{NP}$-complete and remains $\mathcal{NP}$-complete even when the label candidates of each edge are restricted to have the same size and do not overlap (Kakoulis and Tollis, 1996). Similarly to Formann and Wagner (1991) they have reduced from 3SAT.

One of their approaches of labeling edges of graph drawings is restricted to unit height, axis-parallel rectangular labels (Kakoulis and Tollis, 1997). They divide the input drawing of the given graph G into consecutive horizontal strips of equal height. The height of each strip is equal to the height of the labels. Then, the label candidates are determined for each edge inside the strips. Since a label must lie entirely in a horizontal strip and since labels that overlap a vertex or an edge are not considered, each label position overlaps at most one other label position. If two label positions overlap, they are grouped together. A bipartite *matching graph* $G_m = (V_e, V_g, E_m)$ is defined where each edge in G corresponds to a node in V_e and each label group corresponds to a node in V_g. An edge (e, g) in E_m connects a node e in V_e with a node g in V_g if edge e in G has a label position in group g. Furthermore, each edge (e, g) in E_m is assigned a cost corresponding to the cost of labeling edge e with a label position in g. A maximum cardinality minimum weight matching of the matching graph G_m produces a solution to the edge labeling problem of this restricted model. Since the best known algorithm for the computation of a maximum cardinality minimum weight matching takes more than quadratic time with respect to the size of the matching graph (Goldberg and Kennedy, 1995; Tarjan, 1983) the authors have presented an algorithm that solves the maximum cardinality matching problem for their special version of a matching graph in linear time. They have suggested using this algorithm as a fast heuristic for the computation of a maximum cardinality minimum weight matching. Note that the running time of this algorithm is independent of the size of the graph; it depends on the chosen graph drawing, the height, and the number of strips.

The restriction that the label candidates have to lie entirely in horizontal strips is a weakness of this algorithm which causes problems with horizontal edges. The horizontal edges have to be located on the boundary of the strips, otherwise no label candidates would be assigned to such edges. In the given form this approach is not applicable for orthogonal drawings. Furthermore, edges with a small slope generate very few label candidates. Although the maximum cardinality minimum weight matching is solved optimally, nothing can be said about the quality of the labeling with respect to an optimal labeling in a model that allows more label candidates especially for near horizontal edges.

10.5 Graphical Feature Label Placement

10.5.1 Map Labeling Reduced to Graph Problems

Kakoulis and Tollis (1998b) have suggested a unified approach to labeling graphical features that is suitable for labeling all kinds of graphical features with axis parallel rectangular labels of variable size. In particular, their unified approach is also suitable for labeling orthogonal drawings of graphs. Recall that a drawing is called orthogonal if each line segment is either vertical or horizontal. Their approach basically consists of three steps.

Firstly, a large set of label candidates is generated for each graphical feature. Other than in the previous section, the label candidates are no longer restricted to lie in strips. Each label candidate is assigned a cost that reflects the ranking of that label.

Secondly, a graph $G_R = (V_l, E_l)$ is created, where every node l in V_l corresponds to a label candidate. An edge (l_1, l_2) in E_l connects two nodes in V_l, if the corresponding label candidates overlap. Nodes are heuristically removed from G_R until each connected component of G_R is a *clique*. In a clique, each node is adjacent to all other nodes. Since each node in G_R corresponds to a label candidate it follows that the set of label candidates is subdivided into subsets (the cliques) where each label candidate intersectd any other label candidate of that subset. Clearly, at most one label candidate of a clique can be positioned in an overlap-free labeling.

Thirdly, a bipartite matching graph is defined $G_m = (V_f, V_c, E_m)$, where every graphical feature corresponds to a node in V_f and every clique of label candidates corresponds to a node in V_c. An edge (f, c) in E_m connects a node f in V_f with a node c in V_c if feature f has a label candidate in clique c. The edge is associated with the cost of labeling feature f with its label candidate in c. A maximum cardinality minimum weight matching in G_m yields an optimal label assignment with no overlaps with respect to the reduced label set. The overall running time is dominated by the running time of the maximum cardinality minimum weight matching algorithm that is applied. (See Goldberg and Kennedy (1995); Tarjan (1983) for efficient algorithms).

They have further suggested a postprocessing based on local and exhaustive search by exploring the solution space in three ways. They have locally shifted assigned labels to create space for new labels. They have searched the solution space to find if there is enough space to assign a label after repositioning already assigned labels and they have searched the solution space after relaxing the restrictions on the quality of the label assignment by allowing labels to overlap their associated graphical features and/or other labels.

Since the nodes from G_R are removed heuristically until each connected component is a clique, nothing can be said about the quality of the algorithm. Kakoulis and Tollis have not commented on the runtime of their approaches. However, the time required by the matching algorithm of $\mathcal{O}(n^{2.5})$ is certainly a good lower bound. Due to the exponential behavior of their postprocessing steps they allow only a few steps of backtracking in order to keep polynomial runtime of their algorithm.

For the graphical feature labeling problem where each graphical feature can receive more than one label, say k, Kakoulis and Tollis (1998a) have suggested two algorithms. One consists of a loop that executes an extension of (Kakoulis and Tollis, 1997, 1998b) k times. The other reduces the matching graph to a network flow problem, where the capacities are adjusted according to the number of labels per feature.

All their approaches are part of the Graph Layout Toolkit, a Tom Sawyer Software product[3] which is a graph layout and editor toolkit. Several test results are given in their articles, but none of their tests compares their approaches with other edge labeling or general labeling algorithms.

10.5.2 A Combinatorial Framework for Map Labeling

Given is a set of graphical features and a finite set of label candidates for each feature of arbitrary shape. Wagner and Wolff (1998) have designed an algorithm for label number maximization that transforms the label problem into a combinatorial problem that is related to a concept suggested in the artificial intelligence community under the name *constraint satisfaction* (Mackworth and Freuder, 1985; Knuth and Raghunathan, 1992; Freuder and Wallace, 1992).

We chose this approach to be discussed more in detail since all kinds of graphical features and all kinds of label shapes are allowed and since we are convinced, that this approach allows to be extended, improved, or adapted to special cases.

The transformation of the label problem into a constraint satisfaction problem (CSP) is defined as follows: For each graphical feature to be labeled the authors define a variable v_i and a domain D_i and associate the variable with its domain. Each variable can be assigned certain values. Each value corresponds to a label candidate of the graphical feature corresponding to the variable. The set of constraints are binary constraints excluding pairs of variable values, namely those where the corresponding labels intersect.

Other than in the artificial intelligence community, Wagner and Wolff have defined an optimal solution to be a violation free assignment of values to the variables for as many variables as possible. An *m-consistency* algorithm removes all inconsistencies among m of the given n variables.

They have presented rules which, applied exhaustively to a CSP, achieve a weak form of local consistency. They have referred to them as weak since they only prove applying these rules does not destroy an optimal solution. They have not proved that the set of rules they suggest is in any sense complete.

We now give their CSP rules together with a graphical illustration of them. Figures 10.10(a), 10.10(b) and 10.10(c) show typical situations before

[3] `http://www.tomsawyer.com/`

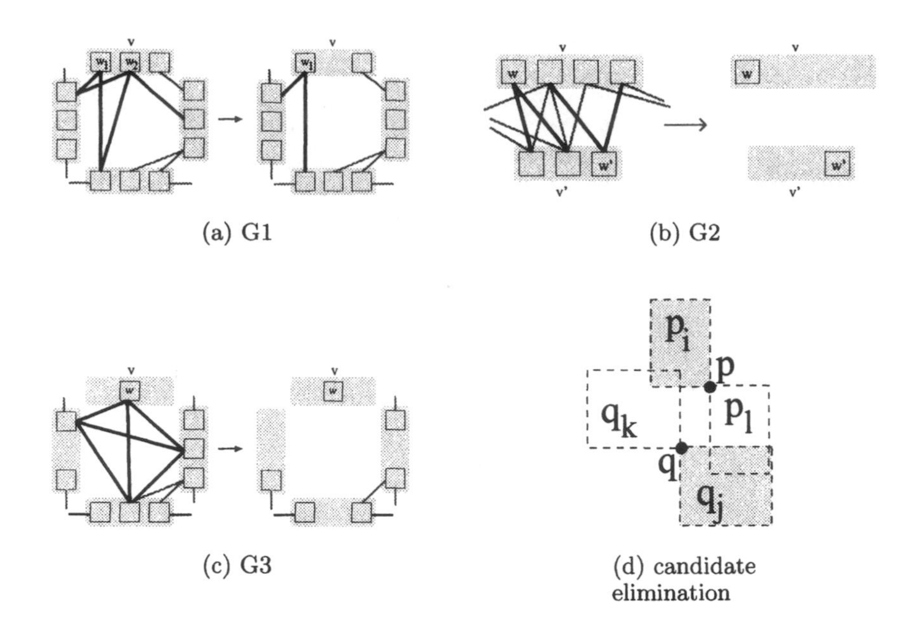

(a) G1

(b) G2

(c) G3

(d) candidate elimination

Fig. 10.10. Consistency rules and candidate elimination.

and after the application of a rule. The domain of a variable is represented by a rectangular shaded area, the values of a variable are squares, and two values are connected by a line segment if their corresponding labels overlap. Bold lines mean that the corresponding constraints are responsible for the application of the depicted rule. Gray lines not ending in a box indicate that the value from which they are emanating might constrain further variables.

G1. If a variable v has two values w_1 and w_2, and all values constrained by w_1 are also constrained by w_2, then set $D_v = D_v - \{w_2\}$, see Figure 10.10(a). *Special case:* If a variable v has a value w without constraints, then set $D_v = \{w\}$.

G2. If there is a subset V of variables $v_1, \ldots, v_l$, each with a value w_i such that w_i only constrains variables in V but does not exclude any w_j for $i \neq j$, then set $D_{v_i} = \{w_i\}$ for $i = 1, \ldots, l$.
Special case: If a variable v has a value w that only constraints a variable v', and v' has a value w' which constrains only v and does not exclude w, then set $D_v = \{w\}$ and $D_{v'} = \{w'\}$, see Figure 10.10(b).

G3. If the domain D_v of a variable v consists only of one value w, and the values $w_1,\ldots, w_l$ excluded by w belong to different variables $v_1,\ldots, v_l$ and pairwise exclude each other (i.e., if the corresponding labels of $w, w_1, \ldots, w_l$ pairwise overlap), then set $D_{v_i} = D_{v_i} - \{w_i\}$ for $i = 1, \ldots, l$, see Figure 10.10(c).

The application of the rules G1, G2, and G3 does not destroy an optimal solution. Since it is $\mathcal{NP}$-hard to decide whether there is a solution that assigns values to all variables, one cannot expect that even an exhaustive application of the above rules immediately gives rise to a solution (which then would be optimal). However, this approach is an effective preprocessing step for heuristics or backtracking, since the search space for an optimal solution can be reduced considerably.

This general concept is then applied to labeling points with axis parallel rectangles. Wagner and Wolff's algorithm consists of two phases. The rules applied in the first phase are restrictions to the more general rules G1, G2, and G3, therefore it is clear that they do not destroy an optimal solution.

In the first phase of the algorithm, the following three constraint satisfaction rules are iteratively applied:

- If p has a candidate p_i without any conflict, then p_i is declared to be part of the solution and all other candidate labels of p are eliminated.
- Let p have a candidate p_i which is only in conflict with some q_k. Let q have a candidate q_j $(j \neq k)$, which is only overlapped by p_l $(l \neq i)$. Then, add p_i and q_j to the solution and eliminate all other candidates of p and q, see Figure 10.10(d).
- Let p be a feature that has only one label candidate p_i. Let p_i overlap k other label candidates that overlap each other (i.e., they form a clique in the candidate conflict graph). Then, p is labeled with p_i and all labels overlapping p_i are eliminated.

Phase II heuristically eliminates label candidates similar to a part of the heuristics in former papers of Wagner and Wolff (1997, 1995b,a). More precisely, for all label features that have a maximum number of candidates they delete the candidate with the maximum number of conflicts among the candidates of that feature. They repeat this process until each feature has at most one candidate left.

The construction of the conflict graph needs $\mathcal{O}(n \log n)$ worst case time in case of n axis parallel rectangular labels. Phase I and Phase II of the algorithm need together at most $\mathcal{O}(n + k^2)$ time, where k is the number of pairs of intersecting label candidates. For very dense problems, i.e., many features are positioned in a small area with respect to the size of the labels, k could be very large and it could dominate the runtime.

Wagner and Wolff have compared the performance of their algorithm in respect to percentage of labeled sites and runtime with a greedy algorithm, similar to the one of van Kreveld et al. (1998) introduced in Section 10.2(b), and an implementation of simulated annealing according to Christensen et al. (1995). Recapitulating their test results, the simulated annealing has outperformed the algorithm of Wagner and Wolff in some example classes by one or two percents, but simulated annealing has needed much longer to achieve a good labeling. For examples with a possible complete labeling their algorithm does not leave more than five percent unlabeled. On the other hand no statement could be made about a constant factor approximation behavior of the algorithm. Additionally, the algorithm is restricted to a finite set of label candidates for one feature, but no longer to only four candidates. As a positive aspect we mention that their algorithm works with arbitrary label shapes.

Recently, Wolff (1999) has been doing research in a new form of local consistency named *r-irreducibility* with application to map labeling. A CSP is r-irreducible if for each variable subset w of cardinality r no value x for a variable v_i exists, such that setting $v_i = x$ reduces the size of an optimal solution for w. He has given an algorithm for 2-irreducibility where the rules are very similar to the rules of the algorithm we described here. He further has shown that an n-irreducible instance gives an optimal solution. Thus, irreducibility introduces a new scale between an unreduced instance and its solution. Due to the hardness of the problem, one cannot expect to find an n-irreducibility algorithm, but applying his 2-irreducibility algorithm to point labeling, Wolff has managed to improve previous results in practice.

10.6 General Optimization Strategies Applied to Map Labeling

Many researchers have attempted to solve map labeling problems using powerful optimization strategies that have their origin in mathematical programming, scientific programming, physical programming, artificial intelligence, etc. To name only the most prominent methods:

Gradient descent. A randomly generated labeling is scored and then improved monotonically by considering all alternative positions for each label (chosen from a discrete set) and making the single label move to the position that most improves the quality of the whole labeling. Since only changes that improve the quality of the labeling are allowed, it is easy to see that this technique gets easily stuck in local optima (Christensen et al., 1993).

Simulated annealing. Simulated annealing is a generalization of gradient descent in which moves that worsen the quality of the labeling are occasionally allowed to avoid getting stuck in local minima (Christensen et al., 1993, 1995; Edmondson et al., 1997).

For a survey on these optimization methods see (Christensen et al., 1995; Edmondson et al., 1997). These methods have the advantage that they are applicable to a wide variety of problems including map labeling. They are easy to implement and at least the simulated annealing approach yields good test results (Christensen et al., 1997, 1995; Edmondson et al., 1997; Wagner and Wolff, 1998). Except for some zero-one integer programming based heuristics, the disadvantages of these methods are the long running time and the lack of quality guarantees. In addition, most general optimization methods do not take advantage of the geometric properties of a map labeling instance.

Acknowledgement. We would like to thank Holger Hennes, Gunnar Klau, Britta Landgraf, Arne Storjohann, and Alexander Wolff for reading this chapter and giving valuable comments.

A. Software Packages

Thomas Willhalm

The theoretical foundations of graph drawing, presented throughout this book, are interesting, if not absorbing. It is, however, even more interesting with the ability to actually draw some graphs. In this appendix, we list some software packages that should enable the reader to try out many of the algorithms that have been presented. In view of the number of the ever growing number of available programs, we are aware that this list is incomplete and will soon be outdated. Research driven software can evolve rapidly, or be abandoned overnight. The list is intended to support first practical steps in graph drawing.

Graph Drawing Server

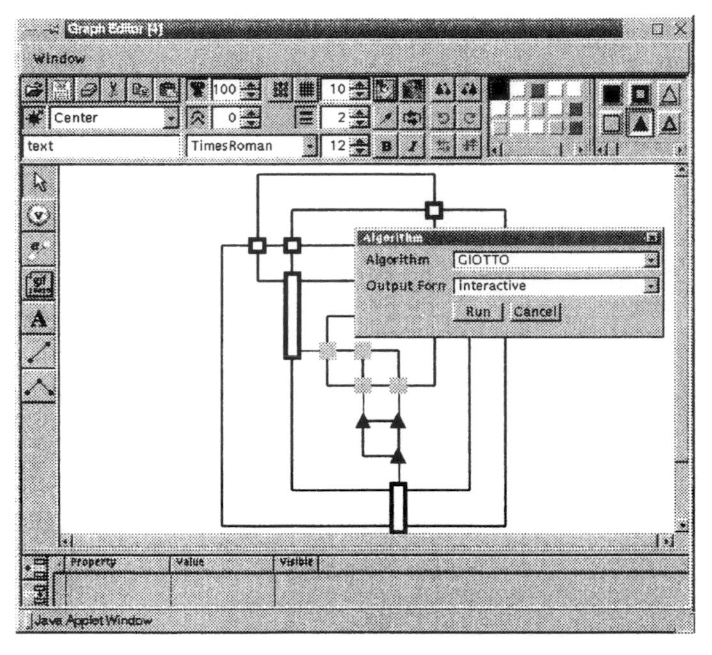

The *Graph Drawing Server* is an Internet service, that returns drawings of graphs, where graphs can be uploaded using either a Java graph editor applet or a Java client program.

Platforms The required software runs on every platform, for which a Java virtual machine exists.

M. Kaufmann and D. Wagner (Eds.): Drawing Graphs, LNCS 2025, pp. 274-281, 2001.

Availability The Web pages at `http://loki.cs.brown.edu:8081/graph-server/` make it easy to access the server. However, bandwidth problems can make this service difficult to use from outside of North America.

Algorithms Currently, most of the algorithms are for orthogonal drawing. Different types of graphs varying from trees to multi-graphs are supported. Furthermore, algorithms for layered graphs are included.

Documentation The Web pages contain explanations for the algorithms and references to publications.

AGD – Algorithms for Graph Drawing

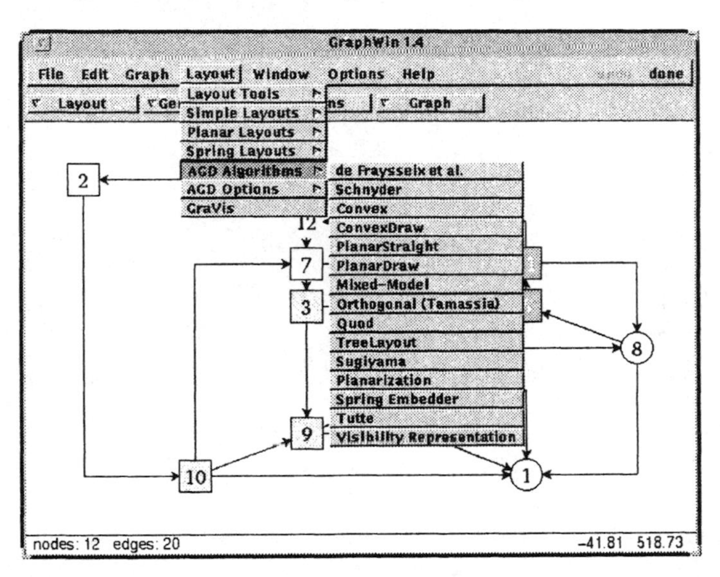

AGD is an object-oriented C++ library of algorithms for handling and drawing graphs.

Platforms AGD has been tested on SunOS (GNU- and SunPro CC compiler), Linux (GNU compiler) and Windows NT (Visual C++ 5.0). For linking, LEDA has to be installed, whereas the use of ABACUS is optional.

Availability For non-commercial purposes, precompiled binaries are available free of charge from `http://www.mpi-sb.mpg.de/AGD/`. Commercial licenses are distributed by Algorithmic-Solutions GmbH, contact `agd@algorithmic-solutions.com`.

Algorithms The library offers a great variety of algorithms for graphical layout of graphs in two dimensions, e.g. methods for drawing planar graphs, hierarchical graphs, or orthogonal graphs. In addition AGD offers planarization methods and other utilities to handle typical subtasks of graph drawing approaches.

Documentation The package includes man page for Unix systems and LaTeX sources for the manual. The Web page also has an online manual.

Graphlet

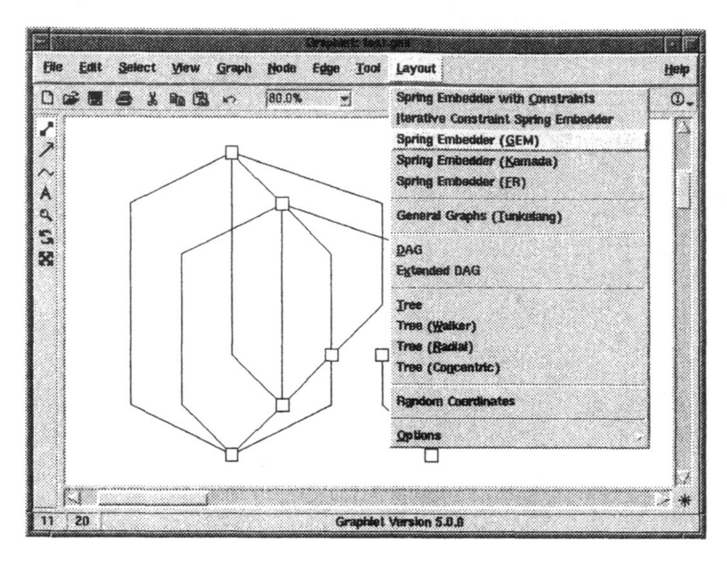

Graphlet is a toolkit for graph editors and graph algorithms, in particular graph drawing algorithms.

Platforms Precompiled binaries are available for Sun Solaris, Linux, and Windows 95/98/NT. The tool is based on GraphScript, an extension of Tcl/Tk with support for graph operations. Algorithms are implemented in C++ and GraphScript.

Availability Graphlet is free for non-commercial use. Binaries are available from `http://www.fmi.uni-passau.de/Graphlet/`, source code upon request.

Algorithms Several variants of the spring embedder, algorithms for layered drawings graphs, and of tree drawing algorithms (among those one for radial layouts).

Documentation No user manual is provided. However, there is a manual for GraphScript, the C++ interface, and sample code showing how to write new algorithms.

GDToolkit – Graph Drawing Toolkit

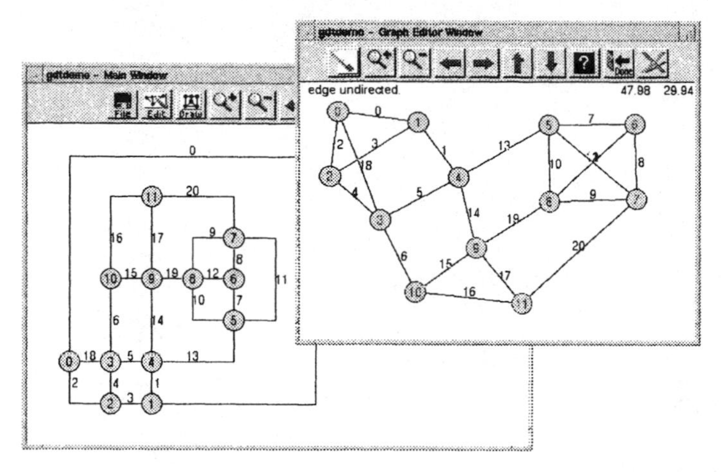

GDToolkit is an object-oriented C++ library for handling and drawing graphs. It includes a graph editor that can be used to call several drawing algorithms and a batch layout generator.

Platforms GDToolkit has been tested on Sun Solaris (GNU compiler), PC Linux (GNU compiler), and Windows 95/98/NT (Borland compiler).

Availability For non-commercial purposes, precompiled binaries are available free of charge from `http://www.dia.uniroma3.it/~gdt/`. For linking, LEDA has to be installed. Commercial licenses are distributed by INTEGRA Sistemi S.r.l. (Italy); contact `Leonforte@Integra-Sistemi.com`.

Algorithms The library offers data structures for several types of graphs such as trees, flow networks, planar graphs, upward planar graphs, and SPQR-trees. It features algorithms for orthogonal drawings, layered drawings, upward planar drawings, and visibility drawings. There are no force-directed or 3D layout algorithms.

Documentation The documentation consists of some online tutorials and the annotated header files.

yFiles

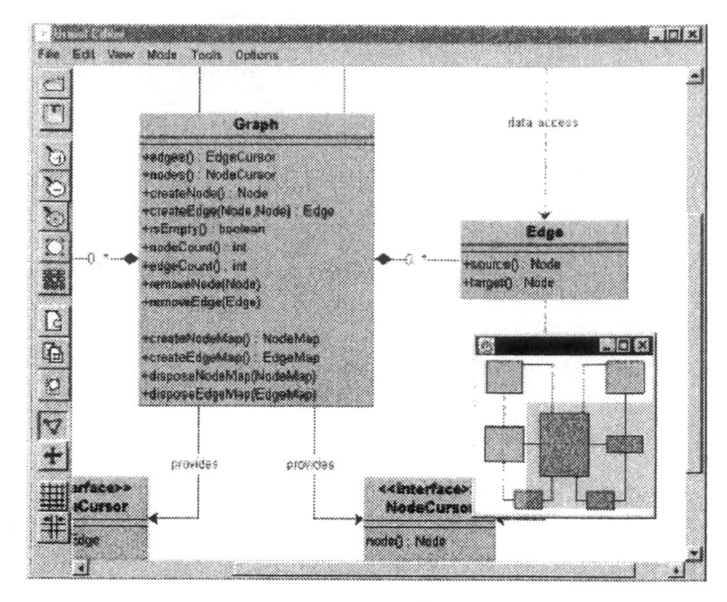

The *yFiles* are Java packages that form a framework for the development of applications that need to visualize graphs. Included are diverse graph layout and labeling algorithms, graph viewer/editor components (2D and 3D) and demo applications.

Platforms All Java 2 platforms, currently including Linux, Solaris, HPUX, Windows, and probably others.

Availability The library can be used free of charge for non-commercial purposes. Java class files are available from `http://www-pr.informatik.uni-tuebingen.de/yfiles/`.

Algorithms Customizable algorithms for layered drawings, force directed drawings, tree drawings, and orthogonal drawings are provided.

Documentation API documentation (javadoc) available.

Other Packages

In addition to the above, some more specialized packages are available for experimentation. They are grouped into categories, but since they often combine several aspects, interesting packages may appear in unexpected categories.

Larger Packages

Pajek `http://vlado.fmf.uni-lj.si/pub/networks/pajek/`
A package for the analysis of large networks for Windows. It supports

hierarchical and clustered graphs, e.g. an interface for genealogy. Several variations of 2D and 3D layout algorithms based on physical models, eigenvalues, and layers are included. Furthermore, various tools for partitioning, numbering, etc. are provided.

daVinci `http://www.tzi.de/~davinci/`
A graph editor and layout engine for directed graphs. It supports dynamic drawings. An API for the graph editor is provided. The graphs are stored in a term representation and can be output in PostScript. Binaries are available for different flavors of Unix. A documentation and a tutorial in HTML are included.

GraphViz `http://www.research.att.com/sw/tools/graphviz/`
A set of graph drawing tools for Unix or MS-Windows (win32). Its use is free of charge for non-commercial purposes, including source code. It supports hierarchical layouts, a spring embedder, and a graph editor. Apart from the Sugiyama algorithm, its speciality is the sophisticated curve drawing procedure which yields smooth edges. Output formats are (among others) PostScript, HPGL, and GIF.

VCG `http://www.cs.uni-sb.de/RW/users/sander/html/gsvcg1.html`
A tool to produce layered drawings of directed graphs. Various Unix platforms are supported and there exists also a Windows 3.1 port. The source code is available (GNU General Public License). The main design goal has been speed. It can be used interactively or as a command line tool.

Commercial Packages

GLT/GET `http://www.tomsawyer.com/`
Two commercial libraries of graph layout algorithms and a graph editor. Universities can participate in a program to obtain the software at a discount rate. There are Java and C++ versions of GLT/GET for Windows and various Unixes.
The GLT offers four different algorithms: circular layout, hierarchical layout (Sugiyama implementation), symmetrical layout (spring embedder implementation), and orthogonal layout (implementation of the 3-phase method).

DataViews `http://www.dvcorp.com/welcome.html`
Commercial tools for data visualization with a C API for Windows and Unix and a C++ API for Windows. Furthermore, Plug-Ins for Web browsers and ActiveX components are supported.

GraphVisualizer3D `http://www.omg.unb.ca/hci/projects/gv3d/`
Visualization of object-oriented code (files, classes, and variables) in three dimensions. Currently, Silicon Graphics and Sun workstations are supported. The upcoming commercial version (`http://www.nvss.nb.ca/`) will also support Linux and Windows.

3D Packages

OrthoPak `http://www.cs.uleth.ac/~wismath/packages/`
Tool for displaying graphs orthogonally in three dimensions. For non-commercial use, binaries and source code are freely available for Solaris and Linux. The C++ code is based on LEDA and outputs VRML files.

3DCube `http://www.dia.uniroma3.it/~patrigna/3dcube/`
A package of recent 3D orthogonal graph drawing algorithms. It is developed in C++ under Unix. VRML and GML output is available and OpenGL output under development. Currently, the program can only be used through a CGI interface.

GEM3D `http://i44s11.info.uni-karlsruhe.de/~frick/gd/gem3Ddraw`
Prototype implementation of Gem3D (spring embedder variant for large data structures in 3D). The program is available for SGI and DEC, further platforms on request. It uses OpenGL.

Java

VGJ `http://www.eng.auburn.edu/department/cse/research/graph_drawing/graph_drawing.html`
Visualizing Graphs with Java. Currently, there is mainly the graph editor distributed under the GNU general public license. The data format is GML. Some basic algorithms for trees as well as a spring embedder are included.

Interactive Graph Drawing
`http://www.cs.rpi.edu/projects/pb/graphdraw/index.html`
A Java applet that provides a graph editor and some drawing algorithms (force directed, hierarchical, circular). Source code is available.

JIGGLE `http://www.cs.cmu.edu/~quixote`
JIGGLE is a Java-based platform for experimenting with numerical optimization approaches to general graph layout. Its features include an implementation of the Barnes-Hut tree code and an optimization procedure on the conjugate gradient method.

GRAPPA `http://www.research.att.com/sw/tools/graphviz/packages/grappa.html`
A GRAPh PAckage written in Java. It provides an application programming interface (API).

LayoutShow `http://www.cs.yorku.ca/~lila/work.html`
LayoutShow is a Java-based multi-threaded application for experimentation with force-directed algorithms and layouts based on eigenvectors. Input and output format is GML.

Other Languages

ffGraph `http://www.fmi.uni-passau.de/~friedric/`
C++ class library to create and display directed graphs. Tcl/tk is used

to display 2 or 3 dimensional drawings for X11 systems. Currently, a 3d spring embedder and a Sugiyama layout algorithm are included, as well as a graph editor.

GraphPlace `ftp://ftp.dcs.warwick.ac.uk/people/Martyn.Amos/packages/graphplace/`
Filter program written in ANSI C that takes a list of nodes and edges and produces a list of coordinates or a PostScript file.

tkgcv `http://www.informatik.uni-stuttgart.de/ipvr/swlab/sopra/tkgcv/tkgcv.html`
Tcl/Tk extension for graph drawing. It requires a C compiler (source code is available). The module has been tested under Linux and HP-Unix. Currently, four algorithms are included.

Mathematica

NetGraph `http://eclectic.ss.uci.edu/linkages/programs/net-graph/netgraph.html`
Visualization packages for Mathematica. It uses symbolic hypotheses to draw graphs.

External Binaries

xdrawgraph `http://rocana.aist-nara.ac.jp/~hayashi/E/graph-tool.html`
A Unix package to edit and draw graphs. The graph drawing algorithms are implemented as external programs.

Graph Editors

angela! `http://www.mpi-sb.mpg.de/~pabst/angela/`
Angela! is a Natural Graph Editor with Layouting Algorithms for Unix. It is written in C and uses Tcl/Tk and Tix.

Ginger `http://www.cs.auc.dk/~normark/Ginger/ginger.html`
Graph editor for Unix. It uses XPM and Elk (optionally).

GraphPanel `http://binger.centre.edu/GraphPanel/`
A simple graph editor written in Java under GNU public license. It can export the graphs to PostScript files.

Bibliography

Abbott, K. R., and Sarin, S. K. (1994). Experiences with workflow management: Issues for the next generation. In *Proceedings of ACM Conference on Computer-Supported Cooperative Work, Workflow and Information Sharing (CSCW'94)*, pages 113–120.

Agarwal, P. K., and Erickson, J. (1997). Geometric range searching and its relatives. Technical Report CS 1997-11, Department of Computer Science, Duke.

Agarwal, P. K., van Kreveld, M., and Suri, S. (1998). Label placement by maximum independent set in rectangles. *Computational Geometry: Theory and Applications*, 11(3-4):209–218.

Aho, A., Hopcroft, J., and Ullman, J. (1974). *The Design and Analysis of Computer Algorithms.* Addison-Wesley.

Ahuja, R. K., Magnanti, T. L., and Orlin, J. B. (1993). *Network Flows: Theory, Algorithms, and Applications.* Prentice Hall.

Alpert, C. J., and Kahng, A. B. (1995). Recent directions in netlist partitioning: a survey. *INTEGRATION, the VLSI Journal*, 19:1–81.

Amtrup, H. H. J., and Jost, U. (1996). What's in a word graph? Evaluation and enhancement of word lattices. Technical Report Verbmobil-Report 186, University Hamburg, Germany.

Andreev, E. M. (1970a). On convex polyhedra in Lobacevskii spaces. *Math. USSR-Sb.*, 10:413–440.

Andreev, E. M. (1970b). On convex polyhedra of finite volume in Lobacevskii space. *Math. USSR-Sb.*, 12:255–259.

Auslander, L., and Parter, S. V. (1961). On imbedding graphs in the plane. *Journal of Mathematics and Mechanics*, 10(3):517–523.

Baker, K. A., Fishburn, P. C., and Roberts, F. S. (1971). Partial orders of dimension 2. *Networks*, 2:11–28.

Batagelj, V., Kerzic, D., and Pisanski, T. (1992). Automatic clustering of languages. *Computational Linguistics*, 18(3):339–352.

Berge, C. (1993). *Graphs.* North Holland, Amsterdam, 3rd edition.

Berger, B., and Shor, P. (1990). Approximation algorithms for the maximum acyclic subgraph problem. In *Proceedings of the 1st ACM-SIAM Symposium on Discrete Algorithms (SODA'90)*, pages 236–243.

Bertolazzi, P., Cohen, R. F., Di Battista, G., Tamassia, R., and Tollis, I. G. (1994a). How to draw a series-parallel digraph. *International Journal of Computational Geometry and Applications*, 4:385–402.

Bertolazzi, P., Di Battista, G., and Didimo, W. (1997). Computing orthogonal drawings with the minimum number of bends. In *Proceedings of the 5th Workshop on Algorithms and Data Structures (WADS'97)*, Spinger LNCS 1272, pages 331–344.

Bertolazzi, P., Di Battista, G., Liotta, G., and Mannino, C. (1994b). Upward drawings of triconnected digraphs. *Algorithmica*, 6(12):476–497.

Bertolazzi, P., Di Battista, G., Mannino, C., and Tamassia, R. (1993). Optimal upward planarity testing of single-source digraphs. In *Proceedings of the 1st European Symposium on Algorithms (ESA '93)*, Springer LNCS 726, pages 37–48.

Bertsekas, D. P. (1998). *Network Optimization: Continuous and Discrete Models*. Athena Scientific.

Biedl, T., Shermer, T., Whitesides, S., and Wismath, S. (1999). Bounds for orthogonal 3D graph drawing. *Journal of Graph Algorithms and Applications*, 3(4):63–79.

Biedl, T. C. (1997). *Orthogonal Graph Visualization: The Three-Phase Method with Applications*. PhD thesis, Rutgers University.

Biedl, T. C. (1998). Three approaches to 3D-orthogonal box-drawings. In *Proceedings of the 6th International Symposium on Graph Drawing (GD'98)*. Springer LNCS 1547, pages 30–43.

Biedl, T. C., and Kant, G. (1994). A better heuristic for orthogonal graph drawing. In *Proceedings of the 2nd European Symposium on Algorithms (ESA '94)*, Springer LNCS 855, pages 24–35.

Biedl, T. C., and Kaufmann, M. (1997). Area-efficient static and incremental graph drawings. In *Proceedings of the 5th European Symposium on Algorithms (ESA '97)*, Springer LNCS 1284, pages 37–52.

Biedl, T. C., Madden, B. P., and Tollis, I. G. (1997a). The three-phase method: A unified approach to orthogonal graph drawing. In *Proceedings of the 5th International Symposium on Graph Drawing (GD'97)*. Springer LNCS 1353, pages 391–402.

Biedl, T. C., Shermer, T., Whitesides, S., and Wismath, S. (1997b). Orthogonal 3D graph drawing. In *Proceedings of the 5th International Symposium on Graph Drawing (GD'97)*. Springer LNCS 1353, pages 76–86.

Blythe, J., McGrath, C., and Krackhardt, D. (1996). The effect of graph layout on inference from social network data. In *Proceedings of the 3rd International Symposiom on Graph Drawing (GD'95)*. Springer LNCS 1027, pages 40–51.

Böhringer, K.-F., and Paulisch, F. N. (1990). Using constraints to achieve stability in automatic graph layout algorithms. In *Proceedings of the ACM Human Factors in Computing Systems Conference (CHI'90)*, pages 43–51.

Booth, K. S., and Lueker, G. S. (1976). Testing for the consecutive ones property, interval graphs and graph planarity using PQ-tree algorithms. *Journal of Computer and System Sciences*, 13:335–379.

Borgida, A., Brachman, R., McGuinness, D., and Resnick, L. (1989). CLASSIC: A structural data model for objects. In *Proceedings of the 1989 ACM-SIGMOD International Conference on Management of Data*, pages 59–67.

Bose, P., Gomez, F., Ramos, P., and Toussaint, G. (1996). Drawings nice projections of objects in space. In *Proceedings of the 3rd International Symposium on Graph Drawing (GD'95)*, Springer LNCS 1027, pages 52–63.

Brandenburg, F. J., Himsolt, M., and Rohrer, C. (1996). An experimental comparison of force-directed and randomized graph drawing algorithms. In *Proceedings of the 3rd International Symposium on Graph Drawing (GD'95)*, Springer LNCS 1027, pages 76–87.

Brandes, U. (1999). *Layout of Graph Visualizations.* PhD thesis, University of Konstanz. `http://www.ub.uni-konstanz/kops/volltexte/1999/255/`.

Brandes, U., Kenis, P., Raab, J., Schneider, V., and Wagner, D. (1999). Explorations into the visualization of policy networks. *Journal of Theoretical Politics*, 11(1):75–106.

Brandes, U., and Wagner, D. (1997). A Bayesian paradigm for dynamic graph layout. In *Proceedings of the 5th International Symposium on Graph Drawing (GD'97)*, Springer LNCS 1353, pages 236–247.

Brandes, U., and Wagner, D. (1998a). Dynamic grid embedding with few bends and changes. In *Proceedings of the 9th Annual International Symposium on Algorithms and Computation (ISAAC'98)*, Springer LNCS 1533, pages 89–98.

Brandes, U., and Wagner, D. (1998b). Using graph layout to visualize train interconnection data. In *Proceedings of the 6th International Symposium on Graph Drawing (GD'98)*, Springer LNCS 1547, pages 44–56.

Branke, J., Bucher, F., and Schmeck, H. (1997). A genetic algorithm for drawing undirected graphs. In *Proceedings of the 3rd Nordic Workshop on Genetic Algorithms and their Applications*, pages 193–206.

Bridgeman, S., Di Battista, G., Didimo, W., Liotta, G., Tamassia, R., and Vismara, L. (2000). Turn-regularity and optimal area drawings for orthogonal representations. *Computational Geometry: Theory and Applications*, 16(1):53–93.

Bridgeman, S., Fanto, J., Garg, A., Tamassia, R., and Vismara, L. (1997). INTERACTIVEGIOTTO: An algorithm for interactive orthogonal graph drawing. In *Proceedings of the 5th International Symposium on Graph Drawing (GD'97)*, Springer LNCS 1353, pages 303–308.

Bridgeman, S., and Tamassia, R. (1998). Difference metrics for interactive orthogonal graph drawing algorithms. In *Proceedings of the 6th International Symposium on Graph Drawing (GD'98)*, Springer LNCS 1457, pages 57–71.

Bruß, I., and Frick, A. (1996). Fast interactive 3-D graph visualization. In *Proceedings of the 3rd International Symposium on Graph Drawing (GD'95)*, Springer LNCS 1027, pages 99–110.

Cai, J., Han, X., and Tarjan, R. E. (1993). An $O(m \log n)$-time algorithm for the maximal planar subgraph problem. *SIAM Journal on Computing*, 22:1142–1162.

Carpano, M. J. (1980b). Automatic display of hierarchized graphs for computer aided decision analysis. *IEEE Transactions on Systems, Man, and Cybernetics*, SMC-10(11):705–715.

Catarci, T. (1995). The assignment heuristic for crossing reduction. *IEEE Trans. Syst. Man Cybern.*, 25(3):515–521.

Chaiken, S., and Kleitman, D. J. (1978). Matrix tree theorems. *Journal of Combinatorial Theory, Series A*, 24:377–381.

Chan, T., Goodrich, M. T., Kosaraju, S. R., and Tamassia, R. (1996). Optimizing area and aspect ration in straight-line orthogonal tree drawings. In *Proceedings of the 4th International Symposium on Graph Drawing (GD'96)*. Springer LNCS 1190, pages 63–75.

Chan, T. M. (1999). A near-linear area bound for drawing binary trees. In *Proceedings of the 10th ACM-SIAM Symposium on Discrete Algorithms (SODA'99)*, pages 161–168.

Chiba, N., Nishizeki, T., Abe, S., and Ozawa, T. (1985). A linear time algorithm for embedding planar graphs using PQ-trees. *Journal of Computer and System Sciences*, 30:54–76.

Christensen, J., Friedman, S., Marks, J., and Shieber, S. (1997). Empirical testing of algorithms for variable-sized label placement. In *Proceedings of the 13th Annual ACM Symposium on Computational Geometry*, pages 415–417.

Christensen, J., Marks, J., and Shieber, S. (1993). Algorithms for cartographic label placement. In *Proceedings of the American Congress on Surveying and Mapping 1*, pages 75–89.

Christensen, J., Marks, J., and Shieber, S. (1995). An empirical study of algorithms for point-feature label placement. *ACM Transactions on Graphics*, 14(3):203–232.

Chrobak, M., and Kant, G. (1997). Convex grid drawings of 3-connected planar graphs. *International Journal of Computational Geometry and Applications*, 7(3):211–224.

Chvátal, V. (1983a). *Linear Programming*. W. H. Freeman.

Closson, M., Everett, H., Gartshore, S., and Wismath, S. (1998). Arrangepak, orthopak and vispak 2.0. Technical Report TR-CS-98, University of Lethbridge.

Coffman, E. G., and Graham, R. L. (1972). Optimal scheduling for two processor systems. *Acta Informatica*, 1:200–213.

Cohen, J. D. (1997). Drawing graphs to convey proximity: An incremental arrangement method. *ACM Transactions on Computer-Human Interaction*, 4(3):197–229.

Cohen, R. F., Di Battista, G., Tamassia, R., and Tollis, I. G. (1995). Dynamic graph drawings: Trees, series-parallel digraphs, and planar *st*-digraphs. *SIAM Journal on Computing*, 24(5):970–1001.

Cohen, R. F., Di Battista, G., Tamassia, R., Tollis, I. G., and Bertolazzi, P. (1992). A framework for dynamic graph drawing. In *Proceedings of the 8th ACM Annual Symposium on Computational Geometry (SCG'92)*, pages 261–270.

Colin de Verdière, Y. (1989). Empilements de cercles: convergence d'une methode de point fixe. *Forum Mathematicum*, 1:395–402.

Cormen, T., Leiserson, C., and Rivest, R. (1990). *Introduction to Algorithms*. The MIT Electrical Engineering and Computer Science Series. The MIT Press and McGraw-Hill Book Company.

Crescenzi, P., Di Battista, G., and Piperno, A. (1992). A note on optimal area algorithms for upward drawings of binary trees. *Computational Geometry: Theory and Applications*, 2:187–200.

Crescenzi, P., and Piperno, A. (1995). Optimal-area upward drawings of AVL-trees. In *Proceedings of the DIMACS International Workshop on Graph Drawing (GD'94)*. Springer LNCS 894, pages 307–317.

Cruz, I. F., and Twarog, J. P. (1996). 3D graph drawing with simulated annealing. In *Proceedings of the 3rd International Symposium on Graph Drawing (GD'95)*, Springer LNCS 1027, pages 162–165.

Cunningham, W. H. (1976). A network simplex method. *Mathematical Programming*, 11:105–116.

Czyzowicz, J. (1991). Lattice diagrams with few slopes. *Journal of Combinatorial Theory, Series A*, 56:96–108.

Czyzowicz, J., Pelc, A., and Rival, I. (1990). Drawing orders with few slopes. *Discrete Mathematics*, 82:233–250.

Dai, W. W.-M., and Kuh, E. S. (1987). Global spacing of building-block layout. In *Proceedings of the IFIP International Conference on Very Large Scale Integration (VLSI'87)*, pages 193–205.

Datta, A., Lenhof, H.-P., Schwarz, C., and Smid, M. H. M. (1993). Static and dynamic algorithms for *k*-point clustering problems. In *Proceedings of the 3rd Workshop on Algorithms and Data Structures (WADS'93)*, Springer LNCS 709, pages 265–276.

Davidson, R., and Harel, D. (1996). Drawing graphs nicely using simulated annealing. *ACM Transactions on Graphics*, 15(4):301–331.

de Fraysseix, H., Pach, J., and Pollack, R. (1990). How to draw a planar graph on a grid. *Combinatorica*, 10:41–51.

Di Battista, G., Eades, P., Tamassia, R., and Tollis, I. G. (1994). Algorithms for drawing graphs: An annotated bibliography. *Computational Geometry*, 4:235–282.

Di Battista, G., Eades, P., Tamassia, R., and Tollis, I. G. (1999). *Graph Drawing: Algorithms for the Visualization of Graphs.* Prentice Hall.

Di Battista, G., Liotta, G., and Vargiu, F. (1998a). Spirality and optimal orthogonal drawings. *SIAM Journal on Computing*, 27(6):1764–1811.

Di Battista, G., Liu, W. P., and Rival, I. (1990). Bipartite graphs, upward drawings, and planarity. *Information Processing Letters*, 36(6):317–322.

Di Battista, G., Patrignani, M., and Vargiu, F. (1998b). A split & push approach to 3-D orthogonal drawing. In *Proceedings of the 6th International Symposium on Graph Drawing (GD'98).* Springer LNCS 1547, pages 87–101.

Di Battista, G., and Tamassia, R. (1988). Algorithms for plane representations of acyclic digraphs. *Theoretical Computer Science*, 61(2-3):175–198.

Di Battista, G., and Tamassia, R. (1989). Incremental planarity testing. In *Proceedings of the 30th Symposium on the Foundations of Computer Science (FOCS'89)*, pages 436–441.

Di Battista, G., and Tamassia, R. (1990). On-line graph algorithms with SPQR-trees. In *Proceedings of the 17th International Colloqium on Automata, Languages and Programming (ICALP'90)*, Springer LNCS 443, pages 598–611.

Di Battista, G., and Tamassia, R. (1996). On-line planarity testing. *SIAM Journal on Computing*, 25(5):956–997.

Di Battista, G., and Vismara, L. (1993). Angles of planar triangulated graphs. In *Proceedings of the 25th Annual ACM Symposium on the Theory of Computing (STOC'93)*, pages 431–437.

Di Battista, G., and Vismara, L. (1996). Angles of planar triangulated graphs. *SIAM Journal on Discrete Mathematics*, 9(3):349–359.

Didimo, W., and Liotta, G. (1998). Computing orthogonal drawings in a variable embedding setting. In *Proceedings of the 9th Annual International Symposium on Algorithms and Computation (ISAAC'98)*, Springer LNCS 1533, pages 79–88.

Dietz, P. F., and Sleator, D. D. (1987). Two algorithms for maintaining order in a list. In *Proceedings of the 19th Annual ACM Symposium of Theory of Computing (STOC'87)*, pages 365–372.

Djidjev, H. N. (1995). A linear algorithm for the maximal planar subgraph problem. In *Proceedings of the 4th Workshop on Algorithms and Data Structures (WADS'95).* Springer LNCS 955, pages 369–380.

Doddi, S., Mararthe, M. V., Mirzaian, A., Moret, B. M. E., and Zhu, B. (1999). Map labeling and its generalizations. Technical Report LA-UR-96-2411, Los Alamos National Labatory.

Doddi, S., Marathe, M. V., Mirzaian, A., Moret, B. M. E., and Zhu, B. (1997). Map labeling and its generalizations. In *Proceedings of the 8th ACM-SIAM Symposium on Discrete Algorithms (SODA'97)*, pages 148–157.

Dresbach, S. (1995). A new heuristic layout algorithm for directed acyclic graphs. In *Operations Research Proceedings 1994*, pages 121–126.

Duncan, C. A., Goodrich, M. T., and Kobourov, S. G. (1998). Balanced aspect ratio trees and their use for drawing very large graphs. In *Proceedings of the 6th International Symposium on Graph Drawing (GD'98)*. Springer LNCS 1547, pages 111–124.

Eades, P. (1984). A heuristic for graph drawing. *Congressus Numerantium*, 42:149–160.

Eades, P., Cohen, R. F., and Huang, M. L. (1997a). Online animated graph drawing for Web navigation. In *Proceedings of the 5th International Symposium on Graph Drawing (GD'97)*, Springer LNCS 1353, pages 330–335.

Eades, P., and Feng, Q. W. (1996). Multilevel visualization of clustered graphs. In *Proceedings of the 4th International Symposium on Graph Drawing (GD'96)*. Springer LNCS 1190, pages 101–112.

Eades, P., and Feng, Q. W. (1997). Drawing clustered graphs on an orthogonal grid. In *Proceedings of the 5th International Symposium on Graph Drawing (GD'97)*. Springer LNCS 1353, pages 146–157.

Eades, P., Feng, Q. W., and Lin, X. (1996a). Straight-line drawing algorithms for hierarchical graphs and clustered graphs. In *Proceedings of the 4th International Symposium on Graph Drawing (GD'96)*. Springer LNCS 1190, pages 113–128.

Eades, P., Feng, Q., and Nagamochi, H. (1999). Drawing clustered graphs on an orthogonal grid. *Journal on Graph Algorithms and Applications*, 3(4):3–29.

Eades, P., Huang, M. L., and Wang, J. (1997b). Online animated graph drawing using a modified spring algorithm. Technical Report 97–05, Department of Computer Science and Software Engineering, University of Newcastle.

Eades, P., and Kelly, D. (1986). Heuristics for reducing crossings in 2-layered networks. *Ars Combinatorica*, 21.A:89–98.

Eades, P., Lai, W., Misue, K., and Sugiyama, K. (1991). Preserving the mental map of a diagram. In *Proceedings of Compugraphics '91*, pages 24–33.

Eades, P., and Lin, X. (1995). A new heuristic for the feedback arc set problem. *Australian Journal of Combinatorics*, 12:15–26.

Eades, P., Lin, X., and Smyth, W. F. (1993). A fast and effective heuristic for the feedback arc set problem. *Information Processing Letters*, 47:319–323.

Eades, P., and Marks, J. (1995). Graph drawing contest report. In *Proceedings of the DIMACS International Workshop on Graph Drawing (GD'94)*, Springer LNCS 894, pages 143–146.

Eades, P., and Marks, J. (1996). Graph-drawing contest report. In *Proceedings of the 3rd International Symposium on Graph Drawing (GD'95)*, Springer LNCS 1027, pages 224–233.

Eades, P., Marks, J., and North, S. C. (1996). Graph-drawing contest report. In *Proceedings of the 4th International Symposium on Graph Drawing (GD'96)*, Springer LNCS 1190, pages 129–138.

Eades, P., Marks, J., and North, S. C. (1997c). Graph-drawing contest report. In *Proceedings of the 5th International Symposium on Graph Drawing (GD'97)*, Springer LNCS 1353, pages 438–445.

Eades, P., Marks, J., Mutzel, P., and North, S. C. (1998). Graph drawing contest report. In *Proceedings of the 6th International Symposium on Graph Drawing (GD'98)*, Springer LNCS 1547, pages 423–435.

Eades, P., Nagamochi, H., and Feng, Q. (1998). Straight-line drawing algorithms for hierarchical graphs and clustered graphs. Technical Report 98-03, Department of Computer Science and Software Engineering, University of Newcastle, Australia. Available at `ftp://ftp.cs.newcastle.edu.au/pub/techreports/tr98-03.ps.Z`.

Eades, P., Stirk, C., and Whitesides, S. (1996). The techniques of Kolmogorov and Bardzin for three-dimensional orthogonal graph drawing. *Information Processing Letters*, 60(2):97–103. University.

Eades, P., and Sugiyama, K. (1990). How to draw a directed graph. *Journal of Information Processing*, 13:424–437.

Eades, P., Symvonis, A., and Whitesides, S. (1996b). Two algorithms for three dimensional orthogonal graph drawing. In *Proceedings of the 4th International Symposium on Graph Drawing (GD'96)*. Springer LNCS 1190, pages 139–154.

Eades, P., Symvonis, A., and Whitesides, S. (2000). Three-dimensional orthogonal graph drawing. *Discrete Applied Mathematics*, 103(1-3):55–87.

Eades, P., and Whitesides, S. (1994). Drawing graphs in two layers. *Theoretical Computer Science*, 131(2):361–374.

Eades, P., and Wormald, N. C. (1990). Fixed edge-length graph drawing is NP-hard. *Discrete Applied Mathematics*, 28:111–134.

Eades, P., and Wormald, N. C. (1994). Edge crossings in drawings of bipartite graphs. *Algorithmica*, 11(4):379–403.

Edmondson, S., Christensen, J., Marks, J., and Shieber, S. (1997). A general cartographic labeling algorithm. *Cartographica*, 33(4):13–23.

Eiglsperger, M., Fößmeier, U., and Kaufmann, M. (2000). Orthogonal graph drawing with constraints. In *Proceedings of the 11th ACM-SIAM Symposium on Discrete Algorithms (SODA 2000)*, pages 3–11.

Eppstein, D., and Erickson, J. (1994). Iterated nearest neighbors and finding minimal polytopes. *Discrete Computational Geometry*, 11:321–350.

Even, S. (1979). *Graph Algorithms*. Pitman.

Even, S., and Tarjan, R. E. (1976). Computing an st-numbering. *Theoretical Computer Science*, 2:436–441.

Faria, L., De Figueiredo, C. M. H., and Mendonca, C. F. X. (1998). Splitting number is $\mathcal{NP}$-complete. *Proceedings of the 24th International Workshop on Graph-Theoretic Concepts in Computer Science (WG'98)*, Springer LNCS 1517, pages 285–297.

Fekete, S. P., and Meijer, H. (1999). Rectangle and box visibility graphs in 3D. *International Journal of Computational Geometry and Applications*, 9(1):1–27.

Feng, Q. (1997). *Algorithms for Drawing Clustered Graphs*. PhD thesis, University of Newcastle. `http://www.cs.newcastle.edu.au/Dept/theses.html`.

Feng, Q.-W., Cohen, R. F., and Eades, P. (1995). Planarity for clustered graphs. In *Proceedings of the 3rd European Symposium on Algorithms (ESA'95)*. Springer LNCS 979, pages 213–226.

Fialko, S., and Mutzel, P. (1998). A new approximation algorithm for the planar augmentation problem. In *Proceedings of the 9th Annual ACM-SIAM Symposium on Discrete Algorithms (SODA'98)*, pages 260–269.

Fisk, C. J., Caskey, D. L., and West, L. E. (1967). ACCEL: Automated circuit card etching layout. *Proceedings of the IEEE*, 55(11):1971–1982.

Foley, J. D., van Dam, A., Feiner, S. K., and Hughes, J. F. (1990). *Computer Graphics*, 2nd edition. Addison-Wesley.

Force, A. C. G. I. T. (1996). Application challenges to computational geometry. Technical Report TR-521-96, Princeton University.

Formann, M., Hagerup, T., Haralambides, J., Kaufmann, M., Leighton, F. T., Simvonis, A., Welzl, E., and Woeginger, G. (1990). Drawing graphs in the plane with high resolution. In *Proceedings of the 31st Symposium on the Foundations of Computer Science (FOCS'90)*, pages 86–95.

Formann, M., and Wagner, F. (1991). A packing problem with applications to lettering of maps. In *Proceedings of the 7th Annual Symposium on Computational Geometry (SCG '91)*, pages 281–288.

Formella, A., and Keller, J. (1995). Generalized fisheye views of graphs. In *Proceedings of the 3rd International Symposium on Graph Drawing (GD'95)*. Springer LNCS 1027, pages 242–253.

Fößmeier, U. (1997a). Interactive orthogonal graph drawing: Algorithms and bounds. In *Proceedings of the 5th International Symposium on Graph Drawing (GD'97)*. Springer LNCS 1353, pages 111–123.

Fößmeier, U. (1997b). *Orthogonale Visualisierungstechniken für Graphen*. PhD thesis, Eberhard-Karls-Universität zu Tübingen.

Fößmeier, U., Heß, C., and Kaufmann, M. (1998). On improving orthogonal drawings: The 4M-algorithm. In *Proceedings of the 6th International Symposium on Graph Drawing (GD'98)*. Springer LNCS 1547, pages 125–137.

Fößmeier, U., Kant, G., and Kaufmann, M. (1996). 2-visibility drawings of planar graphs. In *Proceedings of the 4th International Symposium on Graph Drawing (GD'96)*. Springer LNCS 1190, pages 155–168.

Fößmeier, U., and Kaufmann, M. (1995). Drawing high degree graphs with low bend numbers. In *Proceedings of the 3rd International Symposium on Graph Drawing (GD'95)*. Springer LNCS 1027, pages 254–266.

Foulds, L. R., Gibbons, P. B., and Giffin, J. W. (1985). Facilities layout adjacency determination: An experimental comparison of three graph theoretic heuristics *Operations Research*, 33:1091–1106.

Foulds, L. R., and Robinson, D. F. (1978). Graph theoretic heuristics for the plant layout problem. *International Journal of Production Research*, 16:27–37.

Fowler, R. J., Paterson, M. S., and Tanimoto, S. L. (1981). Optimal packing and covering in the plane are $\mathcal{NP}$-complete. *Information Processing Letters*, 12(3):133–137.

Freeman, L. C. (1999a). The social network graphics source. School of Social Science, University of California Irvine. http://eclectic.ss.uci.edu/~lin/gallery.html.

Freeman, L. C. (1999b). Using molecular modeling software in social network analysis: A practicum. School of Social Science, University of California Irvine. http://eclectic.ss.uci.edu/~lin/chem.html.

Freuder, E. C., and Wallace, R. J. (1992). Partial constraint satisfaction. *Artificial Intelligence*, 58(1-3):21–70.

Frick, A. (1997). Upper bounds on the number of hidden nodes in Sugiyama's algorithm. In *Proceedings of the 4th International Symposium on Graph Drawing (GD'96)*. Springer LNCS 1190, pages 169–183.

Frick, A., Ludwig, A., and Mehldau, H. (1995). A fast adaptive layout algorithm for undirected graphs. In *Proceedings of the DIMACS International Workshop on Graph Drawing (GD'94)*. Springer LNCS 894, pages 388–403.

Fruchterman, T. M. J., and Reingold, E. M. (1991). Graph-drawing by force-directed placement. *Software — Practice and Experience*, 21(11):1129–1164.

Gansner, E. R., Koutsofios, E., North, S. C., and Vo, K.-P. (1993). A technique for drawing directed graphs. *IEEE Transactions on Software Engineering*, 19(3):214–230.

Ganter, B., and Wille, R. (1999). *Formal Concept Analysis — Mathematical Foundations*. Springer.

Garey, M. R., and Johnson, D. S. (1983). Crossing number is $\mathcal{NP}$-complete. *SIAM Journal on Algebraic and Discrete Methods*, 4(3):312–316.

Garey, M. R., and Johnson, D. S. (1991). *Computers and Intractability: A Guide to the Theory of $\mathcal{NP}$-Completeness*. W.H. Freeman & Co.

Garg, A., Goodrich, M. T., and Tamassia, R. (1996). Planar upward tree drawings with optimal area. *International Journal Computational Geometry and Applications*, 6:333–356.

Garg, A., and Tamassia, R. (1993). Efficient computation of planar straight-line upward drawings. In *Graph Drawing '93 (Proc. ALCOM Workshop on Graph Drawing)*.

Garg, A., and Tamassia, R. (1994). Planar drawings and angular resolution: Algorithms and bounds. In *Proceedings of the 2nd European Symposium on Algorithms (ESA'94)*. Springer LNCS 855, pages 12–23.

Garg, A., and Tamassia, R. (1995a). On the computational complexity of upward and rectilinear planarity testing. In *Proceedings of the DIMACS International Workshop on Graph Drawing (GD'94)*. Springer LNCS 894, pages 286–297.

Garg, A., and Tamassia, R. (1995b). Upward planarity testing. *Order*, 12:109–133.

Garg, A., and Tamassia, R. (1996a). GIOTTO3D: A system for visualizing hierarchical structures in 3D. In *Proceedings of the 4th International Symposium on Graph Drawing (GD'96)*. Springer LNCS 1190, pages 193–200.

Garg, A., and Tamassia, R. (1996b). A new minimum cost flow algorithm with applications to graph drawing. In *Proceedings of the 4th International Symposium on Graph Drawing (GD'96)*. Springer LNCS 1190, pages 201–216.

Garg, A., and Tamassia, R. (1997). A new minimum cost flow algorithm with applications to graph drawing. In *Proceedings of the 4th International Symposium on Graph Drawing (GD'96)*. Springer LNCS 1190, pages 201–216.

Georgakopoulos, D., Hornick, M., and Sheth, A. (1995). An overview of workflow management: From process modeling to workflow automation infrastructure. *Distributed and Parallel Databases*, 3(2):119–153.

German Research Center for Artificial Intelligence GmbH (1999). The Verbmobil project. `http://www.dfki.de/verbmobil`.

Godehardt, E. (1988). *Graphs as Structural Models*, Advances in System Analysis 4. Vieweg.

Goldberg, A. V., and Kennedy, R. (1995). An efficient cost scaling algorithm for the assignment problem. *Mathematical Programming*, 71:153–178.

Goldstein, A. J. (1963). An efficient and constructive algorithm for testing whether a graph can be embedded in a plane. In *Graph and Combinatorics Conference, Contract No. NONR 1858-(21)*. Princeton University.

Grötschel, M., Jünger, M., and Reinelt, G. (1985). On the acyclic subgraph polytope. *Mathematical Programming*, 33(1):28–42.

Gutwenger, C., and Mutzel, P. (1998). Planar polyline drawings with good angular resolution. In *Proceedings of the 6th International Symposium on Graph Drawing (GD'98)*. Springer LNCS 1547, pages 167–182.

Hayashi, K., Inoue, M., Masuzawa, T., and Fujiwara, H. (1998). A layout adjustment problem for disjoint rectangles preserving orthogonal order. In *Proceedings of the 6th International Symposium on Graph Drawing*, number 1547 in LNCS, pages 183–197.

He, W., and Marriott, K. (1998). Constrained graph layout. *Constraints*, 3(4):289–314.

Herdeg, W., editor (1981). *Diagrams*. Graphis Press Corporation.

Hermansson, K., and Ojamae, L. (1994). MOVIEMOL — An easy-to-use molecular display and animation program. Technical Report UUIC-B19-500, Institute of Chemistry, University of Uppsala.

Hochbaum, D. S. (1995). *Approximation Algorithms for $\mathcal{NP}$-hard Problems.* PWS Publishing Company, Boston.

Hochbaum, D. S., and Maass, W. (1985). Approximation schemes for covering and packing problems in image processing and VLSI. *Journal of the ACM*, 32(1):130–136.

Hong, S.-H., Eades, P., Quigley, A., and Lee, S.-H. (1998). Drawing algorithms for series-parallel digraphs in two and three dimensions. In *Proceedings of the 6th International Symposium on Graph Drawing (GD'98).* Springer LNCS 1547, pages 198–209.

Hong, S.-H., Eades, P., Quigley, A., and Lee, S.-H. (1999a). Drawing series-parallel digraphs symmetrically. To appear in *International Journal of Computational Geometry and Applications.*

Hong, S.-H., Eades, P., Quigley, A., and Lee, S.-H. (1999b). A three dimensional drawing algorithm for series-parallel graphs. Manuscript.

Hopcroft, J., and Tarjan, R. E. (1974). Efficient planarity testing. *Journal of the ACM*, 21:549–568.

Hopcroft, J. E., and Tarjan, R. E. (1973). Dividing a graph into triconnected components. *SIAM Journal on Computing*, 2(3):135–158.

Hsu, W.-L. (1995). A linear time algorithm for finding maximal planar subgraphs. In *Proceedings of the 6th International Symposium on Algorithms and Computation (ISAAC'95).* Springer LNCS 1004, pages 352–361.

Huang, M. L., and Eades, P. (1998a). A fully animated interactive system for clustering and navigating huge graphs. In *Proceedings of the 6th International Symposium on Graph Drawing (GD'98)*, Springer LNCS 1547, pages 374–383.

Hughes, J. G. (1993). *Object-Oriented Databases.* International Series in Computer Science. Prentice-Hall.

Humphrey, W., Dalke, A., and Schulten, K. (1996). VMD — Visual molecular dynamics. *Journal of Molecular Graphics*, 14(1):33–38.

Hutton, M. D., and Lubiw, A. (1991). Upward planar drawing of single source acyclic digraphs. In *Proceedings of the 2nd ACM-SIAM Symposium on Discrete Algorithms (SODA'91)*, pages 203–211.

Imai, H., and Asano, T. (1983). Finding the connected components and a maximum clique of an intersection graph of rectangles in the plane. *Journal of Algorithms*, 4:310–323.

Imai, H., and Asano, T. (1986). Efficient algorithms for geometric graph search problems. *SIAM Journal on Computing*, 15(2):478–494.

Imhof, E. (1962). Die Anordnung der Namen in der Karte. *International Yearbook of Cartography*, 2:93–129.

Imhof, E. (1975). Positioning names on maps. *The American Cartographer*, 2(2):128–144.

Indermark, K., Thomas, W., Huch, F., Leucker, M., and Noll, T. (1999). Various texts about the TRUTH system for modelling concurrent systems,

Lehrstuhl für Informatik II, RWTH Aachen. `http://www-i2.informatik.rwth-aachen.de/Forschung/MCS/`.

Isoda, S., Shimomura, T., and Ono, Y. (1987). VIPS: A visual debugger. *IEEE Software*, 4(3):8–19.

Iturriaga, C., and Lubiw, A. (1997). $\mathcal{NP}$-hardness of some map labeling problems. Technical Report CS-97-18, University of Waterloo.

Jain, A. K., and Dubes, R. C. (1988). *Algorithms for Clustering Data*. Prentice Hall.

Johnson, D. S. (1982). The $\mathcal{NP}$-completeness column: An ongoing guide. *Journal of Algorithms*, 3:89–99.

Jünger, M., Lee, E. K., Mutzel, P., and Odenthal, T. (1997). A polyhedral approach to the multi-layer crossing minimization problem. In *Proceedings of the 5th International Symposium on Graph Drawing (GD'97)*. Springer LNCS 1353, pages 13–24.

Jünger, M., and Mutzel, P. (1996). Maximum planar subgraphs and nice embeddings: Practical layout tools. *Algorithmica*, 16:33–59.

Jünger, M., and Mutzel, P. (1997). 2-Layer straightline crossing minimization: Performance of exact and heuristic algorithms. *Journal on Graph Algorithms and Applications*, 1(1):1–25.

Kakoulis, K. G., and Tollis, I. G. (1996). On the edge label placement problem. In *Proceedings of the 4th International Symposium on Graph Drawing (GD'96)*. Springer LNCS 1190, pages 241–256.

Kakoulis, K. G., and Tollis, I. G. (1997). An algorithm for labeling edges of hierarchical drawings. In *Proceedings of the 5th International Symposium on Graph Drawing (GD'97)*. Springer LNCS 1353, pages 169–180.

Kakoulis, K. G., and Tollis, I. G. (1998a). On the multiple label placement problem. In *Proceedings of the 10th Canadian Conference on Computational Geometry (CCCG'98)*, pages 66–67.

Kakoulis, K. G., and Tollis, I. G. (1998b). A unified approach to labeling graphical features. In *Proceedings of the 14th Annual ACM Symposium on Computional Geometry (SCG'98)*, pages 347–356.

Kamada, T., and Kawai, S. (1988). A simple method for computing general positions in displaying three-dimensional objects. *Computer Vision, Graphics and Image Processing*, 41:43–56.

Kamada, T., and Kawai, S. (1989). An algorithm for drawing general undirected graphs. *Information Processing Letters*, 31:7–15.

Kant, G. (1996). Drawing planar graphs using the canonical ordering. *Algorithmica*, 16:4–32.

Kant, G., and Bodlaender, H. L. (1991). Planar graph augmentation problems. In *Proceedings of the 2nd Workshop on Algorithms and Data Structures (WADS'91)*, Springer LNCS 519, pages 286–298.

Karp, R. (1972). Reducibility among combinatorical problems. In *Complexity of Computer Computations*, pages 85–103. Plenum Press.

Kato, T., and Imai, H. (1988). The $\mathcal{NP}$-completeness of the character placement problem of 2 or 3 degrees of freedom. In *Record of Joint Conference of Electrical and Electronic Engineers in Kyushu*, page 1138.

Keahey, T. A., and Robertson, E. (1996). Techniques for non-linear maginifaction transformations. In *Proceedings of the IEEE Symposium on Information Visualization (InfoVis'96)*, pages 38–45.

Kedem, G., and Watanabe, H. (1984). Graph optimization techniques for IC-layout and compaction. *IEEE Transactions on Computer-Aided Design of Integrated Circuits and Systems*, CAD-3(1):12–20.

Kelly, D., and Rival, I. (1975). Planar lattices. *Canadian Journal of Mathematics*, 27:636–665.

Kenis, P. (1999). Analysing social network data by means of visualisation techniques. Paper presented at the *19th International Conference on Social Network Analysis (Sunbelt XIX)*, Charleston.

Kernighan, B. W., and Lin, S. (1970). An efficient heuristic procedure for partitioning graphs. *The Bell System Technical Journal*, 49(2):291–307.

Kirchhoff, G. R. (1847). Über die Auflösung der Gleichungen, auf welche man bei der Untersuchung der linearen Verteilung galvanischer Ströme geführt wird. *Annalen der Physik und Chemie*, 72:497–508.

Klau, G. W., and Mutzel, P. (1998). Quasi-orthogonal drawing of planar graphs. Technical Report 98-1-013, Max-Planck-Institut für Informatik, Saarbrücken.

Klau, G. W., and Mutzel, P. (1999a). Combining graph labeling and compaction. In *Proceedings of the 7th International Symposium on Graph Drawing (GD'99)*. Springer LNCS 1731, pages 27–37.

Klau, G. W., and Mutzel, P. (1999b). Optimal compaction of orthogonal grid drawings. In *Integer Programming and Combinatorial Optimization (IPCO'99)*. Springer LNCS 1610, pages 304–319.

Knipping, L. (1998). Beschriftung von Linienzügen. Diplomarbeit, Fachbereich Mathematik und Informatik, Freie Universität Berlin.

Knuth, D. E., and Raghunathan, A. (1992). The problem of compatible representatives. *SIAM Journal on Discrete Mathematics*, 5(3):422–427.

Koebe, P. (1936). Kontaktprobleme auf der konformen Abbildung. *Berichte über die Verhandlungen der Sächsischen Akademie der Wissenschaften zu Leipzig, Mathematisch-Physikalische Klasse*, 88:141–164.

Kolmogorov, A. N., and Bardzin, Y. M. (1967). About realization of sets in 3-dimensional space. *Problems in Cybernetics*, pages 261–268.

Kosak, C., Marks, J., and Shieber, S. (1994). Automating the layout of network diagrams with specified visual organization. *IEEE Transactions on Systems, Man and Cybernetics*, 24(3):440–454.

Krackhardt, D., Blythe, J., and McGrath, C. (1994). KrackPlot 3.0: An improved network drawing program. *Connections*, 17(2):53–55.

Kruskal, J. B., and Wish, M. (1978). *Multidimensional Scaling.* Sage University Paper Series on Quantitative Applications in the Social Sciences 07-011.

Kucera, L., Mehlhorn, K., Preis, B., and Schwarzenecker, E. (1993). Exact algorithms for a geometric packing problem. *Proceedings of the 10th Symposium on the Theoretical Aspects of Computer Science (STACS'93).* Springer LNCS 665, pages 317–322.

Kumar, A., and Fowler, R. H. (1994). A spring modelling algorithm to position nodes of an undirected graph in three dimensions. Technical report, Department of Computer Science, University of Texas.

Laguna, M., and Martí, R. (1999). Grasp and path relinking for 2-layer straight line crossing minimization. *INFORMS Journal on Computing,* 11(1):44–52.

Laguna, M., Martí, R., and Valls, V. (1997). Arc crossing minimization in hierarchical digraphs with tabu search. *Computers and Operations Research,* 24(12):1175–1186.

Lam, S., and Sethi, R. (1977). Worst case analysis of two scheduling problems. *SIAM Journal on Computing,* 6:518–536.

LaPaugh, A. S. (1998). VLSI Layout Algorithms. In *Algorithms and Theory of Computation Handbook.* CRC Press.

Leiserson, C. E. (1980). Area-efficient graph layouts (for VLSI). In *Proceedings of the 21st Annual IEEE Symposium on Foundations of Computer Science (FOCS'80),* pages 270–281.

Lempel, A., and Cederbaum, I. (1966). Minimum feedback arc and vertex sets of a directed graph. *IEEE Transactions on Circuit Theory,* CT–13(4):339–403.

Lempel, A., Even, S., and Cederbaum, I. (1967). An algorithm for planarity testing of graphs. In *Theory of Graphs: International Symposium (Rome 1966),* pages 215–232. Gordon and Breach.

Lengauer, T. (1989). Hierarchical planarity testing algorithms. *Journal of the ACM,* 36:474–509.

Lengauer, T. (1990). *Combinatorial Algorithms for Integrated Circuit Layout.* Applicable Theory in Computer Science. B. G. Teubner and John Wiley & Sons.

Leung, J. (1992). A new graph-theoretic heuristic for facility layout. *Management Science,* 38(4):594–605.

Lewis, J. M., and Yannakakis, M. (1980). The node-deletion problem for hereditary properties is $\mathcal{NP}$-complete. *Journal of Computer and System Sciences,* 20(2):219–230.

Liebers, A. (1996). Methods for planarizing graphs - A survey and annotated bibliography. Technical Report Konstanzer Schriften in Mathematik und Informatik Nr. 12, Fakultät für Mathematik und Informatik, Universität Konstanz. ISSN 1430-3558. To appear in *Journal on Graph Algorithms and Applications.*

Lin, X. (1992). *Analysis of Algorithms for Drawing Graphs.* PhD thesis, University of Queensland.

Lino, P., Martí, R., and Valls, V. (1996). A branch and bound algorithm for minimizing the number of crossing arcs in bipartite graphs. *Journal of Operational Research*, 90:303–319.

Lipton, R. J., Rose, D. J., and Tarjan, R. E. (1979). Generalized nested dissection. *SIAM Journal on Numerical Analysis*, 16:346–358.

Lipton, R. J., and Tarjan, R. E. (1970). A seperator theorem for planar graphs. In *Proceedings of the Conference on Theoretical Computer Science*, pages 1–10.

Liu, P. C., and Geldmacher, R. C. (1977). On the deletion of nonplanar edges of a graph. In *Proceedings of the 10th Southeastern Conference on Combinatorics, Graph Theory, and Computing*, pages 727–738.

Lyons, K. A. (1992). Cluster busting in anchored graph drawing. In *Proceedings of the '92 CAS Conference (CASCON'92)*, pages 7–17.

Lyons, K. A., Meijer, H., and Rappaport, D. (1998). Algorithms for cluster busting in anchored graph drawing. *Journal on Graph Algorithms and Applications*, 2(1):1–24.

Mackworth, A. K., and Freuder, E. C. (1985). The complexity of some polynomial network consistency algorithms for constraint satisfaction problem. *Artificial Intelligence*, 25(1):65–74.

Mäkinen, E. (1990). Experiments on drawing 2-level hierarchical graphs. *International Journal of Computer and Mathematics*, 36:175–181.

Mäkinen, E., and Sieranta, M. (1994). Genetic algorithms for drawing bipartite graphs. *Internatonal Journal of Computer Mathematics*, 53:157–166.

Malitz, S., and Papakostas, A. (1992). On the angular resolution of planar graphs. In *Proceedings of the 24th Annual ACM Symposium on the Theory of Computing (STOC'92)*, pages 527–538.

Malitz, S., and Papakostas, A. (1994). On the angular resolution of planar graphs. *SIAM Journal on Discrete Mathematics*, 7(2):172–183.

Manning, J. (1990). *Geometric Symmetry in Graphs.* PhD thesis, Purdue University.

Marks, J., and Shieber, S. (1991). The computational complexity of cartographic label placement. Technical Report TR-05-91, Harvard University Computer Science.

Masuda, S., Kimura, S., Kashiwabara, T., and Fujisawa, T. (1983). On the Manhattan wiring problem. Technical Report CAS 83-20, Institute of Electronics and Communication Engineers of Japan.

Masui, T. (1992). Graphic object layout with interactive genetic algorithms. In *Proceedings of the 1992 IEEE Workshop on Visual Languages (VL'92)*, pages 74–87.

Matuszewski, C., Schönfeld, R., and Molitor, P. (1999). Using sifting for k-layer straightline crossing minimization. *Proceedings of the 7th Symposium on Graph Drawing (GD'99)*. Springer LNCS 1731, pages 217–224.

McGrath, C., Blythe, J., and Krackhardt, D. (1996). Seeing groups in graph layouts. *Connections*, 19(2):22–29.

McGrath, C., and Borgatti, S. P. (1999). *The International Network for Social Network Analysis Homepage.* http://www.heinz.cmu.edu/project/INSNA/.

Mehlhorn, K. (1984). *Data Structures and Algorithms. Volume 2: Graph Algorithms and $\mathcal{NP}$-Completeness.* EATCS Monographs on Theoretical Computer Science. Springer.

Mehlhorn, K., and Näher, S. (1999). *The* LEDA *Platform of Combinatorial and Geometric Computing.* Cambridge University Press. Project home page at http://www.mpi-sb.mpg.de/LEDA/.

Messinger, E. B., Rowe, L. A., and Henry, R. H. (1991). A divide-and-conquer algorithm for the automatic layout of large directed graphs. *IEEE Transactions on Systems, Man, and Cybernetics*, SMC-21(1):1–12.

Miriyala, K., Hornik, S. W., and Tamassia, R. (1993). An incremental approach to aesthetic graph layout. In *Proceedings of the 6th International Workshop on Computer-Aided Software Engineering (CASE'93)*, pages 297–308.

Misue, K., Eades, P., Lai, W., and Sugiyama, K. (1995). Layout adjustment and the mental map. *Journal of Visual Languages and Computing*, 6:183–210.

Moen, S. (1990). Drawing dynamic trees. *IEEE Software*, 7:21–28.

Monien, B., Ramme, F., and Salmen, H. (1995). A parallel simulated annealing algorithm for generating 3D layouts of undirected graphs. In *Proceedings of the 3rd International Symposium on Graph Drawing (GD'95).* Springer LNCS 1027, pages 396–408.

Monien, B., Ramme, F., and Salmen, H. (1996). A parallel simulated annealing algorithm for generating 3d layouts of undirected graphs. In *Proceedings of the 3rd International Symposium on Graph Drawing (GD'95).* Springer LNCS 1027, pages 396–408.

MTA (1999). MTA New York City subway map. http://www.mta.nyc.ny.us/nyct/images/sub1a.gif and http://www.mta.nyc.ny.us/nyct/images/sub2a.gif.

Mukherjea, S., Foley, J., and Hudson, S. (1994). Interactive clustering for navigating in hypermedia systems. In *Proceedings of the ACM European Conference on Hypermedia Tehcnologie.*

Mutzel, P. (1994). *The Maximum Planar Subgraph Problem.* PhD thesis, Universität zu Köln.

Mutzel, P. (1995). A polyhedral approach to planar augmentation and related problems. In *Proceedings of the 3rd European Symposium on Algorithms (ESA'95).* Springer LNCS 979, pages 494–507.

Mutzel, P. (1997). An alternative method to crossing minimization on hierarchical graphs. In *Proceedings of the 4th International Symposium on Graph Drawing (GD'96).* Springer LNCS 1190, pages 318–333.

Nakano, S., Rahman, M. S., and Nishizeki, T. (1997). A linear-time algorithm for four-partitioning four-connected planar graphs. In *Proceedings of the 5th International Symposium on Graph Drawing (GD'97)*. Springer LNCS 1353, pages 334 - 344.

Nemhauser, G. L., and Sigismondi, G. (1992). A strong cutting plane/branch-and-bound algorithm for node packing. *Journal of the Operational Research Society*, 43:443–457.

Nielsen, J. (1990). The art of navigating throuh hypertext. *Communications of the ACM*, 33(3):296–310.

Nishizeki, T., and Chiba, N. (1988). *Planar Graphs: Theory and Algorithms.* North-Holland Mathematics Studies 140/32.

North, S. C. (1996). Incremental layout with DynaDag. In *Proceedings of the 3rd International Symposium on Graph Drawing (GD'95)*. Springer LNCS 1027, pages 409–418.

Oerder, M., and Ney, H. (1993). Word graphs: An efficient interface between continuous-speech recognition and language understanding. In *Proceedings of the International Conference on Acoustics, Speech and Signal Processing (ICASSP'93)*, volume II, pages 119–122.

Ostry, D. (1996). Some three-dimensional graph drawing algorithms. Master's thesis, University of Newcastle.

Otten, R. H. J. M., and van Wijk, J. G. (1978). Graph representation in interactive layout design. In *Proceedings of the IEEE International Symposium on Circuits and Systems*, pages 914–918.

Papakostas, A. (1995). Upward planarity testing of outerplanar dags. In *Proceedings of the DIMACS International Workshop on Graph Drawing (GD'94)*. Springer LNCS 894, pages 298–306.

Papakostas, A., Six, J. M., and Tollis, I. G. (1996). Experimental and theoretical results in interactive orthogonal graph drawing. In *Proceedings of the 4th International Symposium on Graph Drawing (GD'96)*. Springer LNCS 1190, pages 371–386.

Papakostas, A., and Tollis, I. G. (1997a). Incremental orthogonal graph drawing in three dimensions. In *Proceedings of the 5th International Symposium on Graph Drawing (GD'97)*. Springer LNCS 1353, pages 52–63.

Papakostas, A., and Tollis, I. G. (1997b). Incremental orthogonal graph drawing in three-dimensions. Technical Report UTDCS-02-97, Dept. of Computer Sciencs, University of Texas at Dallas.

Papakostas, A., and Tollis, I. G. (1997c). Orthogonal drawing of high degree graphs with small area and few bends. In *Proceedings of the 5th Workshop on Algorithms and Data Structures (WADS'97)*. Springer LNCS 1272, pages 354–367.

Papakostas, A., and Tollis, I. G. (1997d). A pairing technique for area-efficient orthogonal drawings. In *Proceedings of the 4th International Symposium on Graph Drawing (GD'96)*. Springer LNCS 1190, pages 354–370.

Papakostas, A., and Tollis, I. G. (1998). Interactive orthogonal graph drawing. *IEEE Transactions on Computers*, 47(11):1297–1309.

Patrignani, M. (1999a). On the complexity of orthogonal compaction. Technical Report RT–DIA–39–99, Dipartimento di Informatica e Automazione, Università degli Studi di Roma Tre.

Patrignani, M. (1999b). On the complexity of orthogonal compaction. *Proceedings of the 6th Workshop on Algorithms and Data Structures (WADS'99)*. Springer LNCS 1663, pages 56–61.

Patrignani, M., and Vargiu, F. (1997). 3DCube: A tool for the three dimensional graph drawing. In *Proceedings of the 5th International Symposium on Graph Drawing (GD'97)*. Springer LNCS 1353, pages 284–290.

Paulish, F. N. (1993). *The Design of an Extendible Graph Editor.* Springer LNCS 704.

Platt, C. (1976). Planar lattices and planar graphs. *Journal of Combinatorial Theory, Series B*, 21:30–39.

Poon, C. K., Zhu, B., and Chin, F. (1998). A polynomial time solution for labeling a rectilinear map. *Information Processing Letters*, 65:201–207.

Poutré, J. A. L. (1994). Alpha-algorithms for incremental planarity testing. In *Proceedings of the 26th Annual ACM Symposium on the Theory of Computation (STOC'94)*, pages 706–715.

Purchase, H. C. (1997). Which aesthetic has the greatest effect on human understanding? In *Proceedings of the 5th International Symposium on Graph Drawing (GD'97)*. Springer LNCS 1353, pages 248–261.

Purchase, H. C., Cohen, R. F., and James, M. (1996). Validating graph drawing aesthetics. In *Proceedings of the 3rd International Symposium on Graph Drawing (GD'95)*. Springer LNCS 1027, pages 435–446.

Purchase, H. C., Cohen, R. F., and James, M. (1997). An experimental study of the basis for graph drawing algorithms. *ACM Journal of Experimental Algorithmics*, 2(4).

Quinn, N. R., and Breuer, M. A. (1979). A force directed component placement procedure for printed circuit boards. *IEEE Transactions on Circuits and Systems*, 26(6):377–388.

Reeves, C. M. (1995). *Modern Heuristic Techniques for Combinatorial Problems.* McGraw-Hill.

Reggiani, M. G., and Marchetti, F. E. (1988). A proposed method for representing hierarchies. *IEEE Transactions on Systems, Man, and Cybernetics*, 18(1):2–8.

Reinelt, G. (1985). *The linear ordering problem: algorithms and applications.* Research and Exposition in Mathematics 8, Heldermann.

Reingold, E. M., and Tilford, J. S. (1981). Tidier drawings of trees. *IEEE Transactions on Software Engineering*, 7(2):223–228.

Rival, I. (1985). The diagram. In *Graphs and Order*, NATO ASI Series, pages 103–133. Reidel Publishing.

Robertson, G. G., Mackinlay, J. D., and Card, S. K. (1993). Cone trees: Animated 3*d* visualizations of hierarchical information. In *Proceedings of the ACM Conference on Human Factors in Computing Systems*, pages 189–193.

Rosenstiehl, P., and Tarjan, R. E. (1986). Rectilinear planar layouts of planar graphs and bipolar orientations. *Discrete & Computational Geometry*, 1(4):342–351.

Roxborough, T., and Sen, A. (1997). Graph clustering using multiway ratio cut. In *Proceedings of the 5th International Symposium on Graph Drawing (GD'97)*. Springer LNCS 1353, pages 291–296.

Rudell, R. (1993). Dynamic variable ordering for ordered binary decision diagrams. In *Proceedings of the IEEE/ACM International Conference on Computer-Aided Design (ICCAD'93)*, pages 42–47.

Sablowski, R., and Frick, A. (1996). Automatic graph clustering. In *Proceedings of the 4th International Symposium on Graph Drawing (GD'96)*. Springer LNCS 1190, pages 395–400.

Sander, G. (1994). Graph layout through the VCG tool. Technical Report A03/94, Universität des Saarlandes.

Sander, G. (1996a). A fast heuristic for hierarchical Manhattan layout. In *Proceedings of the 3rd International Symposium on Graph Drawing (GD'95)*. Springer LNCS 1027, pages 447–458.

Sander, G. (1996b). Graph layout for applications in compiler construction. Technical Report A/01/96, FB 14 Informatik, Universität des Saarlandes.

Sarkar, M., and Brown, M. H. (1994). Graphical fisheye views. *Communications of the ACM*, 37(12):73–84.

Schlag, M., Liao, Y.-Z., and Wong, C. K. (1983). An algorithm for optimal two-dimensional compaction of VLSI layouts. *Integration, the VLSI Journal*, 1:179–209.

Schnyder, W. (1990). Embedding planar graphs on the grid. In *Proceedings of the 1st ACM-SIAM Symposium on Discrete Algorithms (SODA'90)*, pages 138–148.

Schrijver, A. (1986). *Theory of Linear and Integer Programming*. Wiley-Interscience.

Sedgewick, R. (1988). *Algorithms*, pages 438–441. Addison–Wesley, 2nd edition.

Shiloach, Y. (1976). *Arrangements of planar graphs on the planar lattice*. PhD thesis, Weizmann Institute of Science.

Sim, S. (1996). Automatic graph drawing algorithms. Manuscript, available at `http://www.cs.toronto.edu/~simsuz/papers/grafdraw.ps.gz`.

Six, J. M., Kakoulis, K. G., and Tollis, I. G. (1998). Refinement of orthogonal graph drawings. In *Proceedings of the 6th International Symposium on Graph Drawing (GD'98)*. Springer LNCS 1547, pages 302–315.

Strijk, T., and van Kreveld, M. (1999). Labeling a rectilinear map more efficiently. *Information Processing Letters*, 69(1):25–30.

Strijk, T., and Wolf, A. (1999). Labeling points with circles. Technical Report TR-99-08, Institut für Informatik, Freie Universität Berlin.

Stumme, G., and Wille, R. (1995). A geometrical heuristic for drawing concept lattices. In *Proceedings of the DIMACS International Workshop on Graph Drawing (GD'94)*. Springer LNCS 894, pages 452–460.

Sugiyama, K. (1987). A cognitive approach for graph drawing. *Cybernetic Systems*, 18(6):447–488.

Sugiyama, K., and Misue, K. (1991). Visualisation of structural information: Automatic drawing of compound digraphs. *IEEE Transactions on Systems, Man, and Cybernetics*, 21(4):876–892.

Sugiyama, K., and Misue, K. (1995). A simple and unified method for drawing graphs: Magnetic-spring algorithm. In *Proceedings of the DIMACS International Workshop on Graph Drawing (GD'94)*. Springer LNCS 894, pages 364–375.

Sugiyama, K., Tagawa, S., and Toda, M. (1981). Methods for visual understanding of hierarchical system structures. *IEEE Transactions on Systems, Man, and Cybernetics*, 11(2):109–125.

Supowit, K. J., and Reingold, E. M. (1983). The complexity of drawing trees nicely. *Acta Informatica*, 18:377–392.

Tamassia, R. (1987). On embedding a graph in the grid with the minimum number of bends. *SIAM Journal on Computing*, 16(3):421–444.

Tamassia, R. (1998). Constraints in graph drawing algorithms. *Constraints*, 3(1):87–120.

Tamassia, R., Di Battista, G., and Batini, C. (1988). Automatic graph drawing and readability of diagrams. *IEEE Transactions on Systems, Man, and Cybernetics*, 18(1):61–79.

Tamassia, R., and Tollis, I. G. (1986). A unified approach to visibility representations of planar graphs. *Discrete & Computational Geometry*, 1(4):321–341.

Tamassia, R., and Tollis, I. G. (1989). Planar grid embedding in linear time. *IEEE Transactions on Circuits and Systems*, 36(9):1230–1234.

Tamassia, R., Tollis, I. G., and Vitter, J. S. (1991). Lower bounds for planar orthogonal drawings of graphs. *Information Processing Letters*, 39(1):35–40.

Tanenbaum, A. S. (1995). *Distributed Operating Systems*. Prentice Hall.

Tarjan, R. E. (1983). *Data structures and network algorithms* CBMS-NSF Regional Conference Series in Applied Mathematics 44, SIAM.

Thomassen, C. (1980). Planarity and duality of finite and infinite planar graphs. *Journal of Combinatorial Theory, Series B*, 29:244–271.

Thompson, C. D. (1980). *A Complexity Theory for VLSI*. PhD thesis, Carnegie Mellon University.

Tunkelang, D. (1994). A practical approach to drawing undirected graphs. Technical Report CMU-CS-94-161, School of Computer Science, Carnegie Mellon University.

Tutte, W. T. (1960). Convex representations of graphs. *Proceedings of the London Mathematical Society, Third Series*, 10:304–320.

Tutte, W. T. (1963). How to draw a graph. *Proceedings of the London Mathematical Society, Third Series*, 13:743–768.

Ullman, J. (1989). *Principles of Database and Knowledgebase Systems*, volume 1. Computer Science Press.

Utech, J., Branke, J., Schmeck, H., and Eades, P. (1998). An evolutionary algorithm for drawing directed graphs. In *Proceedings of the International Conference on Imaging Science, Systems, and Technology*, pages 154–160.

Valdes, J., Tarjan, R. E., and Lawler, E. L. (1982). The recognition of series parallel digraphs. *SIAM Journal on Computing*, 11:298–313.

Valiant, L. (1981). Universality considerations in VLSI circuits. *IEEE Transactions on Computers*, C-30(2):135–140.

van Kreveld, M., Strijk, T., and Wolff, A. (1998). Point set labeling with sliding labels. In *Proceedings of the 14th Annual ACM Symposium on Computational Geometry (SCG'98)*, pages 337–346.

Verweij, B., and Aardal, K. (1999). An optimisation algorithm for maximum independent set with applications in map labelling. In *Proceedings of the 7th European Symposium on Algorithms (ESA'99)*. Springer LNCS 1643, pages 426–437.

Vogt, F. (1996). *Formale Begriffsanalyse mit C++: Datenstrukturen und Algorithmen*. Springer.

Vogt, F., and Wille, R. (1995). TOSCANA — a graphical tool for analyzing and exploring data. In *Proceedings of the DIMACS International Workshop on Graph Drawing (GD'94)*. Springer LNCS 894, pages 226–233.

Vossen, G. (1991). *Datenbankmodelle, Datenbanksprachen und Datenbankmanagement-Systeme*. Addison-Wesley.

Wagner, F. (1994). Approximate map labeling is in $\Omega(n \log n)$. *Information Processing Letters*, 52(3):161–165.

Wagner, F., and Wolff, A. (1995a). An efficient and effective approximation algorithm for the map labeling problem. In *Proceedings of the 3rd European Symposium on Algorithms (ESA'95)*. Springer LNCS 979, pages 420–433.

Wagner, F., and Wolff, A. (1995b). Map labeling heuristics: Provably good and practically useful. In *Proceedings of the 11th Annual ACM Symposium on Computational Geometry (SCG'95)*, pages 109–118.

Wagner, F., and Wolff, A. (1997). A practical map labeling algorithm. *Computational Geometry: Theory and Applications*, 7:387–404.

Wagner, F., and Wolff, A. (1998). A combinatorial framework for map labeling. In *Proceedings of the 6th International Symposium on Graph Drawing (GD'98)*. Springer LNCS 1547, pages 316–331.

Wang, X., and Miyamoto, I. (1995). Generating cunstomized layouts. In *Proceedings of the 3rd International Symposium on Graph Drawing (GD'95)*. Springer LNCS 1027, pages 504–515.

Wang, X., and Miyamoto, I. (1996). Generating customized layouts. In *Proceedings of the 3rd International Symposium on Graph Drawing (GD'95)*. Springer LNCS 1027, pages 504–515.

Warfield, J. (1977). Crossing theory and hierarchy mapping. *IEEE Transactions on Systems, Man, and Cybernetics*, SMC-7(7):502–523.

Warnke, V., Kompe, R., Niemann, H., and Nöth, E. (1997). Integrated dialog act segmentation and classification using prosodic features and language models. Technical Report Verbmobil-Report 218, Lehrstuhl für Mustererkennung 5, Universität Erlangen-Nürnberg.

Wasserman, S., and Faust, K. (1994). *Social Network Analysis: Methods and Applications*. Cambridge University Press.

Watanabe, H. (1984). *IC Layout Generation and Compaction Using Mathematical Optimization*. PhD thesis, University of Rochester.

Watanabe, T., Ae, T., and Nakamura, A. (1983). On the $\mathcal{NP}$-hardness of edge-deletion and -contraction problems. *Discrete Applied Mathematics*, 6:63–78.

Webber, R. (1997). Finding the best viewpoints for three-dimensional graph drawings. In *Proceedings of the 5th International Symposium on Graph Drawing (GD'97)*. Springer LNCS 1353, pages 87–98.

Webber, R. (1998). *Finding the Best Viewpoint for Three-Dimensional Graph Drawings*. PhD thesis, University of Newcastle. `http://www.cs.mu.oz.au/~rwebber/research/thesis/`.

Wei, Y.-C., and Cheng, C.-K. (1991). Ratio cut partitioning for hierarchical designs. *IEEE Transactions on Computer-Aided Design*, 10(7):911–921.

West, D. (1996). *Introduction to Graph Theory*. Prentice Hall.

White, D. (1999). Pgraph of Canaan genealogy made by Pajek program. Manuscript. `http://eclectic.ss.uci.edu/~drwhite/pgraph/p-graphs.html`.

Wiese, R., and Kaufmann, M. (1998). Adding constraints to an algorithm for orthogonal graph drawing. In *Proceedings of the 6th International Symposium on Graph Drawing (GD'98)*. Springer LNCS 1547, pages 462–463.

Wille, R. (1989). Lattices in data analysis: How to draw them with a computer. In *Algorithms and Order*, NATO ASI Series, pages 33–58. Kluwer Academic Publishers.

Wille, R. (1997). Introduction to formal concept analysis. In *Modelli e modellizzazione. Models and modelling.* Consiglio Nazionale delle Ricerche, Instituto di Studi sulli Ricerca e Documentazione Scientifica, Roma, pages 39–51.

Winter, A., and Schürr, A. (1997). Modules and updatable graph views for programmed graph rewriting systems. Technical Report AIB 97-3, Lehrstuhl für Informatik III, RWTH Aachen.

Wolff, A. (1999). *Map Labeling in Theory and Practice*. PhD thesis, Freie Universität Berlin.

Wolff, A., Knipping, L., van Kreveld, M., Strijk, T., and Agarwal, P. K. (1999). A simple and efficient algorithm for high-quality line labeling. In *Proceedings of GISRUK'99*.

Wood, D. (1998a). An algorithm for three-dimensional orthogonal graph drawing. In *Proceedings of the 6th International Symposium on Graph Drawing (GD'98)*. Springer LNCS 1547, pages 332–346.

Wood, D. (1998b). Two-bend three-dimensional orthogonal grid drawing of maximum degree five graphs. Technical Report 98/03, Monash University.

Wood, D. R. (1999a). Multi-dimensional orthogonal graph drawing in the general position model. Technical Report 99/38, Monash University.

Wood, D. R. (1999b). A new algorithm and open problems in three-dimensional orthogonal graph drawing. In *Proceedings of the 10th Australasian Workshop on Combinatorical Algorithms (AWOCA'99)*, pages 157–167.

Wood, D. R. (2000). Three-Dimensional Orthogonal Graph Drawing. PhD thesis, Monash University.

Yannakakis, M. (1978). Node- and edge-deletion $\mathcal{NP}$-complete problems. In *Proceedings 10th Annual ACM Symposium on the Theory of Computing (STOC'78)*, pages 253–264.

Yoeli, P. (1972). The logic of automated map lettering. *The Cartographic Journal*, 9:99–108.

Zeller, A., and Lütkehaus, D. (1996). DDD — A free graphical front-end for UNIX debuggers. *ACM SIGPLAN Notices*, 31(1):22–27.

Zoraster, S. (1986). Integer programming applied to the map label placement problem. *Cartographica*, 23(3):16–27.

Zoraster, S. (1990). The solution of large 0-1 integer programming problems encountered in automated cartography. *Operations Research*, 38(5):752–759.

Index

Lecture Notes in Computer Science

For information about Vols. 1–1953
please contact your bookseller or Springer-Verlag

Vol. 1991: F. Dignum, C. Sierra (Eds.), Agent Mediated Electronic Commerce. VIII, 241 pages. 2001. (Subseries LNAI).

Vol. 1992: K. Kim (Ed.), Public Key Cryptography. Proceedings, 2001. XI, 423 pages. 2001.

Vol. 1993: E. Zitzler, K. Deb, L. Thiele, C.A.Coello Coello, D. Corne (Eds.), Evolutionary Multi-Criterion Optimization. Proceedings, 2001. XIII, 712 pages. 2001.

Vol. 1995: M. Sloman, J. Lobo, E.C. Lupu (Eds.), Policies for Distributed Systems and Networks. Proceedings, 2001. X, 263 pages. 2001.

Vol. 1997: D. Suciu, G. Vossen (Eds.), The World Wide Web and Databases. Proceedings, 2000. XII, 275 pages. 2001.

Vol. 1998: R. Klette, S. Peleg, G. Sommer (Eds.), Robot Vision. Proceedings, 2001. IX, 285 pages. 2001.

Vol. 1999: W. Emmerich, S. Tai (Eds.), Engineering Distributed Objects. Proceedings, 2000. VIII, 271 pages. 2001.

Vol. 2000: R. Wilhelm (Ed.), Informatics: 10 Years Back, 10 Years Ahead. IX, 369 pages. 2001.

Vol. 2001: G.A. Agha, F. De Cindio, G. Rozenberg (Eds.), Concurrent Object-Oriented Programming and Petri Nets. VIII, 539 pages. 2001.

Vol. 2002: H. Comon, C. Marché, R. Treinen (Eds.), Constraints in Computational Logics. Proceedings, 1999. XII, 309 pages. 2001.

Vol. 2003: F. Dignum, U. Cortés (Eds.), Agent Mediated Electronic Commerce III. XII, 193 pages. 2001. (Subseries LNAI).

Vol. 2004: A. Gelbukh (Ed.), Computational Linguistics and Intelligent Text Processing. Proceedings, 2001. XII, 528 pages. 2001.

Vol. 2006: R. Dunke, A. Abran (Eds.), New Approaches in Software Measurement. Proceedings, 2000. VIII, 245 pages. 2001.

Vol. 2007: J.F. Roddick, K. Hornsby (Eds.), Temporal, Spatial, and Spatio-Temporal Data Mining. Proceedings, 2000. VII, 165 pages. 2001. (Subseries LNAI).

Vol. 2009: H. Federrath (Ed.), Designing Privacy Enhancing Technologies. Proceedings, 2000. X, 231 pages. 2001.

Vol. 2010: A. Ferreira, H. Reichel (Eds.), STACS 2001. Proceedings, 2001. XV, 576 pages. 2001.

Vol. 2011: M. Mohnen, P. Koopman (Eds.), Implementation of Functional Languages. Proceedings, 2000. VIII, 267 pages. 2001.

Vol. 2012: D.R. Stinson, S. Tavares (Eds.), Selected Areas in Cryptography. Proceedings, 2000. IX, 339 pages. 2001.

Vol. 2013: S. Singh, N. Murshed, W. Kropatsch (Eds.), Advances in Pattern Recognition – ICAPR 2001. Proceedings, 2001. XIV, 476 pages. 2001.

Vol. 2015: D. Won (Ed.), Information Security and Cryptology – ICISC 2000. Proceedings, 2000. X, 261 pages. 2001.

Vol. 2018: M. Pollefeys, L. Van Gool, A. Zisserman, A. Fitzgibbon (Eds.), 3D Structure from Images – SMILE 2000. Proceedings, 2000. X, 243 pages. 2001.

Vol. 2020: D. Naccache (Ed.), Topics in Cryptology – CT-RSA 2001. Proceedings, 2001. XII, 473 pages. 2001

Vol. 2021: J. N. Oliveira, P. Zave (Eds.), FME 2001: Formal Methods for Increasing Software Productivity. Proceedings, 2001. XIII, 629 pages. 2001.

Vol. 2022: A. Romanovsky, C. Dony, J. Lindskov Knudsen, A. Tripathi (Eds.), Advances in Exception Handling Techniques. XII, 289 pages. 2001

Vol. 2024: H. Kuchen, K. Ueda (Eds.), Functional and Logic Programming. Proceedings, 2001. X, 391 pages. 2001.

Vol. 2025: M. Kaufmann, D. Wagner (Eds.), Drawing Graphs. XIV, 312 pages. 2001.

Vol. 2026: F. Müller (Ed.), High-Level Parallel Programming Models and Supportive Environments. Proceedings, 2001. IX, 137 pages. 2001.

Vol. 2027: R. Wilhelm (Ed.), Compiler Construction. Proceedings, 2001. XI, 371 pages. 2001.

Vol. 2028: D. Sands (Ed.), Programming Languages and Systems. Proceedings, 2001. XIII, 433 pages. 2001.

Vol. 2029: H. Hussmann (Ed.), Fundamental Approaches to Software Engineering. Proceedings, 2001. XIII, 349 pages. 2001.

Vol. 2030: F. Honsell, M. Miculan (Eds.), Foundations of Software Science and Computation Structures. Proceedings, 2001. XII, 413 pages. 2001.

Vol. 2031: T. Margaria, W. Yi (Eds.), Tools and Algorithms for the Construction and Analysis of Systems. Proceedings, 2001. XIV, 588 pages. 2001.

Vol. 2033: J. Liu, Y. Ye (Eds.), E-Commerce Agents. VI, 347 pages. 2001. (Subseries LNAI).

Vol. 2034: M.D. Di Benedetto, A. Sangiovanni-Vincentelli (Eds.), Hybrid Systems: Computation and Control. Proceedings, 2001. XIV, 516 pages. 2001.

Vol. 2035: D. Cheung, G.J. Williams, Q. Li (Eds.), Advances in Knowledge Discovery and Data Mining – PAKDD 2001. Proceedings, 2001. XVIII, 596 pages. 2001. (Subseries LNAI).

Vol. 2037: E.J.W. Boers et al. (Eds.), Applications of Evolutionary Computing. Proceedings, 2001. XIII, 516 pages. 2001.

Vol. 2038: J. Miller, M. Tomassini, P.L. Lanzi, C. Ryan, A.G.B. Tettamanzi, W.B. Langdon (Eds.), Genetic Programming. Proceedings, 2001. XI, 384 pages. 2001.

Vol. 2039: M. Schumacher, Objective Coordination in Multi-Agent System Engineering. XIV, 149 pages. 2001. (Subseries LNAI).

Vol. 2040: W. Kou, Y. Yesha, C.J. Tan (Eds.), Electronic Commerce Technologies. Proceedings, 2001. X, 187 pages. 2001.

Vol. 2044: S. Abramsky (Ed.), Typed Lambda Calculi and Applications. Proceedings, 2001. XI, 431 pages. 2001.

Vol. 2045: B. Pfitzmann (Ed.), Advances in Cryptology – EUROCRYPT 2001. Proceedings, 2001. XII, 545 pages. 2001.

Vol. 2053: O. Danvy, A. Filinski (Eds.), Programs as Data Objects. Proceedings, 2001. VIII, 279 pages. 2001.

Vol. 2054: A. Condon, G. Rozenberg (Eds.), DNA Computing. Proceedings, 2000. X, 271 pages. 2001.